Networked Control Systems with Intermittent Feedback

AUTOMATION AND CONTROL ENGINEERING
A Series of Reference Books and Textbooks

Series Editors

FRANK L. LEWIS, Ph.D.,
Fellow IEEE, Fellow IFAC
Professor
The Univeristy of Texas Research Institute
The University of Texas at Arlington

SHUZHI SAM GE, Ph.D.,
Fellow IEEE
Professor
Interactive Digital Media Institute
The National University of Singapore

STJEPAN BOGDAN
Professor
Faculty of Electrical Engineering
and Computing
University of Zagreb

RECENTLY PUBLISHED TITLES

Networked Control Systems with Intermittent Feedback,
Domagoj Tolić and Sandra Hirche

Doubly Fed Induction Generators: Control for Wind Energy,
Edgar N. Sanchez; Riemann Ruiz-Cruz

Optimal Networked Control Systems with MATLAB®, *Jagannathan Sarangapani; Hao Xu*

Cooperative Control of Multi-agent Systems: A Consensus Region Approach,
Zhongkui Li; Zhisheng Duan

Nonlinear Control of Dynamic Networks,
Tengfei Liu; Zhong-Ping Jiang; David J. Hill

Modeling and Control for Micro/Nano Devices and Systems,
Ning Xi; Mingjun Zhang; Guangyong Li

Linear Control System Analysis and Design with MATLAB®, Sixth Edition,
Constantine H. Houpis; Stuart N. Sheldon

Real-Time Rendering: Computer Graphics with Control Engineering,
Gabriyel Wong; Jianliang Wang

Anti-Disturbance Control for Systems with Multiple Disturbances,
Lei Guo; Songyin Cao

Tensor Product Model Transformation in Polytopic Model-Based Control,
Péter Baranyi; Yeung Yam; Péter Várlaki

Fundamentals in Modeling and Control of Mobile Manipulators, *Zhijun Li; Shuzhi Sam Ge*

Optimal and Robust Scheduling for Networked Control Systems, *Stefano Longo; Tingli Su; Guido Herrmann; Phil Barber*

Advances in Missile Guidance, Control, and Estimation, *S.N. Balakrishna; Antonios Tsourdos; B.A. White*

End to End Adaptive Congestion Control in TCP/IP Networks,
Christos N. Houmkozlis; George A Rovithakis

Quantitative Process Control Theory, *Weidong Zhang*

Classical Feedback Control: With MATLAB® and Simulink®, Second Edition, *Boris Lurie; Paul Enright*

Intelligent Diagnosis and Prognosis of Industrial Networked Systems, *Chee Khiang Pang; Frank L. Lewis; Tong Heng Lee; Zhao Yang Dong*

Synchronization and Control of Multiagent Systems, *Dong Sun*

Subspace Learning of Neural Networks, *Jian Cheng; Zhang Yi; Jiliu Zhou*

Reliable Control and Filtering of Linear Systems with Adaptive Mechanisms, *Guang-Hong Yang; Dan Ye*

Reinforcement Learning and Dynamic Programming Using Function Approximators, *Lucian Busoniu; Robert Babuska; Bart De Schutter; Damien Ernst*

Modeling and Control of Vibration in Mechanical Systems, *Chunling Du; Lihua Xie*

Analysis and Synthesis of Fuzzy Control Systems: A Model-Based Approach, *Gang Feng*

Lyapunov-Based Control of Robotic Systems, *Aman Behal; Warren Dixon; Darren M. Dawson; Bin Xian*

System Modeling and Control with Resource-Oriented Petri Nets, *MengChu Zhou; Naiqi Wu*

Deterministic Learning Theory for Identification, Recognition, and Control, *Cong Wang; David J. Hill*

Sliding Mode Control in Electro-Mechanical Systems, Second Edition, *Vadim Utkin; Juergen Guldner; Jingxin Shi*

Linear Control Theory: Structure, Robustness, and Optimization, *Shankar P. Bhattacharyya; Aniruddha Datta; Lee H. Keel*

Intelligent Systems: Modeling, Optimization, and Control, *Yung C. Shin; Myo-Taeg Lim; Dobrila Skataric; Wu-Chung Su; Vojislav Kecman*

Optimal Control: Weakly Coupled Systems and Applications, *Zoran Gajic*

Intelligent Freight Transportation, *Petros A. Ioannou*

Modeling and Control of Complex Systems, *Petros A. Ioannou; Andreas Pitsillides*

Optimal and Robust Estimation: With an Introduction to Stochastic Control Theory, Second Edition, *Frank L. Lewis; Lihua Xie; Dan Popa*

Feedback Control of Dynamic Bipedal Robot Locomotion, *Eric R. Westervelt; Jessy W. Grizzle; Christine Chevallereau; Jun Ho Choi; Benjamin Morris*

Wireless Ad Hoc and Sensor Networks: Protocols, Performance, and Control, *Jagannathan Sarangapani*

Stochastic Hybrid Systems, *Christos G. Cassandras; John Lygeros*

Hard Disk Drive: Mechatronics and Control, *Abdullah Al Mamun; GuoXiao Guo; Chao Bi*

Autonomous Mobile Robots: Sensing, Control, Decision Making and Applications, *Shuzhi Sam Ge; Frank L. Lewis*

Neural Network Control of Nonlinear Discrete-Time Systems, *Jagannathan Sarangapani*

Fuzzy Controller Design: Theory and Applications, *Zdenko Kovacic; Stjepan Bogdan*

Quantitative Feedback Theory: Fundamentals and Applications, Second Edition, *Constantine H. Houpis; Steven J. Rasmussen; Mario Garcia-Sanz*

Chaos in Automatic Control, *Wilfrid Perruquetti; Jean-Pierre Barbot*

Differentially Flat Systems, *Hebertt Sira-Ramírez; Sunil K. Agrawal*

Robot Manipulator Control: Theory and Practice, *Frank L. Lewis; Darren M. Dawson; Chaouki T. Abdallah*

Robust Control System Design: Advanced State Space Techniques, *Chia-Chi Tsui*

Linear Control System Analysis and Design: Fifth Edition, Revised and Expanded, *Constantine H. Houpis; Stuart N. Sheldon; John J. D'Azzo; Constantine H. Houpis; Stuart N. Sheldon*

Nonlinear Control Systems, *Zoran Vukic; Ljubomir Kuljaca; Donlagic; Sejid Tesnjak*

Actuator Saturation Control, *Vikram Kapila; Karolos Grigoriadis*

Sliding Mode Control In Engineering, *Wilfrid Perruquetti; Jean-Pierre Barbot*

Modern Control Engineering, *P.N. Paraskevopoulos*

Advanced Process Identification and Control, *Enso Ikonen; Kaddour Najim*

Optimal Control of Singularly Perturbed Linear Systems and Applications, *Zoran Gajic; Myo-Taeg Lim*

Robust Control and Filtering for Time-Delay Systems, *Magdi S. Mahmoud*

Self-Learning Control of Finite Markov Chains, *A.S. Poznyak; Kaddour Najim; E. Gomez-Ramirez*

Nonlinear Control of Electric Machinery, *Darren M. Dawson; Jun Hun; Timothy C. Burg*

AUTOMATION AND CONTROL ENGINEERING SERIES

Networked Control Systems with Intermittent Feedback

Domagoj Tolić   Sandra Hirche

CRC Press
Taylor & Francis Group
Boca Raton   London   New York

CRC Press is an imprint of the
Taylor & Francis Group, an **informa** business

CRC Press
Taylor & Francis Group
6000 Broken Sound Parkway NW, Suite 300
Boca Raton, FL 33487-2742

© 2017 by Taylor & Francis Group, LLC
CRC Press is an imprint of Taylor & Francis Group, an Informa business

No claim to original U.S. Government works

Printed on acid-free paper

International Standard Book Number-13: 978-1-4987-5634-1 (Hardback)

This book contains information obtained from authentic and highly regarded sources. Reasonable efforts have been made to publish reliable data and information, but the author and publisher cannot assume responsibility for the validity of all materials or the consequences of their use. The authors and publishers have attempted to trace the copyright holders of all material reproduced in this publication and apologize to copyright holders if permission to publish in this form has not been obtained. If any copyright material has not been acknowledged please write and let us know so we may rectify in any future reprint.

Except as permitted under U.S. Copyright Law, no part of this book may be reprinted, reproduced, transmitted, or utilized in any form by any electronic, mechanical, or other means, now known or hereafter invented, including photocopying, microfilming, and recording, or in any information storage or retrieval system, without written permission from the publishers.

For permission to photocopy or use material electronically from this work, please access www.copyright.com (http://www.copyright.com/) or contact the Copyright Clearance Center, Inc. (CCC), 222 Rosewood Drive, Danvers, MA 01923, 978-750-8400. CCC is a not-for-profit organization that provides licenses and registration for a variety of users. For organizations that have been granted a photocopy license by the CCC, a separate system of payment has been arranged.

Trademark Notice: Product or corporate names may be trademarks or registered trademarks, and are used only for identification and explanation without intent to infringe.

Library of Congress Cataloging-in-Publication Data

Names: Tolic, Domagoj, author. | Hirche, Sandra, author.
Title: Networked control systems with intermittent feedback / Domagoj Tolic and Sandra Hirche.
Description: Boca Raton : Taylor & Francis, CRC Press, 2017. |
Series: Automation and control engineering series. |
Includes bibliographical references and index.
Identifiers: LCCN 2016045227| ISBN 9781498756341 (hardback : alk. paper) | ISBN 9781498756358 (ebook)
Subjects: LCSH: Feedback control systems. | Computer networks.
Classification: LCC TJ216 .T64 2017 | DDC 629.8/9--dc23
LC record available at https://lccn.loc.gov/2016045227

Visit the Taylor & Francis Web site at
http://www.taylorandfrancis.com

and the CRC Press Web site at
http://www.crcpress.com

*To my family –
Verica, Petar, Ivan, Hrvoje,
and Ivana.
(DT)*

*To Goya ...
(SH)*

Contents

Preface — xiii

List of Figures — xvii

List of Tables — xxi

Contributors — xxiii

Symbols and Abbreviations — xxv

1 Introduction — 1
- 1.1 Why Study Intermittent Feedback? — 5
- 1.2 Historical Aspects and Related Notions — 8
- 1.3 Open Problems and Perspectives — 9
- 1.4 Notation — 10

I PLANT-CONTROLLER APPLICATIONS — 13

2 MATIs with Time-Varying Delays and Model-Based Estimators — 15
- 2.1 Motivation, Applications and Related Works — 16
- 2.2 Impulsive Delayed Systems and Related Stability Notions — 17
- 2.3 Problem Statement: Stabilizing Transmission Intervals and Delays — 18
- 2.4 Computing Maximally Allowable Transfer Intervals — 22
  - 2.4.1 $\mathcal{L}_p$-Stability with Bias of Impulsive Delayed LTI Systems — 23
  - 2.4.2 Obtaining MATIs via the Small-Gain Theorem — 25
- 2.5 Numerical Examples: Batch Reactor, Planar System and Inverted Pendulum — 27
  - 2.5.1 Batch Reactor with Constant Delays — 27
  - 2.5.2 Planar System with Constant Delays — 31
  - 2.5.3 Inverted Pendulum with Time-Varying Delays — 35
- 2.6 Conclusions and Perspectives — 40
- 2.7 Proofs of Main Results — 40

2.7.1	Proof of Lemma 2.1	40
2.7.2	Proof of Theorem 2.1	44
2.7.3	Proof of Theorem 2.2	46
2.7.4	Proof of Corollary 2.1	48
2.7.5	Proof of Proposition 2.1	48

3 Input-Output Triggering 51

- 3.1 Motivation, Applications and Related Works 52
  - 3.1.1 Motivational Example: Autonomous Cruise Control . 52
  - 3.1.2 Applications and Literature Review 54
- 3.2 Impulsive Switched Systems and Related Stability Notions . 57
- 3.3 Problem Statement: Self-Triggering from Input and Output Measurements 60
- 3.4 Input-Output Triggered Mechanism 62
  - 3.4.1 Why $\mathcal{L}_p$-gains over a Finite Horizon? 63
  - 3.4.2 Proposed Approach 64
  - 3.4.3 Design of Input-Output Triggering 65
    - 3.4.3.1 Cases 3.1 and 3.2 66
    - 3.4.3.2 Case 3.3 68
  - 3.4.4 Implementation of Input-Output Triggering 68
- 3.5 Example: Autonomous Cruise Control 70
- 3.6 Conclusions and Perspectives 73
- 3.7 Proofs of Main Results 76
  - 3.7.1 Properties of Matrix Functions 76
  - 3.7.2 Proof of Theorem 3.1 77
  - 3.7.3 Proof of Theorem 3.2 78
  - 3.7.4 Proof of Results in Section 3.4.3 80
    - 3.7.4.1 $\mathcal{L}_p$ property over an arbitrary finite interval with constant δ 80
    - 3.7.4.2 Extending bounds to (an arbitrarily long) finite horizon 81
    - 3.7.4.3 Proof of Theorem 3.3 82
    - 3.7.4.4 Proof of Theorem 3.4 82

4 Optimal Self-Triggering 85

- 4.1 Motivation, Applications and Related Works 85
- 4.2 Problem Statement: Performance Index Minimization 87
- 4.3 Obtaining Optimal Transmission Intervals 89
  - 4.3.1 Input-Output-Triggering via the Small-Gain Theorem 89
  - 4.3.2 Dynamic Programming 90
  - 4.3.3 Approximate Dynamic Programming 91
  - 4.3.4 Approximation Architecture 91
    - 4.3.4.1 Desired Properties 92

		4.3.5 Partially Observable States	93
	4.4	Example: Autonomous Cruise Control (Revisited)	94
	4.5	Conclusions and Perspectives	95

5 Multi-Loop NCSs over a Shared Communication Channels 99

	5.1	Motivation, Applications and Related Works	100
		5.1.1 Medium Access Control	101
	5.2	Markov Chains and Stochastic Stability	104
		5.2.1 Markov Chains	104
		5.2.2 Stochastic Stability	105
	5.3	Problem Statement: Scheduling in Multi-Loop NCS	107
	5.4	Stability and Performance	109
		5.4.1 Event-Based Scheduling Design	109
		5.4.2 Stability Analysis	112
		5.4.3 Performance and Design Guidelines	114
		5.4.4 Scheduling in the Presence of Channel Imperfections	116
	5.5	Decentralized Scheduler Implementation	117
	5.6	Empirical Performance Evaluation	121
		5.6.1 Optimized Thresholds λ for the Bi-Character Scheduler	121
		5.6.2 Comparison for Different Scheduling Policies	122
		5.6.3 Performance of the Decentralized Scheduler	123
		5.6.4 Performance with Packet Dropouts	125
	5.7	Conclusions and Perspectives	126
	5.8	Proofs and Derivations of Main Results	127
		5.8.1 Proof of Theorem 5.3	127
		5.8.2 Proof of Corollary 5.1	130
		5.8.3 Proof of Theorem 5.4	131
		5.8.4 Proof of Proposition 5.2	132
		5.8.5 Proof of Proposition 5.3	135

II MULTI-AGENT APPLICATIONS 143

6 Topology-Triggering of Multi-Agent Systems 145

	6.1	Motivation, Applications and Related Works	146
	6.2	Initial-Condition-(In)dependent Multi-Agent Systems and Switched Systems	149
		6.2.1 Switched Systems and Average Dwell Time	150
		6.2.2 Graph Theory	151
	6.3	Problem Statement: Transmission Intervals Adapting to Underlying Communication Topologies	151
	6.4	Topology-Triggering and Related Performance vs. Lifetime Trade-Offs	153
		6.4.1 Designing Broadcasting Instants	154

		6.4.2	Switching Communication Topologies	158
			6.4.2.1 Switching without Disturbances	158
			6.4.2.2 Switching with Disturbances	160
	6.5	Example: Output Synchronization and Consensus Control with Experimental Validation		161
		6.5.1	Performance vs. Lifetime Trade-Offs	163
		6.5.2	Experimental Setup	164
		6.5.3	Energy Consumption	166
		6.5.4	Experimental Results	167
	6.6	Conclusions and Perspectives		168
	6.7	Proofs and Derivations of Main Results		170
		6.7.1	From Agent Dynamics to Closed-Loop Dynamics	170
		6.7.2	Introducing Intermittent Data Exchange	171
		6.7.3	Proof of Proposition 6.1	172
		6.7.4	Proof of Theorem 6.2	172
		6.7.5	Proof of Theorem 6.3	174
		6.7.6	Proof of Theorem 6.4	175

7 Cooperative Control in Degraded Communication Environments   177

	7.1	Motivation, Applications and Related Works	178
	7.2	Impulsive Delayed Systems	179
	7.3	Problem Statement: Stabilizing Transmission Intervals and Delays	180
	7.4	Computing Maximally Allowable Transfer Intervals	182
		7.4.1 Interconnecting the Nominal and Error System	183
		7.4.2 MASs with Nontrivial Sets $\mathcal{B}$	183
		7.4.3 Computing Transmission Intervals τ	184
	7.5	Example: Consensus Control with Experimental Validation	185
	7.6	Conclusions and Perspectives	190
	7.7	Proofs of Main Results	193
		7.7.1 Proof of Theorem 7.1	193
		7.7.2 Proof of Corollary 7.1	194

8 Optimal Intermittent Feedback via Least Square Policy Iteration   197

	8.1	Motivation, Applications and Related Works	198
	8.2	Problem Statement: Cost-Minimizing Transmission Policies	199
	8.3	Computing Maximally Allowable Transfer Intervals	201
		8.3.1 Stabilizing Interbroadcasting Intervals	201
		8.3.2 (Sub)optimal Interbroadcasting Intervals	204
	8.4	Example: Consensus Control (Revisited)	208
	8.5	Conclusions and Perspectives	209

Bibliography	**213**
Index	**231**

Preface

Networked Control Systems (NCSs) are spatially distributed systems for which the communication between sensors, actuators and controllers is realized by a shared (wired or wireless) communication network. NCSs offer several advantages, such as reduced installation and maintenance costs and greater flexibility, as compared to conventional control systems in which parts of control loops exchange information via dedicated point-to-point connections. At the same time, NCSs generate imperfections (such as sampled, corrupted, delayed and lossy data) that impair the control system performance and can even lead to instability. In order to reduce data loss (i.e., packet collisions) among uncoordinated NCS links and owing to limited channel capacities, scheduling protocols are employed to govern the communication medium access. Since the aforementioned *network-induced phenomena* occur simultaneously, the investigation of their cumulative adverse effects on the NCS performance is of particular interest. This investigation opens the door to various trade-offs while designing NCSs. For instance, dynamic scheduling protocols, model-based estimators or smaller transmission intervals can compensate for greater delays at the expense of increased implementation complexity/costs.

When multi-agent applications are of interest, emerging technologies provide networks of increasingly accomplished agents that may be mobile and may possess significant (but limited) processing, sensing, communication as well as memory and energy storage capabilities. Multi-Agent Networks (MANs) that coexist side by side sharing the same physical environment may need to share information (e.g., measurements, intentions) as well as resources and services (e.g., storage space, energy supplies, processing power, Internet or routing services) in order to realize their full potential and prolong their mission. However, this inter-network interaction must not compromise the objectives of each individual network.

That being said, a number of recent control community efforts tackle the question: "How often should information between systems/agents or different parts of control systems be exchanged in order to meet a desired performance?" Under the term *intermittent information*, we refer to both *intermittent feedback* (a user-designed property) and *intrinsic properties* of control systems due to packet collisions, hardware-dependent lower bounds on transmission periods and processing time, network throughput, scheduling protocols, lossy communication channels, occlusions of sensors or a limited communication/sensing range. User-designed intermittent feedback is motivated by rational use of expensive resources at hand in an effort to decrease energy consumption as well as processing and sensing requirements. In addition, intermittent feedback allows *multitasking* by not utilizing resources all the time for a sole task.

Stabilizing transmission intervals can be computed *offline*, for the worst-case scenario, which leads to periodic transmissions, or *online*, depending on the actual control performance, which typically results in aperiodic data exchange. Common "on the fly" transmission strategies are event-triggering and self-triggering. In event-triggered approaches, one defines a desired performance, and a transmission of up-to-date information is triggered when an event representing the unwanted performance occurs. In self-triggered approaches, the currently available (but outdated) information is used to determine the next transmission instant, that is, to predict the occurrence of the triggering event. In comparison with event-triggering, where sensor readings are constantly obtained and analyzed in order to detect events (even though the control signals are updated only upon event detection), self-triggering relaxes requirements posed on sensors and processors in embedded systems.

There has been a tremendous amount of theoretical achievements accompanied by numerical simulations comparing different intermittent feedback paradigms in the past several years. However, experimental validations and applicability discussions are typically not found therein. Hence, we believe that application-oriented works, which should attract the attention of hardware-oriented researchers and practicing engineers in industries and bring benefits to the theorists as well, are in order now. Accordingly, this book is written in an application-oriented and rather straightforward manner, leaving technicalities for the proofs, and several of our theoretical results are verified experimentally. The utilized experimental testbed includes wireless sensor networks and aerial robots. Nevertheless, it should not be forgotten that our methodology is couched in a rich and potent theoretical background (e.g., $\mathcal{L}_p$-stability, impulsive and delayed systems, switched systems, small-gain theorem, Lyapunov–Razumikhin techniques, dynamic programming, Markov processes, etc.). The theoretical results presented in this book can readily be adapted by other fields of engineering (e.g., sensor fusion, computer science and signal processing) and toward different application areas (e.g., aerospace, biomedical, power systems and manufacturing).

The principal goal of this book is to present a coherent and versatile framework applicable to various settings investigated by the authors over the last several years. We strongly believe that a significant number of potential readers exist for this book among the professionals working in the engineering fields, which includes practicing engineers working in industries, researchers and scientists in research institutes as well as academics working in universities. Our intention is to broaden the community attracted to intermittent feedback by demonstrating that the existing theory is mature enough to be exploited/applied in real life. Certainly, the intermittent feedback paradigm can only benefit from the potential synergy. The salient features of this book are:

Preface

- a comprehensive exposition of a versatile and coherent intermittent feedback methodology,
- a broad spectrum of applications and settings,
- experimental validation with a detailed treatment of implementation issues,
- it is an educational and entirely self-contained book so that potential readers not working in related fields can appreciate our theoretical considerations and apply them effectively, and
- it is concise enough to be used for self-study.

Our intention is to demonstrate that the presented results can be used

- for resource management (e.g., multitasking and lifetime vs. performance trade-offs),
- when designing control laws, and
- when acquiring an experimental testbed (i.e., purchasing new sensors, actuators, communication devices, mobile robotic platforms, etc.).

Chapter 1 introduces and motivates the concepts of intermittent information and intermittent feedback in control systems. Our methodology in its most general form is presented in Chapter 2 and the remainder of the book, except for Chapter 5, which is reserved for event-based scheduling in multi-loop NCSs, which unfolds from this chapter. Basically, Chapters 3–8 (save for Chapter 5) devise additional results and apply the methodology of Chapter 2 to several plant-controller and MAN problems illustrating its versatility and effectiveness. Chapters 2–5 are dedicated to plant-controller settings, while multi-system settings are reserved for Chapters 6–8. In case the reader is interested merely in one of the problems from Chapters 3–8, we suggest to first read Chapters 1 and 2 before moving to the corresponding chapter. In order to follow the results in Chapter 5, the reader may even skip Chapter 2.

We wish to express our greatest debt to many colleagues and students who helped us to realize this book. Our special thanks go to our collaborators: Rafael Fierro (Chapters 3 and 4), Ricardo Sanfelice (Chapter 3), Silvia Ferrari (Chapter 4), Stjepan Bogdan (Chapter 7), Vedran Bilas (Chapter 6), Ivana Palunko (Chapters 7 and 8), Adam Molin (Chapter 5), and Mohammad H. Mamduhi (Chapter 5). We would also like to express our gratitude to Nora Konopka, Vakili Jessica and Lyndholm Kyra from CRC Press. This list would not be complete without thanking the University of New Mexico, University of Zagreb, University of Dubrovnik, and Technical University of Munich for providing stimulating research environments.

Dubrovnik, Croatia   Domagoj Tolić
Munich, Germany   Sandra Hirche
                                                                                                     September 2016

List of Figures

1.1	Illustration of event-triggering.	3
1.2	Illustration of self-triggering.	4
2.1	A diagram of a control system with the plant and controller interacting over a communication network with intermittent information updates. The two switches indicate that the information between the plant and controller are exchanged (complying with some scheduling protocol among the NCS links) at discrete time instants belonging to a set $\mathcal{T}$. The communication delays in each NCS link are time varying and, in general, different.	20
2.2	Interconnection of the nominal system Σ_n and error impulsive system Σ_e.	26
2.3	Numerically obtained MATIs for different delay values $d \geq 0$ in scenarios with and without estimation: (a) RR and (b) TOD.	34
2.4	Numerically obtained MATIs for various constant delay values $d \geq 0$ in scenarios with and without estimation: (a) RR and (b) TOD.	38
2.5	Numerically obtained MATIs for various time-varying delays $d(t)$ such that $d(t) \leq \check{d}$ and $|\dot{d}(t)| \leq \check{d}_1 = 0.5$ in scenarios with and without estimation: (a) RR and (b) TOD.	39
3.1	An illustration of the trajectory tracking problem considered in this chapter.	53
3.2	A diagram of a control system with the plant and controller interacting over a communication network with intermittent information updates. The three switches indicate that the information between the plant and controller are exchanged at discrete time instants belonging to a set $\mathcal{T}$.	55
3.3	A comparison of the intersampling intervals τ_i's obtained for different notions of $\mathcal{L}_p$-gains. The abbreviation UG stands for "Unified Gain". Red stems indicate time instants when changes in $\hat{\omega}_p$ occur. The solid blue line indicates τ_i's generated via the methodology devised in this chapter. Definitions of the said notions appear in Sections 3.2 and 3.4.	55

xvii

3.4	Interconnection of the nominal switched system Σ_n^δ and the output error impulsive switched system Σ_e^δ.	66
3.5	A realistic scenario illustrating input-output triggering: (a) states x of the tracking system, (b) norm of (x, e), (c) values of intersampling intervals τ_i's between two consecutive transmissions. Red stems indicate time instants when changes in δ happen, and (d) a detail from Figure 3.5(c).	74
3.6	A realistic scenario illustrating input-output triggering using the unified gains: (a) states x of the tracking system, (b) norm of (x, e), and (c) values of intersampling intervals τ_i's between two consecutive transmissions. Red stems indicate time instants when changes in δ happen.	75
4.1	Diagram of a plant and controller with discrete transmission instants and communication channels giving rise to *intermittent feedback*.	88
4.2	Approximation $\hat{V}^*(\hat{x})$ of the optimal value function $V^*(x)$ for $\omega_p = (1, 1)$ depicted as a function of $\hat{x}_1 \in [-70, 70]$ and $\hat{x}_2 \in [-70, 70]$ when $\hat{x}_3 = 0$.	96
4.3	Illustration of the optimal input-output-triggering: (a) state x of the tracking system, (b) norm of (x, e), (c) values of sampling period τ_i between two consecutive transmissions. Red stems indicate time instants when changes in ω_p happen, and (d) a detail from Figure 4.3(c).	97
5.1	Multi-loop NCS operated over a shared communication channel.	102
5.2	Two sub-systems compete for channel access by randomly choosing their waiting times ν_k^1 and ν_k^2, according to their corresponding local probability mass functions with error-dependent means $\mathbb{E}[\nu_k^1]$ and $\mathbb{E}[\nu_k^2]$) (purple arrows). The blue arrows illustrate a possible realization of the waiting times, here sub-system 1 won the competition as its waiting time is lower than for sub-system 2.	119
5.3	Architecture for the decentralized implementation of the scheduler.	120
5.4	Error variance over local error thresholds for multi-loop NCSs with different number of control loops.	122
5.5	Comparison of the average error variance vs. the number of control loops for different scheduling policies.	124
5.6	Comparison of the average error variance vs. the number of control loops for different scheduling policies.	125
5.7	Comparison of the average error variance vs. the number of control loops for different scheduling policies with packet dropout.	126

List of Figures xix

6.1 Block scheme of an agent indicating information and energy flows. 147
6.2 Two decentralized co-located MANs. 147
6.3 The graph partition $\mathcal{P}_\rho^1 = \{1,3\}$, $\mathcal{P}_\rho^2 = \{2,5\}$ and $\mathcal{P}_\rho^3 = \{4\}$ obtained via Algorithm 1. Accordingly, $T_\rho = 3$. In order not to clutter this figure with a battery for each node, only Node 3 is connected to a battery. 155
6.4 An illustration of the considered TDMA scheduling for the partition depicted in Figure 6.3. The abbreviation TX stands for *transmission* while RX stands for *reception*. Apparently, our TDMA scheduling prevents limitations (i)–(iv) from Section 6.3 for a sufficiently large τ_ρ, i.e., $\tau_\rho \geq \max\{t_{TX}, t_{RX}\}$, for each $\rho \in \mathcal{P}$. On the other hand, τ_ρ has to be sufficiently small in order to preserve closed-loop stability. 156
6.5 Interconnection of the nominal and the error dynamics. . . . 158
6.6 eZ430-RF2500 WSN node used in the experimental setup. . 165
6.7 A sequence showing a node listening to the communication medium, then receiving a message and immediately transmitting another message. The packet has a 2 B payload (total 24 B on the physical layer due to the stack and radio overhead). Transitions between operating modes present a significant overhead in time and energy. Notice that the power consumption of the radio when merely listening is almost the same (and even slightly higher!) as when actually receiving a packet. This finding also advocates TDMA scheduling employing agent partitions. 167
6.8 Experimental results that verify the theoretical exposition of Section 6.4: (a) expected lifetime of the battery, (b) time to converge into ϵ-vicinity of the consensus for $\epsilon = 0.4$, and (c) states of the agents for $\tau_\rho = 0.033$ s. 169

7.1 A snapshot of our experimental setup with three quadcopters. 179
7.2 An illustration of communication links among four agents characterized with the same propagation delays and sampling instants. 181
7.3 Experimental verification of the identified transfer function for an AR.Drone Parrot quadcopter in the x-axis. These responses are obtained for the pulse input with duration of 1.75 s and amplitude of 0.2. This 0.2 is the normalized roll angle that is fed to the low-level controller and corresponds to the actual pitch angle of 9°. Since our agents are homogeneous, we write $x[m]$ instead of $x_i[m]$. 186
7.4 Experimental data for one square maneuver and $K = 2$. The computed marginal frequency is 12 Hz while the experimentally obtained one is 9 Hz. 191

7.5	Experimental data for one square maneuver and $K=4$. The computed marginal frequency is 24 Hz while the experimentally obtained one is 20 Hz.	192
8.1	Illustration of the RR protocol.	204
8.2	Numerically obtained data: (a) agents' positions, (b) agents' velocities, and (c) the corresponding (sub)optimal interbroadcasting intervals for each agent.	210
8.3	The process of learning α_κ for each agent.	211

List of Tables

2.1	Comparison between our methodology and [68] for UGAS. All delays d and MATIs $\bar{\tau}$ are measured in milliseconds.	30
2.2	Comparison of our methodology and [68] for $\mathcal{L}_2$-stability. Delays d and MATIs $\bar{\tau}$ are expressed in milliseconds.	30
5.1	Optimal error thresholds λ and the number of collisions depending on the number of control loops.	124
6.1	Current consumption and oscillator characteristics of MSP430 microcontroller.	165
6.2	Current consumption of the eZ430-RF2500 node in different operating modes with a 3 V supply.	166
7.1	Comparison of the normalized $\|\xi[t_0, t_{\text{end}}]\|_{2,\mathcal{B}}$ for different gains K and sampling frequencies. The consensus controller is turned on at t_0 while it is switched off at t_{end}. The difference $t_{\text{end}} - t_0$ is about 40 s for each experiment.	190

Contributors

Vedran Bilas
University of Zagreb
Zagreb, Croatia

Stjepan Bogdan
University of Zagreb
Zagreb, Croatia

Silvia Ferrari
Duke University
Durham, North Carolina

Rafael Fierro
University of New Mexico
Albuquerque, New Mexico

Mohammad H. Mamduhi
Technical University of Munich
Munich, Germany

Adam Molin
KTH Royal Institute of Technology
Stockholm, Sweden

Ivana Palunko
University of Dubrovnik
Dubrovnik, Croatia

Ricardo Sanfelice
University of California, Santa Cruz
Santa Cruz, California

Symbols and Abbreviations

Symbol Description

x, y, z	vectors	$\lvert x \rvert$	absolute value of scalar x
A, B, C	matrices	(x, y)	vector concatenation
$A^\top$	transpose of A	$\overline{x}$	vector operator yielding $(\lvert x_1\rvert, \lvert x_2\rvert, \ldots, \lvert x_n\rvert)$
$\mathcal{A}, \mathcal{B}, \mathcal{C}$	sets		
$\lvert \mathcal{A} \rvert$	set cardinality	$x \preceq y$	partial order indicating $x_i \leq y_i \ \forall i \in \{1, \cdots, n\}$
$\mathbb{N}$	natural numbers		
$\mathbb{N}_0$	natural numbers including zero	$\mathbf{0}_n$	n-dimensional vector with all zero entries
$\mathbb{R}^n$	n-dimensional Euclidean space	$\lVert f[a,b] \rVert_p$	$\mathcal{L}_p$-norm of f on interval $[a,b]$, $p \in [1, \infty]$
$\mathbb{R}^n_+$	nonnegative n-dimensional orthant	$\mathcal{L}_p[a,b]$	set of functions with finite $\mathcal{L}_p$-norm on $[a,b]$
$\lVert A \rVert$	induced matrix 2-norm	$\mathcal{L}_p$	set of functions with bounded $\mathcal{L}_p$-norm on $\mathbb{R}$
$\lambda_i(A)$	eigenvalues of A		
$\sigma_i(A)$	singular values of A	$\mathcal{L}_\infty$	set of essentially bounded measurable functions on $\mathbb{R}$
$\mathrm{Ker}(A)$	kernel of A		
$\mathcal{G}(A)$	geometric multiplicity of zero eigenvalue of A	$f(t^-)$	left-hand limit of f in t
		$f(t^+)$	right-hand limit of f in t
$\mathcal{A}(A)$	algebraic multiplicity of zero eigenvalue of A	$C([\cdot,\cdot],\cdot)$	continuous functions on $[\cdot,\cdot]$
$A(i).A(j)$	element-wise product of ith and j^{th} column of A	$PC([\cdot,\cdot],\cdot)$	piece-wise continuous functions on $[\cdot,\cdot]$ with certain properties
$\mathbf{0}_{n\times m}$	n by m matrix with all zero entries	$\tilde{\mathbf{0}}_{n_x}$	zero element of $PC([-d,0], \mathbb{R}^{n_x})$
I_n	n-dimensional identity matrix	$\mathcal{B}^c$	complementary space of space $\mathcal{B}$
$\mathcal{A}_n$	$n \times n$ matrices		
$\mathcal{A}_n^+$	symmetric matrices with nonnegative entries	$\otimes$	Kronecker product
		$O(f)$	dominant growth rate of function f
n_x	dimension of x		
x_t	translation operator acting on trajectory $x(\cdot)$ and defined by $x_t(\theta) := x(t+\theta)$ for $-d \leq \theta \leq 0$	$\mathbb{E}_x$	expectation w.r.t. random variable x
		$\mathbb{E}[x\lvert y]$	conditional expectation of random variable x given event y
$\lVert x \rVert$	Euclidean norm of x		

xxv

Abbreviations

ACC	Autonomous Cruise Control	MATI	Maximally Allowable Transmission Interval
ADP	Approximate Dynamic Programming	MLP	Multilayer Perceptron
CSMA	Carrier Sense Multiple Access	NCS	Networked Control System
		NN	Neural Network
CSMA-CA	Carrier Sense Multiple Access with Collision Avoidance	ODE	Ordinary Differential Equation
		PF	Particle Filter(ing)
DP	Dynamic Programming	RL	Reinforcement Learning
FDMA	Frequency Division Multiple Access	RR	Round Robin
		RX	reception
ISS	Input-to-State Stability	TDMA	Time Division Multiple Access
KF	Kalman Filter(ing)		
LMI	Linear Matrix Inequality	TOD	Try-Once-Discard
LMSS	Lyapunov Mean Square Stability	TX	transmission
		UGAS	Uniform Global Asymptotic Stability
LSP	Lyapunov Stability in Probability	UGES	Uniform Global Exponential Stability
LSPI	Least Square Policy Iteration	UGS	Uniform Global Stability
LTI	Linear Time-Invariant	VI	Value Iteration
MAC	Medium Access Control	w.r.t.	with respect to
MAN	Multi-Agent Network	WSN	Wireless Sensor Network
MAS	Multi-Agent System	ZOH	Zero-Order-Hold

1

Introduction

CONTENTS

1.1	Why Study Intermittent Feedback?	5
1.2	Historical Aspects and Related Notions	8
1.3	Open Problems and Perspectives	9
1.4	Notation	10

The motto of the Olympic Games is "citius, altius, fortius", which is Latin for "faster, higher, stronger". The same driving force is found in almost every field of human activity and interest. Accordingly, engineering communities are constantly looking for ways to further improve the state of the art by continually pushing technology limits and advancing theoretical foundations. For instance, both manufacturing improvements and theoretical findings are providing more reliable, precise, smaller and faster processing units, sensors and communication devices, as well as enhanced energy storage units. However, the full potential of these resources ought to be exploited in order to satisfy the insatiable needs of the industrial and civil sector. Perhaps the same processor can serve several control loops in case one control loop is not onerous enough. Similarly, one sensor could provide feedback information for several control loops, rather than being idle most of the time while serving a single control loop. That being said, the control system community has recently put under scrutiny its fundamental concept—feedback. Essentially, the following question is tackled: "How often should information between systems be exchanged in order to meet a desired performance?" The desired performance can be estimation quality, disturbance rejection, stability or performance index minimization/maximization, which brings us to the subject of this book. Our efforts naturally fall within the scope of Networked Control Systems (NCSs).

NCSs are spatially distributed systems for which the communication between sensors, actuators and controllers is realized by a shared (wired or wireless) communication network [73, 18, 196, 111, 52]. NCSs offer several advantages, such as reduced installation and maintenance costs as well as greater flexibility, over conventional control systems in which parts of control loops exchange information via dedicated point-to-point connections. At the same time, NCSs generate imperfections (such as sampled, corrupted, delayed and lossy data) that impair the control system performance and can even lead to instability. In order to reduce data loss (i.e., packet collisions) among uncoordinated NCS links and owing to limited channel capacities (i.e., bandwidth), scheduling protocols are employed to govern the communication medium access. For instance, it may not be possible to simultaneously send both position and velocity data of an agent to its neighbors (e.g., underwater applications), but either position or velocity data can be sent at any time instance. Likewise, many communication devices cannot receive and send information simultaneously (see Chapter 6 for more). Since the aforementioned network-induced phenomena occur simultaneously, the investigation of their cumulative adverse effects on the NCS performance is of particular interest. This investigation opens the door to various trade-offs while designing NCSs. For instance, dynamic scheduling protocols (refer to [68] and [117]), model-based estimators [48] or smaller transmission intervals can compensate for greater delays at the expense of increased implementation complexity/costs [35].

In particular, this book examines the notion of *intermittent information*. Under the term intermittent information, we refer to both *intermittent feedback* (a user-designed property as in [169, 5, 6, 103, 200, 67]) and *intrinsic properties* of control systems due to packet collisions, hardware-dependent lower bounds on transmission periods and processing time, network throughput, scheduling protocols, lossy communication channels, occlusions of sensors or a limited communication/sensing range. User-designed intermittent feedback is motivated by rational use of expensive resources at hand in an effort to decrease energy consumption as well as processing and sensing requirements. In addition, intermittent feedback allows *multitasking* by not utilizing resources all the time for a sole task.

Stabilizing transmission intervals can be computed *offline* for the worst-case scenario (Chapters 2 and 7), which leads to periodic transmissions, or *online* depending on the actual control performance, which typically results in aperiodic data exchange (Chapters 3, 4, 5, 6 and 8). Common "on the fly" transmission strategies are *event-triggering* and *self-triggering* and are illustrated in Figures 1.1 and 1.2, respectively. In event-triggered approaches, one defines a desired performance, and a transmission of up-to-date information is triggered when an event representing the unwanted performance occurs [169, 103, 200, 203, 149]. In self-triggered approaches (see Chapters 3 and 6), the currently available (but outdated) information is used to determine the next transmission instant, that is, to predict the occurrence of the triggering event [191, 192, 6, 170]. In comparison with event-triggering, where sensor readings are constantly obtained and analyzed in order to detect events (even though the control signals are updated only upon event detection), self-triggering relaxes requirements posed on sensors and processors in embedded systems.

Owing to their importance in recent years, let us dedicate one more paragraph to event- and self-triggering as depicted in Figures 1.1 and 1.2. Say that one's goal is to keep the red signal below the prespecified green signal by means of "interventions" (e.g., transmissions of up-to-date data) that reset the red signal to zero. Using the event-triggered paradigm, one would con-

Introduction

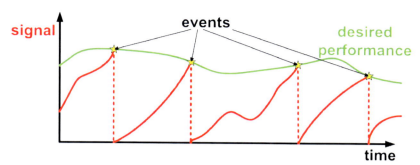

FIGURE 1.1
Illustration of event-triggering.

tinuously monitor both signals and make an "intervention" whenever the red signal becomes equal to the green signal. Clearly, the equality of the red and green signals represents the events in this scenario. Using the self-triggered paradigm, one would need to know the prespecified green signal as well as dynamics of the red signal. Based on this knowledge and the information available at an "intervention" instant, one would compute/predict the occurrence of events (e.g., via a worst-case analysis), commit to this prediction, and preclude events by making another "intervention" at the computed/predicted instant. This commitment to the computed/predicted time instants introduces some conservativeness in "intervention" instants, but eliminates the need for continuous monitoring of both signals, which may not be implementable in real life. In other words, event-triggering typically yields less frequent "intervention" instants at the expense of continuous monitoring and inability to foresee events and act accordingly. Basically, the self-triggered paradigm is a combination of event-triggering and time-triggering. Lastly, observe that the resets of the red signal give rise to impulsive system modeling, which entails Zeno behavior analyses as demonstrated in this book.

Throughout the book, we focus on *emulation-based approaches*. In emulation-based approaches, one first designs a controller without taking into account a communication network, and then, in the second step, one determines how often control and sensor signals have to be transmitted over the network so that the closed-loop system still delivers a desired performance (i.e., stability, disturbance rejection level, etc.). If the reader is interested in *network-aware control design* or *co-design*, which are out of scope of this book, the reference [162] is a good starting point. However, as emulation-based design and network-aware design/co-design are quite intertwined, the present results may be utilized as the first step toward network-aware design/co-design.

The principal requirement of our framework is $\mathcal{L}_p$-stability of the closed-loop system. In other words, if a certain controller does not yield the closed-loop system $\mathcal{L}_p$-stable, one can seek another controller. Hence, unlike some of

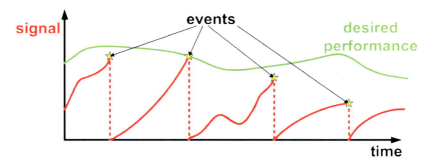

FIGURE 1.2
Illustration of self-triggering.

the related works, our requirements are on the closed-loop systems and not on the plant and controller per se, as illustrated throughout this book.

Altogether, this book presents a coherent and versatile framework applicable to various settings investigated by the authors over the last several years. Our framework is applicable to

- nonlinear time-varying dynamic plants and controllers with delayed dynamics;

- a large class of static, dynamic, probabilistic and priority-oriented scheduling protocols (i.e., Uniformly Globally Exponentially Stable and bi-character prioritizing protocols);

- delayed, noisy, lossy and intermittent information exchange;

- decentralized initial-condition-independent and initial-condition-dependent control problems (e.g., consensus and output synchronization) of heterogeneous agents with time-varying directed (not necessarily balanced) communication topologies;

- state and output feedback;

- offline and online intermittent feedback;

- optimal intermittent feedback through Approximate Dynamic Programming (ADP) and Reinforcement Learning (RL); and

- control systems with exogenous disturbances and modeling uncertainties.

We aim at demonstrating that the presented results can be used

- for resource management (e.g., multitasking and lifetime vs. performance trade-offs),

- when designing control laws, and

- when acquiring an experimental testbed (i.e., purchasing new sensors, actuators, communication devices, mobile robotic platforms, etc.).

In order to improve book readability, we make each chapter as self-contained as possible while developing the selected applications and solving the engineering problems of interest. Hence, when two derivations in different chapters are similar, but not identical, we have the luxury to keep both derivations, rather than referring the reader to the other chapter with different settings and forcing her/him to fill in the omitted details.

1.1 Why Study Intermittent Feedback?

Suppose that one is to develop an Autonomous Cruise Control (ACC) system using laser-based or radar-based sensors (depicted in Figure 3.1). ACC allows a vehicle to slow down when approaching another vehicle and accelerate to the desired speed when traffic allows. The sampling periods of ACC loops are typically fixed and designed for the worst-case scenario (e.g., fast and heavy traffic). Furthermore, these fixed sampling periods are often determined experimentally and are based on the traditional rules of thumb (e.g., 20 times the time constant of the dominant pole [158]). Intuitively, the sampling periods of ACC loops should not remain constant as the desired speed, distance between the cars, the environment (urban on non-urban), and paths (straight or turns) change. Our work quantifies this intuition. In addition, say that merely low-end (e.g., with high noise levels and with low update frequencies) sensors are at one's disposal. A natural question is whether such sensors are able to deliver the desired performance level (e.g., a prespecified overshoot, settling time and disturbance rejection level) for the vehicle of interest (e.g., bounds on linear and angular velocity as well as on turning radius of the vehicle) given a preferred control law. In case this possibility is ruled out by the theoretical findings, one should consider improving the software (e.g., advanced filtering and processing of noisy measurements, model-based estimators, etc.), designing a different control law or purchasing new hardware (e.g., sensors, processors or communication devices) to decrease transmission intervals, delays and noise level. The exact amount of time, effort and money that should be invested in these enhancements, which is often the crucial question in many real-life projects, can be inferred from the results presented in this book. Alternatively, a number of exhausting trial-and-error experiments need to be carried out risking vehicle damage or (potentially) overly powerful, yet economically unjustifiable, hardware components need to be acquired. An example of ACC is thoroughly analyzed in Chapter 3.

Essentially, this book studies the robustness of closed-loop systems with respect to the aforementioned network-induced phenomena (e.g., scheduling protocols, sampled, corrupted, delayed and lossy data, etc.). Hence, the plant and controller (designed via any existing method) are given and we analyze robustness of the resulting control system. We are interested to know whether the hardware/equipment at hand can deliver a desired control performance for a given control system. If not, one can modify the controller parameters, controller design or purchase more advanced hardware components. Accordingly, we want quick and easy transferability of control systems to different communication/sensing networks with the slightest interventions in the control law (e.g., change of controller parameters). One is often given a sensing and communication infrastructure along with inherent delays and needs to make sure in advance (i.e., in order to preclude testbed damage) whether this set-up can yield the required transmission intervals. In fact, it is common practice in industrial applications to design the control laws without having a particular communication/network infrastructure in mind (refer to Chapters 2, 3, 6 and 7).

Moreover, as illustrated throughout this book, the up-to-date data, which is about to arrive at the controller, may not be informative enough with respect to the out-of-date data from the last transmission instant. In the extreme case, where the up-to-date data is exactly the same as the outdated data, there is no justification for a novel transmission. Consequently, in an effort to further improve control system performance (i.e., "faster, higher, stronger"), it may not be beneficial to purchase a more expensive sensor that yields more frequent measurements.

Intrinsically, continuous information flows in real-life applications involving wireless communication (see Chapter 6 for more) are often not achievable for the following reasons:

(i) *Digital technology might be employed:* Every operation within a digital control system takes a certain amount of time for completion. Consequently, the hardware-specific *minimal time* to broadcast/receive a packet by a wireless radio transceiver determines the *maximal frequency* of communication.

(ii) *Communication devices cannot broadcast and receive simultaneously:* Unless there are two parallel communication channels (one only for transmitting and one only for receiving), a wireless device can either broadcast or receive at a particular moment. Since a double communication channel yields a big overhead in terms of complexity and energy consumption of devices, it is not desirable in most applications.

(iii) *A radio transceiver cannot receive messages when inactive:* In order for a radio to receive a message, it has to *listen* to the communication medium and switch to the *active* operation mode. Because the power consumption of the radio while actually receiving and while merely listening is practically the same (see [154, 194, 99] and Section 6.5), it is desirable to replace portions of listening intervals with idle intervals when no incoming information is expected.

(iv) *Message collision:* When several messages are being received at a receiver's end simultaneously, the associated data are lost. To avoid such collisions, scheduling protocols are often used.

The emergence of new technologies provides networks of increasingly accomplished agents. Such agents may be mobile and may possess significant processing, sensing, communication as well as memory and energy storage capabilities (refer to Figure 6.1). At the same time, aspirations to satisfy evergrowing demands of the industrial and civil sector bring about novel engineering paradigms such as Cyber-Physical Systems [148] and the Internet of Things [125]. The essence of these paradigms is fairly similar—to extend even further the concepts of heterogeneity, safety, decentralization, scalability, reconfigurability, and robustness of Multi-Agent Networks (MANs) by laying more burden on the agents. Everyday examples of MANs are cooperative multi-robot systems [159] and Wireless Sensor Networks (WSNs) [1]. Agents usually have limited and costly resources at their disposal which renders agent resource management of critical importance. An avenue toward agent resource management can be decentralized event- and self-triggered control/sensing/communication (Chapters 6, 7 and 8). In addition, in scenarios where somebody could eavesdrop on inter-agent communication, it may be desirable to decrease the rate at which information among agents is exchanged.

Intermittent feedback can be utilized to reduce requirements posed on communication, sensing, processing and energy resources in MANs without compromising objectives of the MANs (Chapters 6, 7 and 8). Consequently, the hardware expenses and energy consumption are driven down whilst MAN multitasking, such as inter-network collaboration, is facilitated. Scenarios in which agents of one network need to achieve a common goal, call for the study of decentralized cooperative control schemes. In order to accomplish this goal in an uncertain and noisy setting and to detect changes in the communication topology, agents have to exchange information. Because each transmission and reception of information necessitates energy, communication should be induced only when the goal completion can no longer be guaranteed in order to prolong the MAN mission.

Naturally, our analysis methodology can be a starting point for synthesis, which we often employ when selecting controller parameters. For instance, in the quadrotor experiments from Chapter 7, we chose controller parameters in order to obtain Maximally Allowable Transmission Intervals (MATIs) greater than 20 ms as that was the limit of our hardware setup. Essentially, as it is the case in real-life problems, merely the plant is given a priori and cannot be altered. Everything else (e.g., scheduling protocol, controller, delays, estimator) is either selected by the control system designer or stems from the hardware limitations. The control system designer can typically choose a different controller or estimator at no cost, but the choice of "better" hardware often entails cost. Thus, selection of a different controller or estimator is almost always preferred.

Altogether, the intermittent feedback investigations can be utilized in a number of ways (e.g., for choosing the controller structure and parameters, designing the network and scheduling protocol, designing the estimator, purchasing hardware, analyzing the influence of noise and disturbances on the control performance, etc.).

1.2 Historical Aspects and Related Notions

Even though feedback implementations traditionally connote periodicity owing to its theoretical and applicational convenience [158, 98], this periodicity has been questioned on and off ever since the dawn of digital technology [58, 106, 16, 185, 36, 126]. Suitable introductory reads regarding time- vs. event-triggering are found in [136] and [137]. However, as opposed to our work, the work presented in [136] and [137] designs the integrated chip architecture for time- and event-triggering. Hence, the concepts of stability, control performance (in terms of $\mathcal{L}_p$-gains), optimal transmission intervals or upper bounds on transmission intervals are not covered in [136] and [137] at all.

Among the more recent works that led to resurgence of interest in intermittent feedback, let us pinpoint [157, 10]. Our book shares the viewpoint of [157, 10] and falls within the recent body of literature regarding intermittent feedback, such as [169, 5, 6, 103, 200, 67, 67, 102, 203, 66, 149] to name a few. In fact, the entire book brings together a number of recent references regarding intermittent feedback making it impossible to state all of them in this paragraph. A significant portion of these recent works investigate event- and self-triggering as depicted in Figures 1.1 and 1.2. However, owing to the space limitations of conference and journal publications, the aforementioned works focus merely on a subset of problems/applications investigated in this book and on rather specific assumptions (e.g., noiseless measurements, disturbances not present, so-called small delays, static controllers, time-invariant plants, linear plants and controllers, etc.). As expected from a book, we present a comprehensive and unifying framework applicable to a rather broad range of problems.

As stated in [52], three principal formalisms used to tackle the problems in the area of NCSs and intermittent feedback are *discrete-time systems, impulsive/hybrid systems* and *delay systems*. Save for Chapter 5, where the discrete-time systems formalism is employed, our framework integrates impulsive systems and delay systems formalisms. As a consequence, we have devised novel results regarding stability of impulsive delay systems, which are presented in this book.

As mentioned earlier, the ability to allow for intermittent feedback can be thought of as a robustness property of control systems. Accordingly, the systems that allow for longer intervals of open-loop control, that is, the inter-

Introduction

vals in between two consecutive arrivals of feedback information, are clearly more robust with respect to the intermittent information phenomenon. Taking the utilized robustness tool into account, the recent approaches pertaining to intermittent feedback can be classified as:

(i) small-gain theorem approaches [131], [168];

(ii) dissipativity and passivity-based approaches [200], [201], [203];

(iii) Input-to-State Stability (ISS) approaches [169], [5], [6], [103]; and

(iv) other robustness approaches [47], [104], [170], [149].

We point out that our framework employs the small-gain theorem and $\mathcal{L}_p$-stability.

Lastly, let us clarify the difference between *sampled-data control* and intermittent feedback, which are indeed quite related terms. Besides some conceptual and historical differences stated in [67, 102, 66], the term *sampled-data control* traditionally implies periodic information exchange schemes and absence of communication protocols (i.e., the data among all parts of a control system are exchanged simultaneously).

1.3 Open Problems and Perspectives

Owing to "citius, altius, fortius", we are sure intermittent feedback will attract the attention of researchers and practitioners for years to come. For instance, tomorrow's hardware components will undoubtedly produce smaller delays and smaller transmission intervals. However, that does not mean smaller delays and/or transmission intervals should be utilized routinely. On the one hand, smaller delays may destabilize some systems as discussed in [61], [133, Chapter 1.], [164], and [52, Section 2.3]. On the other, further reductions of transmission intervals may not bring any benefits, as discussed in this book. In fact, smaller delays may allow for greater transmission intervals, that is, greater information intermittency, without compromising the control system performance. If the latter holds, additional feedback intermittency is worthwhile. Hence, the question "Exactly which values of delays and transmission intervals render the same level of performance (e.g., disturbance rejection, convergence speed, etc.)?" is extensively, but certainly not exhaustively, addressed in our book. Furthermore, smaller delays and transmission intervals often do improve the system performance. Nevertheless, the obtained performance improvement may be negligible with respect to the greater use (e.g., wear and tear, sensor exposure) of resources. Essentially, the order of magnitude of tomorrow's delays and transmission intervals will keep decreasing, but almost all concerns pointed out in this chapter will remain.

Since many works regarding event- and self-triggering are still being published, it is clear that this area has not matured yet and further research endeavors are vital. In fact, most of the proposed methodologies are applicable to a rather specific application-dependent setting (e.g., specific plant/agent and controller dynamics as well as specific topology and inter-agent coupling), which often hinders equitable comparisons between methodologies and impedes their transferability to a different setting. In addition, not every event-triggered scheme has its self-triggered counterpart, which may be troublesome in practical implementations owing to the continuous monitoring required by event-triggering. Thus, generalizing and unifying frameworks are needed in order to attract a broader audience. We believe our unifying and comprehensive framework is a step toward achieving this goal.

At the moment, the research community is interested in extending intertransmission intervals as much as possible without taking into account a deterioration in the performance due to intermittent feedback. In applications where energy consumption for using sensors, transmitting the obtained information, and executing control laws is relatively inexpensive compared to the slower convergence and excessive use of control power, extending intersampling intervals is not desirable. For instance, think of an airplane driven by an autopilot system designed to follow the shortest path between two points. Any deviation from the shortest path caused by intermittent feedback increases overall fuel consumption. This increase in fuel consumption is probably more costly than the cost of energy saved due to intermittent feedback. In Chapters 4 and 8, we encode these energy consumption trade-offs using cost functions, and design ADP approaches that yield (sub)optimal intertransmission intervals with respect to the cost function. However, more formal guarantees regarding (sub)optimality are required.

Finally, more detailed accounts of open problems and perspectives are found in the corresponding sections of each chapter.

1.4 Notation

To simplify notation in the book, we use $(x, y) := [x^\top \ y^\top]^\top$. The dimension of a vector x is denoted n_x. Next, let $f : \mathbb{R} \to \mathbb{R}^n$ be a Lebesgue measurable function on $[a, b] \subset \mathbb{R}$. We use

$$\|f[a,b]\|_p := \left(\int_{[a,b]} \|f(s)\|^p \mathrm{d}s \right)^{\frac{1}{p}}$$

to denote the $\mathcal{L}_p$-norm of f when restricted to the interval $[a, b]$. If the corresponding norm is finite, we write $f \in \mathcal{L}_p[a, b]$. In the above expression, $\|\cdot\|$ refers to the Euclidean norm of a vector. If the argument of $\|\cdot\|$ is a matrix

A, then it denotes the induced 2-norm of A. Eigenvalues and singular values of a matrix A are denoted $\lambda_i(A)$ and $\sigma_i(A)$, respectively. The kernel (or null space) of a matrix A is denoted $\text{Ker}(A)$. The dimension of $\text{Ker}(A)$ is denoted $\mathcal{G}(A)$ and equals the geometric multiplicity of the zero eigenvalue [195, Definition B.14]. The algebraic multiplicity of the zero eigenvalue is denoted $\mathcal{A}(A)$. Furthermore, $|\cdot|$ denotes the (scalar) absolute value function. The elementwise product of the ith and j^{th} columns of a matrix A is denoted $A(i).A(j)$. The n-dimensional vector with all zero entries is denoted $\mathbf{0}_n$. Likewise, the n by m matrix with all zero entries is denoted $\mathbf{0}_{n\times m}$. The identity matrix of dimension n is denoted I_n. In addition, $\mathbb{R}_+^n$ denotes the nonnegative orthant. The natural numbers are denoted $\mathbb{N}$ or $\mathbb{N}_0$ when zero is included. The dominant growth rate of a function $f(x) : \mathbb{R}^{n_x} \to \mathbb{R}$ is denoted $O(f)$. Since this dominant growth rate is in line with the big O notation, we adopt the letter O.

Left-hand and right-hand limits are denoted $x(t^-) = \lim_{t' \nearrow t} x(t')$ and $x(t^+) = \lim_{t' \searrow t} x(t')$, respectively. Next, for a set $\mathcal{S} \subseteq \mathbb{R}^n$, let $PC([a,b],\mathcal{S}) = \{\phi : [a,b] \to \mathcal{S} \mid \phi(t) = \phi(t^+) \text{ for every } t \in [a,b), \phi(t^-) \text{ exists in } \mathcal{S} \text{ for all } t \in (a,b] \text{ and } \phi(t^-) = \phi(t) \text{ for all but at most a finite number of points } t \in (a,b]\}$. Observe that $PC([a,b],\mathcal{S})$ denotes the family of right-continuous functions on $[a,b)$ with finite left-hand limits on $(a,b]$ contained in $\mathcal{S}$ and whose discontinuities do not accumulate in finite time. Finally, let $\tilde{\mathbf{0}}_{n_x}$ denote the zero element of $PC([-d,0],\mathbb{R}^{n_x})$.

Given $x \in \mathbb{R}^n$, we define
$$\bar{x} = (|x_1|, |x_2|, \ldots, |x_n|),$$
where $|\cdot|$ denotes the (scalar) absolute value function. Given $x = (x_1, x_2, \ldots, x_n)$ and $y = (y_1, y_2, \ldots, y_n) \in \mathbb{R}^n$, the partial order $\preceq$ is defined as
$$x \preceq y \iff x_i \leq y_i \quad \forall i \in \{1, \cdots, n\}.$$

The set $\mathcal{A}_n$ denotes the set of all $n \times n$ matrices and $\mathcal{A}_n^+$ denotes the subset of all matrices that are symmetric and have nonnegative entries.

Part I
PLANT-CONTROLLER APPLICATIONS

2

Maximally Allowable Transfer Intervals with Time-Varying Delays and Model-Based Estimators

CONTENTS

2.1	Motivation, Applications and Related Works	16
2.2	Impulsive Delayed Systems and Related Stability Notions	17
2.3	Problem Statement: Stabilizing Transmission Intervals and Delays	18
2.4	Computing Maximally Allowable Transfer Intervals	22
	2.4.1 $\mathcal{L}_p$-Stability with Bias of Impulsive Delayed LTI Systems	23
	2.4.2 Obtaining MATIs via the Small-Gain Theorem	25
2.5	Numerical Examples: Batch Reactor, Planar System and Inverted Pendulum	27
	2.5.1 Batch Reactor with Constant Delays	27
	2.5.2 Planar System with Constant Delays	31
	2.5.3 Inverted Pendulum with Time-Varying Delays	35
2.6	Conclusions and Perspectives	40
2.7	Proofs of Main Results	40
	2.7.1 Proof of Lemma 2.1	40
	2.7.2 Proof of Theorem 2.1	44
	2.7.3 Proof of Theorem 2.2	46
	2.7.4 Proof of Corollary 2.1	48
	2.7.5 Proof of Proposition 2.1	48

In this chapter, we present our methodology in its most general form. More precisely, we consider nonlinear plants with delayed dynamics, nonlinear delayed dynamic controllers, external disturbances (and/or modeling uncertainties), scheduling protocols, lossy communication channels as well as sampled, distorted and delayed information. Furthermore, variable transmission intervals and time-varying nonuniform delays possibly greater than the transmission intervals (so-called *large delays*) are considered along with model-based estimators. Several numerical examples are provided to demonstrate the benefits of the proposed approach. The remaining chapters further exemplify the methodology of this chapter and combine it with additional results to solve more specific intermittent feedback problems.

2.1 Motivation, Applications and Related Works

Consider a nonlinear delayed system to be controlled by a nonlinear delayed dynamic controller over a communication network in the presence of exogenous/modeling disturbances, scheduling protocols among lossy NCS links, time-varying signal delays, time-varying transmission intervals and distorted data. Notice that networked control is not the only source of delays and that delays might be present in the plant and controller dynamics as well. Therefore, we use the term *delayed NCSs*. This chapter presents an emulation-based approach for investigating the cumulative adverse effects in NCSs consisting of plants and controllers with delayed dynamics and characterized with *nonuniform* time-varying NCS link *delays*. In other words, different NCS links induce different and nonconstant delays. In addition, the communication delays are allowed to be larger than the transmission intervals (the so-called large delays). In addition, this chapter investigates the aforementioned cumulative effects for a wide class of plant-controller dynamics (i.e., time-varying, nonlinear, delayed and with disturbances) and interconnections (i.e., output feedback) as well as for the variety of scheduling protocols (i.e., UGES protocols) and other network-induced phenomena (i.e., variable delays, lossy communication channels with distortions). For instance, [119] focuses on time-varying nonlinear control affine plants (i.e., no delayed dynamics in the plant nor controller) and state feedback with a constant delay whilst neither exogenous/modeling disturbances, distorted data nor scheduling protocols are taken into account. The authors in [95] and [205] consider linear control systems, impose Zero-Order-Hold (ZOH) sampling and do not consider noisy data or scheduling protocols. In addition, [95] does not take into account disturbances. Similar comparisons can be drawn with respect to other related works (see [189, 73, 68, 181, 119] and the references therein).

In order to account for large delays, our methodology employs impulsive delayed system modeling and Lyapunov–Razumikhin techniques when computing Maximally Allowable Transmission Intervals (MATIs) that provably stabilize NCSs for the class of Uniformly Globally Exponentially Stable (UGES) scheduling protocols (to be defined later on). Besides MATIs that merely stabilize NCSs, our methodology is also capable of designing MATIs that yield a prespecified level of control system performance. As in [68], the performance level is quantified by means of $\mathcal{L}_p$-gains. According to the batch reactor case study provided in Section 2.5.1, MATI conservativeness repercussions of our approach for the small delay case appear to be modest in comparison with [68]. This conservativeness emanates from the complexity of the tools for computing $\mathcal{L}_p$-gains of delayed (impulsive) systems as pointed out in Section 2.5 and, among others, [28]. On the other hand, delayed system modeling (rather than ODE modeling as in [68]) allows for the employment of model-based estimators, which in turn increases MATIs (see Section 2.5 for more). In addition, real-life applications are characterized by corrupted data due to, among others, measurement noise and communication channel distortions. In order to include distorted information (in addition to exogenous/modeling disturbances) into the stability analyses, we propose the notion of $\mathcal{L}_p$-stability with bias.

The remainder of this chapter is organized as follows. Section 2.2 presents the utilized stability notions regarding impulsive delayed systems. Section 2.3 states the problem of finding MATIs for nonlinear delayed NCSs with UGES protocols in the presence of nonuniform communication delays and exogenous/modeling disturbances. A methodology to solve the problem is presented in Section 2.4. Detailed numerical examples are provided in Section 2.5. Conclusions and future challenges are in Section 2.6. The proofs are provided in Section 2.7.

2.2 Impulsive Delayed Systems and Related Stability Notions

Consider nonlinear *impulsive delayed systems*

$$\Sigma \begin{cases} \chi(t^+) = h_\chi(t, \chi_t) & t \in \mathcal{T} \\ \dot{\chi}(t) = f_\chi(t, \chi_t, \omega) \\ y = \ell_\chi(t, \chi_t, \omega) \end{cases} \text{otherwise}, \quad (2.1)$$

where $\chi \in \mathbb{R}^{n_\chi}$ is the state, $\omega \in \mathbb{R}^{n_\omega}$ is the input and $y \in \mathbb{R}^{n_y}$ is the output. The functions f_χ and h_χ are regular enough to guarantee forward completeness of solutions which, given initial time t_0 and initial condition $\chi_{t_0} \in PC([-d, 0], \mathbb{R}^{n_\chi})$, where $d \geq 0$ is the maximum value of all time-varying delay phenomena, are given by right-continuous functions $t \mapsto \chi(t) \in PC([t_0 - d, \infty], \mathbb{R}^{n_\chi})$. Furthermore, χ_t denotes the *translation operator* acting on the trajectory $\chi(\cdot)$ defined by $\chi_t(\theta) := \chi(t + \theta)$ for $-d \leq \theta \leq 0$. In other words, χ_t is the restriction of trajectory $\chi(\cdot)$ to the interval $[t-d, t]$ and translated to $[-d, 0]$. For $\chi_t \in PC([-d, 0], \mathbb{R}^{n_\chi})$, the norm of χ_t is defined by $\|\chi_t\| = \sup_{-d \leq \theta \leq 0} \|\chi_t(\theta)\|$. Jumps of the state are denoted $\chi(t^+)$ and occur at time instants $t \in \mathcal{T} := \{t_1, t_2, \ldots\}$, where $t_i < t_{i+1}$, $i \in \mathbb{N}_0$. The value of the state after a jump is given by $\chi(t^+)$ for each $t \in \mathcal{T}$. For a comprehensive discussion regarding the solutions to (2.1) considered herein, refer to [13, Chapter 2 and 3]. Even though the considered solutions to (2.1) allow for jumps at t_0, we exclude such jumps in favor of notational convenience.

Definition 2.1 (Uniform Global Stability). *For $\omega \equiv \mathbf{0}_{n_\omega}$, the system Σ is said to be Uniformly Globally Stable (UGS) if for any $\epsilon > 0$ there exists $\delta(\epsilon) > 0$ such that, for each $t_0 \in \mathbb{R}$ and each $\chi_{t_0} \in PC([-d,0], \mathbb{R}^{n_\chi})$ satisfying $\|\chi_{t_0}\| < \delta(\epsilon)$, each solution $t \mapsto \chi(t) \in PC([t_0 - d, \infty], \mathbb{R}^{n_\chi})$ to Σ satisfies $\|\chi(t)\| < \epsilon$ for all $t \geq t_0$ and $\delta(\epsilon)$ can be chosen such that $\lim_{\epsilon \to \infty} \delta(\epsilon) = \infty$.*

Definition 2.2 (Uniform Global Asymptotic Stability). *For $\omega \equiv \mathbf{0}_{n_\omega}$, the system Σ is said to be Uniformly Globally Asymptotically Stable (UGAS) if it is UGS and uniformly globally attractive, i.e., for each $\eta, \zeta > 0$ there exists $T(\eta, \zeta) > 0$ such that $\|\chi(t)\| < \eta$ for every $t \geq t_0 + T(\eta, \zeta)$ and every $\|\chi_{t_0}\| < \zeta$.*

Definition 2.3 (Uniform Global Exponential Stability). *For $\omega \equiv \mathbf{0}_{n_\omega}$, the system Σ is said to be Uniformly Globally Exponentially Stable (UGES) if there exist positive constants λ and M such that, for each $t_0 \in \mathbb{R}$ and each $\chi_{t_0} \in PC([-d, 0], \mathbb{R}^{n_\chi})$, each solution $t \mapsto \chi(t) \in PC([t_0 - d, \infty], \mathbb{R}^{n_\chi})$ to Σ satisfies $\|\chi(t)\| \leq M\|\chi_{t_0}\|e^{-\lambda(t-t_0)}$ for each $t \geq t_0$.*

Definition 2.4 ($\mathcal{L}_p$-Stability with Bias b). *Let $p \in [1, \infty]$. The system Σ is $\mathcal{L}_p$-stable with bias $b(t) \equiv b \geq 0$ from ω to y with (linear) gain $\gamma \geq 0$ if there exists $K \geq 0$ such that, for each $t_0 \in \mathbb{R}$ and each $\chi_{t_0} \in PC([-d, 0], \mathbb{R}^{n_\chi})$, each solution to Σ from χ_{t_0} satisfies $\|y[t_0, t]\|_p \leq K\|\chi_{t_0}\| + \gamma\|\omega[t_0, t]\|_p + \|b[t_0, t]\|_p$ for each $t \geq t_0$.*

Definition 2.5 ($\mathcal{L}_p$-Detectability). *Let $p \in [1, \infty]$. The state χ of Σ is $\mathcal{L}_p$-detectable from (y, ω) with (linear) gain $\gamma \geq 0$ if there exists $K \geq 0$ such that, for each $t_0 \in \mathbb{R}$ and each $\chi_{t_0} \in PC([-d, 0], \mathbb{R}^{n_\chi})$, each solution to Σ from χ_{t_0} satisfies $\|\chi[t_0, t]\|_p \leq K\|\chi_{t_0}\| + \gamma\|y[t_0, t]\|_p + \gamma\|\omega[t_0, t]\|_p$ for each $t \geq t_0$.*

Definitions 2.1, 2.2 and 2.3 are motivated by [88], while Definition 2.5 is inspired by [131]. Definition 2.4 is motivated by [131] and [84]. When $b = 0$, we say "$\mathcal{L}_p$-stability" instead of "$\mathcal{L}_p$-stability with bias 0".

2.3 Problem Statement: Stabilizing Transmission Intervals and Delays

Consider a nonlinear control system consisting of a plant with delayed dynamics

$$\dot{x}_p = f_p(t, x_{p_t}, u, \omega_p),$$
$$y = g_p(t, x_{p_t}), \qquad (2.2)$$

and a controller with delayed dynamics

$$\dot{x}_c = f_c(t, x_{c_t}, y, \omega_c),$$
$$u = g_c(t, x_{c_t}), \qquad (2.3)$$

where $x_p \in \mathbb{R}^{n_p}$ and $x_c \in \mathbb{R}^{n_c}$ are the states, $y \in \mathbb{R}^{n_y}$ and $u \in \mathbb{R}^{n_u}$ are the outputs, and $(u, \omega_p) \in \mathbb{R}^{n_u} \times \mathbb{R}^{n_{\omega_p}}$ and $(y, \omega_c) \in \mathbb{R}^{n_y} \times \mathbb{R}^{n_{\omega_c}}$ are the inputs of the plant and controller, respectively, where ω_p and ω_c are external disturbances to (and/or modeling uncertainties of) the plant and controller, respectively. The translation operators x_{p_t} and x_{c_t} are defined in Section 2.2 while the corresponding plant and controller delays are $d_p \geq 0$ and $d_c \geq 0$, respectively. For notational convenience, constant plant and controller delays are considered.

Let us now model the communication network between the plant and controller over which intermittent and realistic exchange of information takes place (see Figure 2.1). The value of u computed by the controller that arrives at the plant is denoted $\hat{u}$. Similarly, the values of y that the controller actually receives are denoted $\hat{y}$. Consequently, we have

$$u = \hat{u}, \qquad y = \hat{y}, \tag{2.4}$$

on the right-hand side of (2.2) and (2.3). In our setting, the quantity $\hat{u}$ is the delayed and distorted input u fed to the plant (2.2) while the quantity $\hat{y}$ is the delayed and distorted version of y received by the controller (2.3). We proceed further by defining the error vector

$$e = \begin{bmatrix} e_y(t) \\ e_u(t) \end{bmatrix} := \begin{bmatrix} \hat{y}(t) - y_t \\ \hat{u}(t) - u_t \end{bmatrix}, \tag{2.5}$$

where y_t and u_t are translation operators and the maximal network-induced delay is $d \geq 0$ (e.g., propagation delays and/or delays arising from protocol arbitration). In addition, in case of a time-invariant static controller with $\omega_c \equiv 0_{n_\omega}$, d can directly embed nontrivial executions times of the control law. The operator (y_t, u_t) in (2.5) delays each component of (y, u) for the respective delay. Essentially, if the ith component of $(y(t), u(t))$, that is $(y(t), u(t))_i$, is transmitted with delay $d_i : \mathbb{R} \to \mathbb{R}_+$, then the ith component of (y_t, u_t), that is $(y_t, u_t)_i$, is in fact $(y(t - d_i(t)), u(t - d_i(t)))_i$. Accordingly, $d := \max\{\sup_{t \in \mathbb{R}} d_1(t), \ldots, \sup_{t \in \mathbb{R}} d_{n_y + n_u}(t)\}$.

Due to intermittent transmissions of the components of y and u, the respective components of $\hat{y}$ and $\hat{u}$ are updated at time instants $t_1, t_2, \ldots, t_i, \ldots \in \mathcal{T}$, i.e.,[1]

$$\left. \begin{array}{l} \hat{y}(t_i^+) = y_t + h_y(t_i, e(t_i)) \\ \hat{u}(t_i^+) = u_t + h_u(t_i, e(t_i)) \end{array} \right\} \quad t_i \in \mathcal{T}, \tag{2.6}$$

where $h_y : \mathbb{R} \times \mathbb{R}^{n_e} \to \mathbb{R}^{n_y}$ and $h_u : \mathbb{R} \times \mathbb{R}^{n_e} \to \mathbb{R}^{n_u}$ model measurement noise, channel distortion and the underlying scheduling protocol. The roles of h_y and h_u are as follows. Suppose that the NCS has l links. Accordingly, the error vector e can be partitioned as $e := (e_1, \ldots, e_l)$. In order to avoid

---

[1] Recall that the jump times at the controller and plant end obey some scheduling scheme.

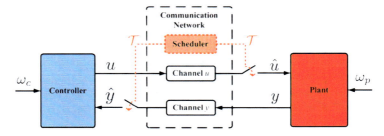

FIGURE 2.1
A diagram of a control system with the plant and controller interacting over a communication network with intermittent information updates. The two switches indicate that the information between the plant and controller are exchanged (complying with some scheduling protocol among the NCS links) at discrete time instants belonging to a set $\mathcal{T}$. The communication delays in each NCS link are time varying and, in general, different.

cumbersome indices, let us assume that each NCS link is characterized by its own delay. Hence, there are merely l (rather than $n_y + n_u$) different delays $d_i : \mathbb{R} \to \mathbb{R}_+$ in (2.5). Besides the already introduced upper bound d on $d_i(t)$'s, we assume that $d_i(t)$'s are differentiable with bounded $|\dot{d}_i(t)|$. As orchestrated by (2.6), if the jth NCS link is granted access to the communication medium at some $t_i \in \mathcal{T}$, the corresponding components of $(\hat{y}(t_i), \hat{u}(t_i))$ jump to the received values. It is to be noted that all other components of $(\hat{y}(t_i), \hat{u}(t_i))$ remain unaltered. Consequently, the related components of $e(t_i)$ reset to the noise $\nu_j(t_i)$ present in the received data, i.e.,

$$e_j(t_i^+) = \nu_j(t_i), \tag{2.7}$$

and we assume that

$$\sup_{t \in \mathbb{R}, j \in \{1,\ldots,l\}} \|\nu_j(t)\| = K_\nu.$$

Noise $\nu_j(t_i)$, which is embedded in h_y and h_u, models any discrepancy between the received values and their actual values at time $t_i - d_j(t)$ (when the jth NCS link of $(y(t), u(t))$ was sampled). As already indicated, this discrepancy can be a consequence of measurement noise and channel distortion. We point out that ν_j has nothing to do with ω_p nor ω_c. Observe that out-of-order packet arrivals, as a consequence of the time-varying delays, are allowed for.

Between transmissions, the values of $\hat{y}$ and $\hat{u}$ need not to be constant as in [68], but can be estimated in order to extend transmission intervals (consult [48] for more). In other words, for each $t \in [t_0, \infty) \setminus \mathcal{T}$ we have

$$\begin{aligned}\dot{\hat{y}} &= \hat{f}_p(t, x_{p_t}, x_{c_t}, \hat{y}_t, \hat{u}_t, \omega_p, \omega_c), \\ \dot{\hat{u}} &= \hat{f}_c(t, x_{p_t}, x_{c_t}, \hat{y}_t, \hat{u}_t, \omega_p, \omega_c),\end{aligned} \tag{2.8}$$

where the translation operators $\hat{y}_t$ and $\hat{u}_t$ are with delay d. The commonly used ZOH strategy is characterized by $\dot{\hat{y}} \equiv \mathbf{0}_{n_y}$ and $\dot{\hat{u}} \equiv \mathbf{0}_{n_u}$.

The following definition of UGES scheduling protocols is extracted from [131] and [68].

Definition 2.6. *Consider the noise-free setting, i.e., $K_\nu = 0$. The protocol given by $h := (h_y, h_u)$ is UGES if there exists a function $W : \mathbb{N}_0 \times \mathbb{R}^{n_e} \to \mathbb{R}_+$ such that $W(i, \cdot) : \mathbb{R}^{n_e} \to \mathbb{R}_+$ is locally Lipschitz (and hence almost everywhere differentiable) for every $i \in \mathbb{N}_0$, and if there exist positive constants $\underline{a}$, $\overline{a}$ and $0 \leq \rho < 1$ such that*

(i) $\underline{a}\|e\| \leq W(i,e) \leq \overline{a}\|e\|$, and

(ii) $W(i+1, h(t_i, e)) \leq \rho W(i, e)$,

for all $(i, e) \in \mathbb{N}_0 \times \mathbb{R}^{n_e}$.

Note that i in $W(i, e)$ counts jumps/transmissions and corresponds to $t_i \in \mathcal{T}$. Additionally, even though the delays could result from protocol arbitration, the delays are not a part of the UGES protocol definition [131, 68]. In addition, $\mathcal{T}$ is not a part of the protocol, but rather a consequence, as it is yet to be designed. Commonly used UGES protocols are the Round Robin (RR) and Try-Once-Discard protocol (TOD) (consult [131, 68, 35]). RR utilizes a cyclic user-defined deterministic scheme to grant the communication medium access while TOD grants the communication medium access to the NCS link with the greatest corresponding error in (2.5). Accordingly, RR is a static whilst TOD is a dynamic scheduling protocol. The corresponding constants are $\underline{a}_{RR} = 1$, $\overline{a}_{RR} = \sqrt{l}$, $\rho_{RR} = \sqrt{(l-1)/l}$ for RR and $\underline{a}_{TOD} = \overline{a}_{TOD} = 1$, $\rho_{TOD} = \sqrt{(l-1)/l}$ for TOD. Explicit expressions of the noise-free $h(t, e)$ for RR and TOD are provided in [131], but are not needed in the context of this chapter.

The properties imposed on the delay NCS in Figure 2.1 are summarized in the following standing assumption.

Assumption 2.1. *The jump times of the NCS links at the controller and plant end obey the underlying UGES scheduling protocol (characterized through h) and occur at transmission instants belonging to $\mathcal{T} := \{t_1, t_2, \ldots, t_i, \ldots\}$, where $\varepsilon \leq t_{i+1} - t_i \leq \tau$ for each $i \in \mathbb{N}_0$ with $\varepsilon > 0$ arbitrarily small. The received data is corrupted by measurement noise and/or channel distortion (characterized through h as well). In addition, each NCS link is characterized by the network-induced delay $d_i(t)$, $i \in \{1, \ldots, l\}$.*

The existence of a strictly positive τ, and therefore the existence of $\varepsilon > 0$, is demonstrated in Remark 2.3.

A typical closed-loop system (2.2)–(2.8) with continuous (yet delayed) information flows in all NCS links might be robustly stable (in the $\mathcal{L}_p$ sense according to (2.16)) only for some sets of $d_i(t)$, $i \in \{1, \ldots, l\}$. We refer to the family of such delay sets as the family of *admissible delays* and denote it $\mathcal{D}$.

Next, given some admissible delays $d_i(t)$, $i \in \{1, \ldots, l\}$, the maximal τ which renders $\mathcal{L}_p$-stability (with a desired gain) of the closed-loop system (2.2)–(2.8) is called MATI and is denoted $\bar{\tau}$. We are now ready to state the main problem studied herein.

Problem 2.1. *Given admissible delays $d_i(t)$, $i \in \{1, \ldots, l\}$, estimator (2.8) and the UGES protocol of interest, determine the MATI $\bar{\tau}$ to update components of $(\hat{y}, \hat{u})$ such that the NCS (2.2)–(2.8) is $\mathcal{L}_p$-stable with bias and a prespecified $\mathcal{L}_p$-gain for some $p \in [1, \infty]$.*

Remark 2.1. *Even though our intuition (together with the case studies provided herein and in [181]) suggests that merely "small enough" delays (including the zero delay) are admissible because the control performance impairs (i.e., the corresponding $\mathcal{L}_p$-gain increases) with increasing delays, this observation does not hold in general [61], [133, Chapter 1.], [164], and [52, Section 2.3]. In fact, "small" delays may destabilize some systems while "large" delays might destabilize others. In addition, even a second-order system with a single discrete delay might toggle between stability and instability as this delay is being decreased. Clearly, the family $\mathcal{D}$ needs to be specified on a case-by-case basis. Hence, despite the fact that the case studies presented herein and in [181] yield MATIs that hold for all smaller delays (including the zero delay) than the delays for which these MATIs are computed, it would be erroneous to infer that this property holds in general.*

2.4 Computing Maximally Allowable Transfer Intervals

Along the lines of [131], we rewrite the closed-loop system (2.2)–(2.8) in the following form amenable for small-gain theorem (see [88, Chapter 5]) analyses:

$$\left.\begin{aligned} x(t^+) &= x(t) \\ e(t^+) &= h(t, e(t)) \end{aligned}\right\} \quad t \in \mathcal{T} \tag{2.9a}$$

$$\left.\begin{aligned} \dot{x} &= f(t, x_t, e, \omega) \\ \dot{e} &= g(t, x_t, e_t, \omega_t) \end{aligned}\right\} \quad \text{otherwise,} \tag{2.9b}$$

where $x := (x_p, x_c)$, $\omega := (\omega_p, \omega_c)$, and functions f, g and h are given by (2.10)–(2.12). Even though (2.10)–(2.12) may seem quite convoluted, we point out that these expressions are obtained via straightforward calculations as illustrated with a number of examples throughout this book. We assume enough regularity on f and g to guarantee existence of the solutions on the interval of interest [13, Chapter 3]. Observe that the differentiability of $d_i(t)$'s and the boundedness of $|\dot{d}_i(t)|$ play an important role in attaining regularity of g. For

MATIs with Time-Varying Delays and Model-Based Estimators   23

the sake of simplicity, our notation does not explicitly distinguish between translation operators with delays d_p, d_c, d or $2d$ in (2.10)–(2.12) and in what follows. In this regard, we point out that the operators x_{p_t} and x_{c_t} are with delays d_p and d_c, respectively, the operators g_{p_t} and g_{c_t} within $\hat{f}_p$ and $\hat{f}_c$ are with delay $2d$ while all other operators are with delay d. In what follows we also use $\overline{d} := 2d + \max\{d_p, d_c\}$, which is the maximum value of all delay phenomena in (2.12).

For future reference, the delayed dynamics

$$x(t^+) = x(t) \quad \} \quad t \in \mathcal{T} \qquad (2.13a)$$
$$\dot{x} = f(t, x_t, e, \omega) \quad \} \quad \text{otherwise,} \qquad (2.13b)$$

are termed the *nominal system* Σ_n, and the impulsive delayed dynamics

$$e(t^+) = h(t, e(t)) \quad \} \quad t \in \mathcal{T} \qquad (2.14a)$$
$$\dot{e} = g(t, x_t, e_t, \omega_t) \quad \} \quad \text{otherwise,} \qquad (2.14b)$$

are termed the *error system* Σ_e. Observe that Σ_n contains delays, but does not depend on h or $\mathcal{T}$ as seen from (2.13). Instead, h and $\mathcal{T}$ constitute the error subsystem Σ_e as seen from (2.14).

The remainder of our methodology interconnects Σ_n and Σ_e using appropriate outputs. Basically, $W(i, e)$ from Definition 2.6 is the output of Σ_e while the output of Σ_n, denoted $H(x_t, \omega_t)$, is obtained from $g(t, x_t, e_t, \omega_t)$ and $W(i, e)$ as specified in Section 2.4.2. Notice that the outputs $H(x_t, \omega_t)$ and $W(i, e)$ are auxiliary signals used to interconnect Σ_n and Σ_e and solve Problem 2.1, but do not exist physically. Subsequently, the small-gain theorem is employed to infer $\mathcal{L}_p$-stability with bias.

2.4.1 $\mathcal{L}_p$-Stability with Bias of Impulsive Delayed LTI Systems

Before invoking the small-gain theorem in the upcoming subsection, let us establish conditions on the transmission interval τ and delay $d(t)$ that yield $\mathcal{L}_p$-stability with bias for a class of *impulsive delayed LTI systems*. Clearly, the results of this subsection are later applied to achieving $\mathcal{L}_p$-stability with bias and an appropriate $\mathcal{L}_p$-gain of Σ_e.

Consider the following class of impulsive delayed LTI systems

$$\dot{\xi}(t) = a\xi(t - d(t)) + \tilde{u}(t), \qquad t \notin \mathcal{T} \qquad (2.15a)$$
$$\xi(t^+) = c\xi(t) + \tilde{\nu}(t), \qquad t \in \mathcal{T}, \qquad (2.15b)$$

where $a \in \mathbb{R}$ and $c \in (-1, 1)$, initialized with some $\xi_{t_0} \in PC([-\check{d}, 0], \mathbb{R})$. In addition, $d(t)$ is a continuous function upper bounded by $\check{d}$ while $\tilde{u}, \tilde{\nu} : \mathbb{R} \to \mathbb{R}$ denote external inputs and $\tilde{\nu} \in \mathcal{L}_\infty$.

$$f(t, x_t, e, \omega) \overset{(2.2),(2.3)}{:=} \begin{bmatrix} f_p\left(t, x_{p_t}, \underbrace{g_{c_t}(t, x_{c_t}) + e_{u_t}(t)}_{=\hat{u}(t) \text{ using } (2.3) \text{ and } (2.5)}, \omega_p(t) \right) \\ f_c\left(t, x_{c_t}, \underbrace{g_{p_t}(t, x_{p_t}) + e_{y_t}(t)}_{=\hat{y}(t) \text{ using } (2.2) \text{ and } (2.5)}, \omega_c(t) \right) \end{bmatrix} =: \begin{bmatrix} f_1(t, x_t, e, \omega) \\ f_2(t, x_t, e, \omega) \end{bmatrix} \tag{2.10}$$

$$h(t, e(t)) := \begin{bmatrix} h_y(t, e(t)) \\ h_u(t, e(t)) \end{bmatrix} \tag{2.11}$$

$$g(t, x_t, e_t, \omega_t)$$
$$\overset{(2.5)}{:=}$$
$$\underbrace{\begin{bmatrix} f_p(t, x_{p_t}, x_{c_t}, g_{p_t}(t, x_{p_t}) + e_{y_t}, g_{c_t}(t, x_{c_t}) + e_{u_t}, \omega(t)) \\ f_c(t, x_{p_t}, x_{c_t}, g_{p_t}(t, x_{p_t}) + e_{y_t}, g_{c_t}(t, x_{c_t}) + e_{u_t}, \omega(t)) \end{bmatrix}}_{\text{model-based estimator } (2.8)} - \underbrace{\begin{pmatrix} \dfrac{\partial g_p}{\partial t} \end{pmatrix}_t (t, x_{p_t})}_{=-\hat{y}_t \text{ using } (2.2) \text{ and } (2.10)} - \underbrace{\begin{pmatrix} \dfrac{\partial g_p}{\partial x_p} \end{pmatrix}_t (t, x_{p_t}) f_{1_t}(t, x_t, e, \omega)}_{} - \underbrace{\begin{pmatrix} \dfrac{\partial g_c}{\partial t} \end{pmatrix}_t (t, x_{c_t})}_{} - \underbrace{\begin{pmatrix} \dfrac{\partial g_c}{\partial x_c} \end{pmatrix}_t (t, x_{c_t}) f_{2_t}(t, x_t, e, \omega)}_{=-\hat{u}_t \text{ using } (2.3) \text{ and } (2.10)} \tag{2.12}$$

Lemma 2.1. Assume $\tilde{u} \equiv 0$, $\tilde{\nu} \equiv 0$ and consider a positive constant r. In addition, let $\lambda_1 := \frac{a^2}{r}$, and $\lambda_2 := c^2$ for $c \neq 0$ or merely $\lambda_2 \in (0,1)$ for $c = 0$. If there exist constants $\lambda > 0$, $M > 1$ such that the conditions

(I) $\tau\left(\lambda + r + \lambda_1 M e^{-\lambda \tau}\right) < \ln M$, and

(II) $\tau\left(\lambda + r + \frac{\lambda_1}{\lambda_2} e^{\lambda \check{d}}\right) < -\ln \lambda_2$

hold, then the system (2.15) is UGES and $\|\xi(t)\| \leq \sqrt{M} \|\xi_{t_0}\| e^{-\frac{\lambda}{2}(t-t_0)}$ for all $t \geq t_0$.

The previous lemma, combined with the work presented in [4], results in the following theorem.

Theorem 2.1. Suppose that the system given by (2.15) is UGES with constants $\lambda > 0$ and $M > 1$ and that $\sup_{t \in \mathbb{R}} \|\tilde{\nu}(t)\| \leq \tilde{K}_\nu$. Then, the system (2.15) is $\mathcal{L}_p$-stable with bias $\frac{\tilde{K}_\nu \sqrt{M}}{e^{\frac{\lambda \varepsilon}{2}} - 1}$ from $\tilde{u}$ to ξ and with gain $\frac{2}{\lambda} \sqrt{M}$ for each $p \in [1, \infty]$.

2.4.2 Obtaining MATIs via the Small-Gain Theorem

We are now ready to state and prove the main result of this chapter. Essentially, we interconnect Σ_n and Σ_e via suitable outputs (i.e., $H(x_t, \omega_t)$ and $W(i,e)$, respectively) as depicted in Figure 2.2, impose the small-gain condition and invoke the small-gain theorem.

Theorem 2.2. Suppose the underlying UGES protocol, $d_1(t), \ldots, d_l(t)$ and $K_\nu \geq 0$ are given. In addition, assume that

(a) there exists a continuous function $H : PC([-\bar{d}, 0], \mathbb{R}^{n_x}) \times PC([-d, 0], \mathbb{R}^{n_\omega}) \to \mathbb{R}^m$ such that the system Σ_n given by (2.13) is $\mathcal{L}_p$-stable from (W, ω) to $H(x_t, \omega_t)$ for some $p \in [1, \infty]$, i.e., there exist $K_H, \gamma_H \geq 0$ such that

$$\|H[t, t_0]\|_p \leq K_H \|x_{t_0}\| + \gamma_H \|(W, \omega)[t, t_0]\|_p, \quad (2.16)$$

for all $t \geq t_0$, and

(b) there exists $L \geq 0$ and $d : \mathbb{R} \to \mathbb{R}_+$, $\sup_{t \in \mathbb{R}} d(t) = \check{d}$, such that for almost all $t \geq t_0$, almost all $e \in \mathbb{R}^{n_e}$ and for all $(i, x_t, \omega_t) \in \mathbb{N}_0 \times PC([-\bar{d}, 0], \mathbb{R}^{n_x}) \times PC([-d, 0], \mathbb{R}^{n_\omega})$ it holds that

$$\left\langle \frac{\partial W(i, e)}{\partial e}, g(t, x_t, e_t, \omega_t) \right\rangle \leq LW(i, e(t - d(t))) + \|H(x_t, \omega_t)\|. \quad (2.17)$$

Then, the NCS (2.9) is $\mathcal{L}_p$-stable with bias from ω to (H, e) for each τ for which there exist $M > 1$ and $\lambda > 0$ satisfying (I), (II) and $\frac{2}{\lambda}\sqrt{M}\gamma_H < 1$ with parameters $a = \frac{\bar{a}}{\underline{a}} L$ and $c = \rho$.

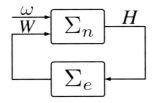

FIGURE 2.2
Interconnection of the nominal system Σ_n and error impulsive system Σ_e.

Remark 2.2. *According to Problem 2.1, condition (a) requires the underlying delays to be admissible, i.e., $\{d_1(t), \ldots, d_l(t)\} \in \mathcal{D}$. Condition (a) implies that the nominal system (i.e., the closed-loop system) is robust with respect to intermittent information and disturbances. Besides $\mathcal{L}_p$-stability, typical robustness requirements encountered in the literature include Input-to-State Stability (ISS) and passivity [184]. Condition (b) relates the current growth rate of $W(i,e)$ with its past values. As shown in Section 2.5, all recommendations and suggestions from [131] and [68] regarding how to obtain a suitable $W(i,e)$ readily apply because $W(i,e)$ characterizes the underlying UGES protocol (and not the plant-controller dynamics).*

Remark 2.3 (Zeno-freeness). *The left-hand side of conditions (I) and (II) from Lemma 2.1 are nonnegative continuous functions of $\tau \geq 0$ and approach ∞ as $\tau \to \infty$. Also, these left-hand sides equal zero for $\tau = 0$. Note that both sides of (I) and (II) are continuous in λ, M, λ_1, λ_2 and $\check{d}$. Hence, for every $\lambda > 0$, $\lambda_1 \geq 0$, $M > 1$, $\lambda_2 \in (0, 1)$ and $\check{d} \geq 0$ there exists $\tau > 0$ such that (I) and (II) are satisfied. Finally, since $\frac{2}{\lambda}\sqrt{M}$ is continuous in λ and M, we infer that for every finite $\gamma_H > 0$, there exists $\tau > 0$ such that $\frac{2}{\lambda}\sqrt{M}\gamma_H < 1$. In other words, for each admissible $d_i(t)$, $i \in \{1, \ldots, l\}$, the unwanted Zeno behavior is avoided and the proposed methodology does not yield continuous feedback, which might be impossible to implement. Notice that each τ yielding $\frac{2}{\lambda}\sqrt{M}\gamma_H < 1$ is a candidate for $\bar{\tau}$. Depending on r, λ_2, λ and M, the maximal such τ is in fact MATI $\bar{\tau}$.*

Remark 2.4. *The right-hand side of (2.17) might not be descriptive enough for many problems of interest. In general, (2.17) should be sought in the form $\left\langle \frac{\partial W(i,e)}{\partial e}, g(t, x_t, e_t, \omega_t) \right\rangle \leq \sum_{k=1}^{m} L_k W(i, e(t - \grave{d}_k(t))) + \|H(x_t, \omega_t)\|$, where $\grave{d}_k : \mathbb{R} \to \mathbb{R}_+$ and $m \geq 1$. As this general form leads to tedious computations (as evident from the proof of Lemma 2.1 in Section 2.7), we postpone its consideration for the future. For the time being, one can intentionally delay the communicated signals in order to achieve a single discrete delay $d(t)$ in (2.17). This idea is often found in the literature and can be accomplished via the time-stamping of data and introduction of buffers at receiver ends (refer to [73] and references therein).*

Remark 2.5. *Noisy measurements can be a consequence of quantization errors. According to [118], feedback control prone to quantization errors cannot yield closed-loop systems with linear $\mathcal{L}_p$-gains. Hence, the bias term in the linear gain $\mathcal{L}_p$-stability with bias result of Theorem 2.2 cannot be removed without contradicting the points in [118]. Further investigations of quantized feedback are high on our future research agenda.*

Remark 2.6. *Let us consider the case of lossy communication channels. If there is an upper bound on the maximum number of successive dropouts, say $N_d \in \mathbb{N}$, simply use $\frac{\tau}{N_d}$ as the transmission interval in order for Theorem 2.2 to hold. Moreover, the transmission instants among NCS links need not be (and often cannot be) synchronized. In this case, each NCS must transmit at a rate smaller than τ_{RR} (instead of $\tau_{RR}l$), where τ_{RR} is the MATI obtained for the RR protocol, in order to meet the prespecified performance requirements. Observe that this leads to asynchronous transmission protocols, which in turn increases the likelihood of packet collisions [117].*

Corollary 2.1. *Assume that the conditions of Theorem 2.2 hold and that x is $\mathcal{L}_p$-detectable from (W, ω, H). Then the NCS (2.9) is $\mathcal{L}_p$-stable with bias from ω to (x, e).*

In the following proposition, we provide conditions that yield UGS and GAS of the interconnection Σ_n and Σ_e. Recall that $\omega \equiv \mathbf{0}_{n_\omega}$ and $K_\nu = 0$ are the disturbance and noise settings, respectively, corresponding to UGS and GAS.

Proposition 2.1. *Assume that the interconnection of systems Σ_n and Σ_e, given by (2.13) and (2.14), is $\mathcal{L}_p$-stable from ω to (x, e). If $p = \infty$, then this interconnection is UGS. When $p \in [1, \infty)$, assume that $f(t, x_t, e, \mathbf{0}_{n_\omega})$ and $g(t, x_t, e_t, \tilde{\mathbf{0}}_{n_\omega})$ are (locally) Lipschitz uniformly in t and that $\|H(x_t, \tilde{\mathbf{0}}_{n_\omega})\| \to 0$ as $\|x_t\| \to 0$. Then, this interconnection is GAS.*

2.5 Numerical Examples: Batch Reactor, Planar System and Inverted Pendulum

2.5.1 Batch Reactor with Constant Delays

According to [68], the batch reactor case study has become a benchmark example in the area of NCSs over the years. Hence, we apply our work to this example and compare it with the approach presented in [68].

Consider the linearized model of an unstable batch reactor given by

$$\dot{x}_p = A_p x_p + B_p u, \qquad y = C_p x_p,$$

where

$$A_p = \begin{bmatrix} 1.38 & -0.2077 & 6.715 & -5.676 \\ -0.5814 & -4.29 & 0 & 0.675 \\ 1.067 & 4.273 & -6.654 & 5.893 \\ 0.048 & 4.273 & 1.343 & -2.104 \end{bmatrix}, B_p = \begin{bmatrix} 0 & 0 \\ 5.679 & 0 \\ 1.136 & -3.146 \\ 1.136 & 0 \end{bmatrix}, C_p = \begin{bmatrix} 1 & 0 & 1 & -1 \\ 0 & 1 & 0 & 0 \end{bmatrix},$$

and a PI controller given by

$$\dot{x}_c = A_c x_c + B_c y, \qquad u = C_c x_c + D_c y,$$

where

$$A_c = \begin{bmatrix} 0 & 0 \\ 0 & 0 \end{bmatrix}, B_c = \begin{bmatrix} 0 & 1 \\ 1 & 0 \end{bmatrix}, C_c = \begin{bmatrix} -2 & 0 \\ 0 & 8 \end{bmatrix}, D_c = \begin{bmatrix} 0 & -2 \\ 5 & 0 \end{bmatrix}.$$

In addition, assume that only the plant outputs y are transmitted via the network; hence, $e = e_y$ and $e_u \equiv \mathbf{0}_{n_u}$. Following the development of Section 2.4, we obtain

$$\dot{x}(t) = \begin{bmatrix} \dot{x}_p(t) \\ \dot{x}_c(t) \end{bmatrix} = \begin{bmatrix} A_p & B_p C_c \\ \mathbf{0}_{2\times 4} & A_c \end{bmatrix} x(t) + \begin{bmatrix} B_p D_c C_p & \mathbf{0}_{4\times 2} \\ B_c C_p & \mathbf{0}_{2\times 2} \end{bmatrix} x(t-d)$$
$$+ \begin{bmatrix} B_p D_c \\ B_c \end{bmatrix} e(t) + \begin{bmatrix} 10 & 0 & 10 & 0 & 0 & 0 \\ 0 & 5 & 0 & 5 & 0 & 0 \end{bmatrix}^\top \omega(t),$$

where the last term including $\omega(t)$ is a posteriori added by the authors in [68], and

$$\dot{e}(t) = \underbrace{\begin{bmatrix} -C_p B_p D_c \end{bmatrix}}_{A_1} e(t-d) + \underbrace{\begin{bmatrix} -C_p A_p & -C_p B_p C_c \end{bmatrix}}_{A_2} x(t-d) +$$
$$+ \underbrace{\begin{bmatrix} -C_p B_p D_c C_p & \mathbf{0}_{2\times 2} \end{bmatrix}}_{A_3} x(t-2d),$$

with A_1 being diagonal

$$A_1 = \begin{bmatrix} 15.73 & 0 \\ 0 & 11.3580 \end{bmatrix}.$$

Since the example in [68] does not consider noisy information, one can set $K_\nu = 0$. In other words, bias $b(t) \equiv b = 0$ and all $\mathcal{L}_2$-stability results in this subsection are without bias. Likewise, in order to provide a better comparison with [68], we assume that the received values of y and u given by $\hat{y}$ and $\hat{u}$, respectively, remain constant between updates (see (2.12)). This strategy is known as the Zero-Order Hold (ZOH) strategy. Let us point out that the ZOH strategy is an integral part of the small-delay approach in [68]; hence, the ZOH assumption in [68] cannot be trivially relaxed.

According to [131] and [68], we select $W_{RR}(i,e) := \|D(i)e\|$ and $W_{TOD}(t,e) := \|e\|$, where $D(i)$ is a diagonal matrix whose diagonal elements are upper bounded by $\sqrt{l}$. Next, we determine L_{RR}, $H_{RR}(x,\omega,d)$, L_{TOD} and $H_{TOD}(x,\omega,d)$ as follows:

$$\left\langle \frac{\partial W_{RR}(i,e)}{\partial e}, \dot{e} \right\rangle \leq \|D(i)\dot{e}\| =$$
$$= \|D(i)\big(A_1 e(t-d) + A_2 x(t-d) + A_3 x(t-2d)\big)\|$$
$$\leq \underbrace{\|A_1\|}_{L_{RR}} \underbrace{\|D(i)e(t-d)\|}_{W_{RR}(i,e(t-d))} + \underbrace{\sqrt{l}\|A_2 x(t-d) + A_3 x(t-2d)\|}_{H_{RR}(x,\omega,d)},$$

$$\left\langle \frac{\partial W_{TOD}(i,e)}{\partial e}, \dot{e} \right\rangle = \frac{1}{W_{TOD}(i,e)} \Big(e(t)^\top A_1 e(t-d) +$$
$$+ e(t)^\top \big(A_2 x(t-d) + A_3 x(t-2d)\big) \Big)$$
$$\leq \frac{1}{W_{TOD}(i,e)} \Big(\|e(t)\|\|A_1\|\|e(t-d)\| +$$
$$+ \|e(t)\|\|A_2 x(t-d) + A_3 x(t-2d)\| \Big)$$
$$= \underbrace{\|A_1\|}_{L_{TOD}} \underbrace{\|e(t-d)\|}_{W_{TOD}(i,e(t-d))} + \underbrace{\|A_2 x(t-d) + A_3 x(t-2d)\|}_{H_{TOD}(x,\omega,d)}.$$

In order to compute γ_H for $p=2$, we use the software HINFN [124]. However, HINFN is not able to handle delayed states in the output (see H_{RR} and H_{TOD}). In order to accommodate H_{RR} and H_{TOD} to HINFN, we first compute an estimate of the $\mathcal{L}_2$-gain, say L_1, from (e,ω) to $y_1 := A_2 x(t)$, and then an estimate of the $\mathcal{L}_2$-gain, say L_2, from (e,ω) to $y_2 := A_3 x(t)$. Afterward, we simply add together those two gains because

$$\|y_1(t)[t_0,\infty) + y_2(t-d)[t_0,\infty)\|_p = \|y_1(t)[t_0,\infty) + y_2(t)[t_0+d,\infty)\|_p$$
$$\leq \|y_1(t)[t_0,\infty)\|_p + \|y_2(t)[t_0,\infty)\|_p$$
$$\leq \underbrace{(L_1 + L_2)}_{\gamma_{H_{TOD}} \leq} \|(e,\omega)[t_0,\infty)\|_p,$$

where we used the fact that the $\mathcal{L}_2$-gain of the time-delay operator $\mathcal{D}_d$ is less than unity (refer to [87] for more) and that $y_2(t-d) = 0$ when $t-d < t_0$. Notice that another d seconds delay added to $\|y_1(t) + y_2(t-d)\|$ is exactly H_{TOD} and the corresponding $\mathcal{L}_p$-gain $\gamma_{H_{TOD}}$ remains unaltered (i.e., $\gamma_{H_{TOD}}$ is less or equal to $L_1 + L_2$). In addition, from the form of H_{RR} we infer that we can use $\gamma_{H_{RR}} = \sqrt{l}\gamma_{H_{TOD}}$. According to HINFN, the maximally admissible delay $\bar{d}$ is 40 ms. In addition, let us choose the controlled output z given by $z(t) = Cx(t)$ where

$$C = \begin{bmatrix} 1 & 0 & 1 & -1 & 0 & 0 \\ 0 & 1 & 0 & 0 & 0 & 0 \end{bmatrix}.$$

One goal of [68] is to determine pairs $(d,\bar{\tau})$ such that the $\mathcal{L}_2$-gain, denoted γ_z,

		UGAS		
		$d=0$	$d=6.3$	$d=40$
our work	RR	$\bar{\tau}=6.5$	$\bar{\tau}=3.6$	$\bar{\tau}=1.3$
	TOD	$\bar{\tau}=7.8$	$\bar{\tau}=5.3$	$\bar{\tau}=2.8$
[68]	RR	$\bar{\tau}_{[68]}=8.9$	N/A	N/A
	TOD	$\bar{\tau}_{[68]}=10.8$	$\bar{\tau}_{[68]}=8.3$	N/A

TABLE 2.1
Comparison between our methodology and [68] for UGAS. All delays d and MATIs $\bar{\tau}$ are measured in milliseconds.

		$\mathcal{L}_2$-gain $\gamma_z=200$	
		$d=10$	$d=40$
our work	RR	$\bar{\tau}=1.6$	$\bar{\tau}=0.18$
	TOD	$\bar{\tau}=2.6$	$\bar{\tau}=0.64$
[68]	RR	N/A	N/A
	TOD	N/A	N/A

TABLE 2.2
Comparison of our methodology and [68] for $\mathcal{L}_2$-stability. Delays d and MATIs $\bar{\tau}$ are expressed in milliseconds.

from ω to z is below a certain value. This goal is achieved in a similar manner as Corollary 2.1.

When one is interested merely in asymptotic stability, then the input to Σ_n is e instead of (e,ω) because, for this particular example, the corresponding γ_H is smaller (i.e., $\bar{\tau}$ is greater). Owing to Proposition 2.1, we infer UGAS. When interested in UGAS, HINFN suggests that $\bar{d}$ is 2.34 s. However, the corresponding $\bar{\tau}$ is extremely small (the order of magnitude is 10^{-200} s for both RR and TOD) and has no practical merit but rather confirms Remark 2.3.

In Tables 2.1 and 2.2, a comparison between the methodology presented herein and in [68] is provided. We consider the case without ω, which renders UGAS, and the case with ω, which renders $\mathcal{L}_2$-stability from ω to z with a prespecified gain γ_z. Notice that the maximal possible $\bar{\tau}_{[68]}$, obtained with TOD for $d=0$ ms and when interested in UGAS, is 10.8 ms. Hence, the maximal theoretical delay that can be considered in [68] for TOD is $d=10.8$ ms while for RR is $d=8.9$ ms. Recall that this corresponds to the small delay case. We point out that our methodology is able to consider delays that are significantly greater than 10.8 ms even when $\mathcal{L}_p$-stability is of interest (refer to Tables 2.1 and 2.2). In fact, provided that Σ_n is $\mathcal{L}_p$-stable for some $d \geq 0$ (recall that each such d is an admissible delay), there exists a stabilizing $\tau > 0$ (see Remark 2.3). As can be concluded from Tables 2.1 and 2.2, those MATIs that can be computed in [68] are about 50% greater than ours for the same d.

2.5.2 Planar System with Constant Delays

The following example is motivated by [186, Example 2.2] and all the results are provided for $p = 2$. It is straightforward to show that [186, Example 2.2] is not $\mathcal{L}_2$-stable when the error vector e is considered as the input. In addition, [186, Example 2.2] does not contain disturbance terms. Therefore, we introduce disturbance ω and modify [186, Example 2.2] to make it (more precisely, its nominal system (2.13)) $\mathcal{L}_2$-stable in the sense of (2.16) with delay-independent $\mathcal{L}_2$-gain. Basically, consider the following nonlinear delayed plant (compare with (2.2))

$$\begin{bmatrix} \dot{x}_{p1}(t) \\ \dot{x}_{p2}(t) \end{bmatrix} = \begin{bmatrix} -0.5x_{p1}(t)+x_{p2}(t)-0.25x_{p1}(t)\sin\left(u(t)x_{p2}(t-d_{p1})\right) \\ x_{p1}(t)\sin\left(u(t)x_{p2}(t-d_{p1})\right)+1.7x_{p2}(t-d_{p2})+u(t)-x_{p2}(t) \end{bmatrix} + \begin{bmatrix} \omega_1(t) \\ \omega_2(t) \end{bmatrix}$$

controlled with (compare with (2.3))

$$u(t) = -2x_{p1}(t) - 2x_{p2}(t).$$

As this controller is without internal dynamics, therefore $x(t) := x_p(t) = (x_{p1}(t), x_{p2}(t))$. Additionally, $\omega(t) := (\omega_1(t), \omega_2(t))$.

Let us consider the NCS setting in which noisy information regarding x_{p1} and x_{p2} are transmitted over a communication network while the control signal is not transmitted over a communication network nor distorted (i.e., $\hat{u} = u$). Anyway, since we are considering a static controller (i.e., no state on its own), a potential NCS link for transmitting u can readily be embedded in NCS links from the plant to the controller. In addition, consider that the information regarding x_{p2} arrives at the controller with delay d while information regarding x_{p1} arrives in timely manner. For the sake of simplicity, let us take $d = d_{p2}$. Apparently, the output of the plant is $y(t) = x_p(t) = x(t)$ and there are two NCS links so that $l = 2$. Namely, x_{p1} is transmitted through one NCS link while x_{p2} is transmitted through the second NCS link. The repercussions of these two NCS links are modeled via the following error vector (compare with (2.5))

$$e = \begin{bmatrix} e_1 \\ e_2 \end{bmatrix} = \hat{y} - \underbrace{\left(\begin{bmatrix} x_{p1}(t) \\ 0 \end{bmatrix} + \begin{bmatrix} 0 \\ x_{p2}(t-d) \end{bmatrix} \right)}_{y_t}.$$

The expressions (2.10) and (2.12) for this example become:

$$\dot{x}(t) = \underbrace{\begin{bmatrix} -0.5 & 1 \\ -2 & -1 \end{bmatrix}}_{A_1} x(t) + \underbrace{\begin{bmatrix} 0 & 0 \\ 0 & -0.3 \end{bmatrix}}_{A_2} x(t-d) +$$

$$+ \underbrace{\begin{bmatrix} -0.25 & 0 \\ 1 & 0 \end{bmatrix}}_{B_1} x(t) N(x_t, e) + \underbrace{\begin{bmatrix} 0 & 0 \\ -2 & -2 \end{bmatrix}}_{B} e(t) + \omega(t), \qquad (2.18)$$

$$\left\langle \frac{\partial W_{RR}(i,e)}{\partial e}, \dot{e} \right\rangle \leq \|D(i)\dot{e}\| \leq \underbrace{\sqrt{l}\|B\|}_{L_{RR}} \underbrace{\|D(i)e(t-d)\|}_{W_{RR}(i,e(t-d))} +$$

$$+ \underbrace{\sqrt{l}\left(\|C_1 x(t) + C_2 x(t-d) + C_3 x(t-2d) + C_6 \omega(t) + C_7 \omega(t-d)\| + \|C_4 x(t)\| + \|C_5 x(t-d)\|\right)}_{H_{RR}(x_t, \omega_t)},$$
(2.19)

$$\left\langle \frac{\partial W_{TOD}(i,e)}{\partial e}, \dot{e} \right\rangle \leq \underbrace{\|B\|}_{L_{TOD}} \underbrace{\|e(t-d)\|}_{W_{TOD}(i,e(t-d))} +$$

$$+ \underbrace{\left(\|C_1 x(t) + C_2 x(t-d) + C_3 x(t-2d) + C_6 \omega(t) + C_7 \omega(t-d)\| + \|C_4 x(t)\| + \|C_5 x(t-d)\|\right)}_{H_{TOD}(x_t, \omega_t)}.$$
(2.20)

$$\dot{e}(t) = \dot{\hat{y}} - Be(t-d) + \underbrace{\begin{bmatrix} 0.5 & -1 \\ 0 & 0 \end{bmatrix}}_{C_1} x(t) + \underbrace{\begin{bmatrix} 0 & 0 \\ 2 & 2 \end{bmatrix}}_{C_2} x(t-d) +$$

$$+ \underbrace{\begin{bmatrix} 0 & 0 \\ 0 & 0.3 \end{bmatrix}}_{C_3} x(t-2d) + \underbrace{\begin{bmatrix} 0.25 & 0 \\ 0 & 0 \end{bmatrix}}_{C_4} x(t) N(x_t, e) +$$

$$+ \underbrace{\begin{bmatrix} 0 & 0 \\ -1 & 0 \end{bmatrix}}_{C_5} x(t-d) N(x_t, e_t) + \underbrace{\begin{bmatrix} -1 & 0 \\ 0 & 0 \end{bmatrix}}_{C_6} \omega(t) + \underbrace{\begin{bmatrix} 0 & 0 \\ 0 & -1 \end{bmatrix}}_{C_7} \omega(t-d),$$

where $N(x_t, e) := \sin\left(\left[-2(x_{p1}(t)+e_1(t))-2(x_{p2}(t-d)+e_2(t))\right]x_{p2}(t-d_{p1})\right)$

and $N(x_t, e_t) := \sin\left(\left[-2(x_{p1}(t-d)+e_1(t-d))-2(x_{p2}(t-2d)+e_2(t-d))\right]x_{p2}(t-d_{p1}-d)\right)$.

According to [131] and [68], we select $W_{RR}(i,e) := \|D(i)e\|$ and $W_{TOD}(t,e) := \|e\|$, where $D(i)$ is a diagonal matrix whose diagonal elements are lower bounded by 1 and upper bounded by $\sqrt{l}$. Next, we determine L_{RR}, $H_{RR}(x,\omega,d)$, L_{TOD} and $H_{TOD}(x,\omega,d)$ from Theorem 2.2 for the ZOH strategy (i.e., $\dot{\hat{y}} \equiv \mathbf{0}_{n_y}$) obtaining (2.19) and (2.20). In order to estimate γ_H, we utilize Lyapunov–Krasovskii functionals according to [29, Chapter 6] and [38]. Basically, if there exist $\gamma \geq 0$ and a Lyapunov–Krasovskii functional $V(x_t)$ for the nominal system (2.13), that is (2.18), with the input (W, ω) and the output H such that its time derivative along the solution of (2.13) with a zero initial condition satisfies

$$\dot{V}(x_t) + H^\top H - \gamma^2 (W,\omega)^\top (W,\omega) \leq 0, \qquad \forall x_t \in C([-\overline{d}, 0], \mathbb{R}^{n_x}), \quad (2.21)$$

$$\begin{bmatrix} A_1^\top C + CA_1 + E + N(x_t,e)(B_1^\top C + CB_1) + C_1^\top C_1 & CA_2 + C_1^\top C_2 & CB & C + C_1^\top C_6 \\ A_2^\top C + C_2^\top C_1 & -E + C_2^\top C_2 & 0_{2\times 2} & C_2^\top C_6 \\ B^\top C & 0_{2\times 2} & -\gamma^2 I_2 & 0_{2\times 2} \\ C + C_6^\top C_1 & C_6^\top C_2 & 0_{2\times 2} & -\gamma^2 I_2 + C_6^\top C_6 \end{bmatrix} \leq 0, \quad (2.23)$$

then the corresponding $\mathcal{L}_2$-gain γ_H is less than γ. Refer to [133] for more regarding time-derivatives of functionals along solutions. The functional used herein is

$$V(x_t) = x(t)^\top C x(t) + \int_{-d}^{0} x(t+\theta)^\top E x(t+\theta) d\theta, \quad (2.22)$$

where C and E are positive-definite symmetric matrices.

Next, we illustrate the steps behind employing (2.22). Let us focus on TOD (i.e., the input is (e, ω)) and the output $C_1 x(t) + C_2 x(t-d) + C_6 \omega(t)$. The same procedure is repeated for the remaining terms of $H_{RR}(x_t, \omega_t)$ and $H_{TOD}(x_t, \omega_t)$. For the Lyapunov–Krasovskii functional in (2.22), the expression (2.21) boils down to the Linear Matrix Inequality (LMI) (see [29] for more) given by (2.23). For a comprehensive discussion regarding LMIs, refer to [29]. Notice that the above LMI has to hold for all $N(x_t, e) \in [-1, 1]$. Using the LMI Toolbox in MATLAB®, we find that the minimal γ for which (2.23) holds is in fact γ_H. For our TOD example and the specified output, we obtain $\gamma_H = 18.7051$. This γ_H holds for all $d \geq 0$. In other words, any $d \geq 0$ is an admissible delay and belongs to the family $\mathcal{D}$. For RR, simply multiply γ_H by $\sqrt{2}$.

Detectability of x from (W, x, H), which is a condition of Corollary 2.1, is easily inferred by taking $x(t)$ to be the output of the nominal system and computing the respective $\mathcal{L}_2$-gain γ_d. Next, let us take the output of interest to be x and find MATIs that yield the desired $\mathcal{L}_p$-gain from ω to x to be $\gamma_{\text{des}} = 50$. Combining (2.56) with γ_d leads to the following condition

$$\gamma_W \gamma_H < 1 - \frac{\gamma_d}{\gamma_{\text{des}}}$$

that needs to be satisfied (by changing γ_W through changing MATIs) in order to achieve the desired gain γ_{des}. In addition, observe that the conditions of Proposition 2.1 hold (and the closed-loop system is an autonomous system) so that we can infer UGAS when $\omega \equiv 0_{n_\omega}$ and $K_\nu = 0$.

Let us now introduce the following estimator (compare with (2.8))

$$\dot{\hat{y}} = B\hat{y}(t-d) = B\left(e(t-d) + \begin{bmatrix} 1 & 0 \\ 0 & 0 \end{bmatrix} x(t-d) + \begin{bmatrix} 0 & 0 \\ 0 & 1 \end{bmatrix} x(t-2d)\right), \quad (2.24)$$

which can be employed when one is interested in any of the three performance objectives (i.e., UGAS, $\mathcal{L}_p$-stability or $\mathcal{L}_p$-stability with a desired gain).

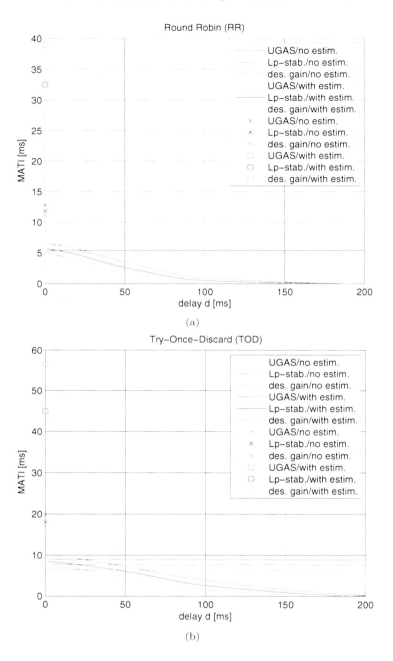

FIGURE 2.3
Numerically obtained MATIs for different delay values $d \geq 0$ in scenarios with and without estimation: (a) RR and (b) TOD.

Figure 2.3 provides evidence that the TOD protocol results in greater MATIs (at the expense of additional implementation complexity/costs) and that the model-based estimators significantly prolong MATIs, when compared with the ZOH strategy, especially as d increases. We point out that a different estimator (such as $\hat{y} = kB\hat{y}(t-d)$ for some $k \in \mathbb{R}$) can be employed as d approaches zero (because the estimator slightly decreases the MATIs as seen in Figure 2.3) to render greater MATIs in comparison with the scenarios without estimation. However, we do not modify our estimator as d approaches zero for the sake of clarity. In addition, notice that the case $d = 0$ boils down to ODE modeling so that we can employ less conservative tools for computing $\mathcal{L}_2$-gains. Accordingly, the 4×4 LMI given by (2.23) becomes a 3×3 LMI resulting in a smaller γ_H. Furthermore, the constant a in Theorem 2.2 becomes L, rather than $\frac{\bar{a}}{a}L$, which in turn decreases γ_W for the same τ. Apparently, MATIs pertaining to UGAS are greater than the MATIs pertaining to $\mathcal{L}_p$-stability from ω to (x, e) and these are greater than the MATIs pertaining to $\mathcal{L}_p$-stability from ω to x with $\gamma_{\text{des}} = 50$.

For completeness, we provide the gains used to obtain Figure 2.3: $\gamma_{H,TOD} = 9.6598$ for UGAS with ZOH and $d = 0$; $\gamma_{H,TOD} = 4.3344$ for UGAS with estimation and $d = 0$; $\gamma_{H,TOD} = 22.3631$ for UGAS with ZOH and $d > 0$; $\gamma_{H,TOD} = 27.3659$ for UGAS with estimation and $d > 0$; $\gamma_{H,TOD} = 10.8958$ for $\mathcal{L}_p$-stability with ZOH and $d = 0$; $\gamma_{H,TOD} = 5.3258$ for $\mathcal{L}_p$-stability with estimation and $d = 0$; $\gamma_{H,TOD} = 26.4601$ for $\mathcal{L}_p$-stability with ZOH and $d > 0$; $\gamma_{H,TOD} = 31.7892$ for $\mathcal{L}_p$-stability with estimation and $d > 0$; $\gamma_d = 3.5884$ for $d = 0$; and $\gamma_d = 7.9597$ for $d > 0$. Recall that $\gamma_{H,RR} = \sqrt{2}\gamma_{H,TOD}$. For $\gamma_{H,RR}$, simply multiply $\gamma_{H,TOD}$ with $\sqrt{2}$.

2.5.3 Inverted Pendulum with Time-Varying Delays

The following example is taken from [170, 183] and the results are provided for $p = 2$. Consider the inverted pendulum (compare with (2.2)) given by

$$\dot{x}_{p1} = x_{p2} + \omega_1$$
$$\dot{x}_{p2} = \frac{1}{L}(-g\cos(x_{p1}) + u) + \omega_2,$$

where $g = 9.8$ and $L = 2$, controlled with

$$u = -L\lambda x_{p2} + g\cos(x_{p1}) - K(x_{p2} + \lambda x_{p1}),$$

where $K = 50$ and $\lambda = 1$. Clearly, the control system goal is to keep the pendulum at rest in the upright position. Because this controller is without internal dynamics, $x(t) := x_p(t) = (x_{p1}(t), x_{p2}(t))$. Additionally, $\omega(t) := (\omega_1(t), \omega_2(t))$.

Consider the NCS setting in which noisy information regarding x_{p1} and x_{p2} are transmitted over a communication network while the control signal is not transmitted over a communication network nor distorted (i.e., $\hat{u} = u$). In addition, consider that the information regarding x_{p2} arrives at the controller with delay $d(t) \leq \check{d}$ and $|\dot{d}(t)| \leq \check{d}_1$ while information regarding x_{p1} arrives instantaneously. Apparently, the output of the plant is $y(t) = x_p(t) = x(t)$ and there are two NCS links so that $l = 2$. Namely, x_{p1} is transmitted through one NCS link while x_{p2} is transmitted through the second NCS link. The repercussions of these two NCS links are modeled via the following error vector (compare with (2.5))

$$e = \begin{bmatrix} e_1 \\ e_2 \end{bmatrix} = \hat{y} - \underbrace{\left(\begin{bmatrix} x_{p1}(t) \\ 0 \end{bmatrix} + \begin{bmatrix} 0 \\ x_{p2}(t-d(t)) \end{bmatrix} \right)}_{y_t}.$$

The expressions (2.10) and (2.12) for this example become

$$\dot{x}(t) = \underbrace{\begin{bmatrix} 0 & 1 \\ \frac{-K\lambda}{L} & -1 \end{bmatrix}}_{A_1} x(t) + \underbrace{\begin{bmatrix} 0 & 0 \\ 0 & \frac{-K}{L} - \lambda L \end{bmatrix}}_{A_2} x(t-d(t)) +$$

$$+ \begin{bmatrix} 0 \\ n(x_1(t), e_1(t)) \end{bmatrix} + \underbrace{\begin{bmatrix} 0 & 0 \\ \frac{-K\lambda}{L} & \frac{-K}{L} - \lambda L \end{bmatrix}}_{B} e(t) + \omega(t),$$

$$\dot{e}(t) = \dot{\hat{y}} + \underbrace{\begin{bmatrix} 0 & -1 \\ 0 & 0 \end{bmatrix}}_{B_1} x(t) + \underbrace{\begin{bmatrix} -1 & 0 \\ 0 & 0 \end{bmatrix}}_{C_1} \omega(t) + \Bigg(-Be(t-d(t)) +$$

$$+ \underbrace{\begin{bmatrix} 0 & 0 \\ \frac{K\lambda}{L} & 0 \end{bmatrix}}_{B_2} x(t-d(t)) - \begin{bmatrix} 0 \\ n(x_1(t-d(t)), e_1(t-d(t))) \end{bmatrix} +$$

$$+ \underbrace{\begin{bmatrix} 0 & 0 \\ 0 & -1 \end{bmatrix}}_{C_2} \omega(t-d(t)) + \underbrace{\begin{bmatrix} 0 & 0 \\ 0 & \frac{K}{L} + \lambda L \end{bmatrix}}_{B_3} x(t-2d(t)) \Bigg)(1-\dot{d}(t)),$$

where $n(x_1(t), e_1(t)) = \frac{-2g}{L} \sin\left(\frac{e_1(t)+2x_1(t)}{2}\right) \sin\left(\frac{e_1(t)}{2}\right)$.

According to [131] and [68], we select $W_{RR}(i, e) := \|D(i)e\|$ and $W_{TOD}(t, e) := \|e\|$, where $D(i)$ is a diagonal matrix whose diagonal elements are lower bounded by 1 and upper bounded by $\sqrt{l}$. Next, we determine L_{RR}, $H_{RR}(x, \omega, d)$, L_{TOD} and $H_{TOD}(x, \omega, d)$ from Theorem 2.2 for the ZOH strategy (i.e., $\dot{\hat{y}} \equiv \mathbf{0}_{n_y}$) obtaining (2.25) and (2.26).

The Lyapunov–Krasovskii functional used for the pendulum example is

$$V(t, x_t, \dot{x}_t) = x(t)^\top P x(t) + \int_{t-\check{d}}^{t} x(s)^\top S x(s) \mathrm{d}s +$$

$$+ \check{d} \int_{-\check{d}}^{0} \int_{t+\theta}^{t} \dot{x}(s)^\top R \dot{x}(s) \mathrm{d}s \mathrm{d}\theta + \int_{t-d(t)}^{t} x(s)^\top Q x(s) \mathrm{d}s,$$

MATIs with Time-Varying Delays and Model-Based Estimators

$$\left\langle \frac{\partial W_{RR}(i,e)}{\partial e}, \dot{e} \right\rangle \leq \|D(i)\dot{e}\| \leq \underbrace{\sqrt{l}(1+\check{d}_1)\left(\|B\| + \frac{g}{L}\right)}_{L_{RR}} \underbrace{\|D(i)e(t-d(t))\|}_{W_{RR}(i,e(t-d(t)))} +$$

$$+ \sqrt{l}\underbrace{\Big(\|B_1 x(t) + C_1 \omega(t) + (1-\dot{d}(t))(B_2 x(t-d(t)) + B_3 x(t-2d(t)) + C_2 \omega(t-d(t)))\|\Big)}_{H_{RR}(x_t,\omega_t)},$$
(2.25)

$$\left\langle \frac{\partial W_{TOD}(i,e)}{\partial e}, \dot{e} \right\rangle \leq \underbrace{(1+\check{d}_1)\left(\|B\| + \frac{g}{L}\right)}_{L_{TOD}} \underbrace{\|e(t-d(t))\|}_{W_{TOD}(i,e(t-d(t)))} +$$

$$+ \underbrace{\Big(\|B_1 x(t) + C_1 \omega(t) + (1-\dot{d}(t))(B_2 x(t-d(t)) + B_3 x(t-2d(t)) + C_2 \omega(t-d(t)))\|\Big)}_{H_{TOD}(x_t,\omega_t)},$$
(2.26)

where P is a positive-definite symmetric matrix while S, R and Q are positive-semidefinite symmetric matrices. Next, we take the output of interest to be x and find MATIs that yield the desired $\mathcal{L}_p$-gain from ω to x to be $\gamma_{\text{des}} = 15$. In addition, observe that the conditions of Proposition 2.1 hold (and the closed-loop system is an autonomous system) so that we can infer UGAS when $\omega \equiv \mathbf{0}_{n_\omega}$ and $K_\nu = 0$.

We use the following estimator (compare with (2.8))

$$\dot{\hat{y}} = B\hat{y}(t-d(t))(1-\dot{d}(t)) = B\left(e(t-d(t)) + \begin{bmatrix} 1 & 0 \\ 0 & 0 \end{bmatrix} x(t-d(t)) + \right.$$

$$\left. + \begin{bmatrix} 0 & 0 \\ 0 & 1 \end{bmatrix} x(t-2d(t)) \right)(1-\dot{d}(t)),$$

which can be employed in any of the three performance objectives (i.e., UGAS, $\mathcal{L}_p$-stability or $\mathcal{L}_p$-stability with a desired gain) provided $d(t)$ is known. One can use the ideas from Remark 2.4 toward obtaining known delays.

Figures 2.4 and 2.5 provide evidence that the TOD protocol results in greater MATIs (at the expense of additional implementation complexity/costs) and that the model-based estimators significantly prolong MATIs, when compared with the ZOH strategy. In addition, notice that the case $\check{d} = 0$ boils down to ODE modeling, so that we can employ less conservative tools for computing $\mathcal{L}_2$-gains. Apparently, MATIs pertaining to UGAS are greater than the MATIs pertaining to $\mathcal{L}_p$-stability from ω to (x, e) and these are greater than the MATIs pertaining to $\mathcal{L}_p$-stability from ω to x with $\gamma_{\text{des}} = 15$. As expected, time-varying delays upper bounded with some $\check{d}$ lead to smaller MATIs when compared to constant delays $\check{d}$. It is worth mentioning that $\check{d} = 33$ ms is the maximal value for which we are able to establish condition (a) of Theorem 2.2. Consequently, the delays from Figures 2.4 and 2.5 are instances

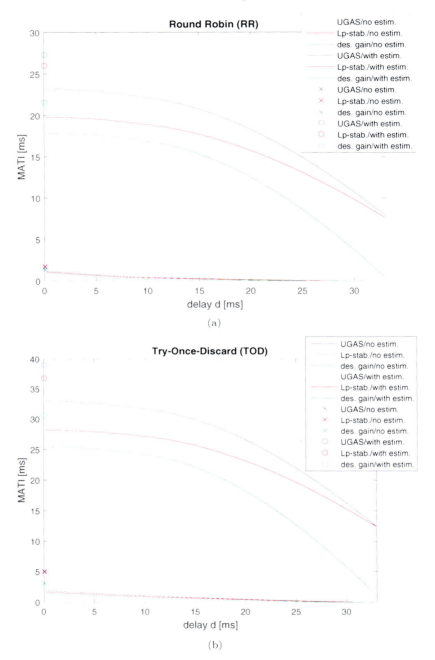

FIGURE 2.4
Numerically obtained MATIs for various constant delay values $d \geq 0$ in scenarios with and without estimation: (a) RR and (b) TOD.

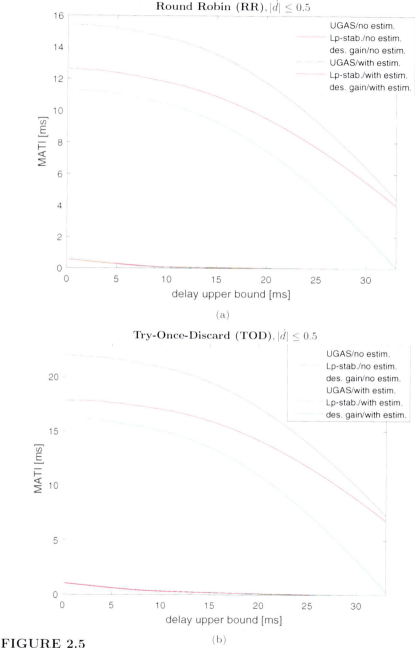

FIGURE 2.5
Numerically obtained MATIs for various time-varying delays $d(t)$ such that $d(t) \leq \check{d}$ and $|\dot{d}(t)| \leq \check{d}_1 = 0.5$ in scenarios with and without estimation: (a) RR and (b) TOD.

of admissible delays. The exhaustive search for admissible delays is an open problem that is out of the scope of this chapter.

2.6 Conclusions and Perspectives

In this chapter, we studied how much information exchange between a plant and controller can become intermittent (in terms of MATIs) such that the performance objectives of interest are not compromised. Depending on the noise and disturbance setting, the performance objective can be UGAS or $\mathcal{L}_p$-stability (with a prespecified gain and toward the output of interest). Our framework incorporates time-varying delays and transmission intervals that can be smaller than the delays, nonlinear plants/controllers with delayed dynamics, external disturbances (or modeling uncertainties), UGES scheduling protocols (e.g., the RR and TOD protocols), distorted data and model-based estimators. As expected, the TOD protocol results in greater MATIs than the RR protocol. Likewise, estimation (rather than the ZOH strategy) between two consecutive transmission instants extends the MATIs.

Regarding potential future research avenues, one can devise conditions rendering $\mathcal{L}_p$-stability of the error dynamics involving several time-varying delays (see Remark 2.4). In addition, in light of [44], we plan to design event- and self-triggering for delayed NCS with scheduling protocols. Observe that Chapter 3 does not consider delays or scheduling protocols.

2.7 Proofs of Main Results

2.7.1 Proof of Lemma 2.1

This proof follows the exposition in [208]. The following two definitions regarding (2.1) are utilized in this proof and are taken from [208].

Definition 2.7 (Lyapunov Function). *The function $V : [t_0, \infty) \times \mathbb{R}^{n_\xi} \to \mathbb{R}^+$ is said to belong to the class ν_0 if we have the following:*

1. V is continuous in each of the sets $[t_{k-1}, t_k) \times \mathbb{R}^{n_\xi}$, and for each $\xi \in \mathbb{R}^{n_\xi}$ and each $t \in [t_{k-1}, t_k)$, where $k \in \mathbb{N}$, the limit $\lim_{(t,y) \to (t_k^-, x)} V(t, y) = V(t_k^-, x)$ exists;

2. $V(t, \xi)$ is locally Lipschitz in all $\xi \in \mathbb{R}^{n_\xi}$; and

3. $V(t, 0) \equiv 0$ for all $t \geq t_0$.

Definition 2.8 (Upper Dini Derivative). *Given a function $V : [t_0, \infty) \times \mathbb{R}^{n_\xi} \to \mathbb{R}^+$, the upper right-hand derivative of V with respect to system (2.1) is defined by $D^+V(t, \xi(t)) = \limsup_{\delta \searrow 0} \frac{1}{\delta}[V(t + \delta, \xi(t + \delta)) - V(t, \xi(t))]$.*

Proof. We prove this theorem employing mathematical induction. Consider the following Lyapunov function for (2.15) with $\tilde{u} \equiv 0$, $\tilde{\nu} \equiv 0$:

$$V(t, \xi(t)) = r\xi(t)^2. \tag{2.27}$$

Using $2ab \leq a^2 + b^2$, $a, b \in \mathbb{R}$ in what follows, we obtain

$$D^+V(t, \xi(t)) \leq 2r\xi(t)a\xi(t - d(t)) \leq r^2\xi(t)^2 + a^2\xi(t - d(t))^2$$
$$\leq rV(t, \xi(t)) + \lambda_1 V(t - d(t), \xi(t - d(t))), \tag{2.28}$$

along the solutions of (2.15) with $\tilde{u}, \tilde{\nu} \equiv 0$, for each $t \notin \mathcal{T}$. In what follows, we are going to show that

$$V(t, \xi(t)) \leq rM\|\xi_{t_0}\|^2 e^{-\lambda(t - t_0)}, \quad \forall t \geq t_0, \tag{2.29}$$

where

$$M > e^{\lambda\tau} \geq e^{\lambda(t_1 - t_0)}. \tag{2.30}$$

One can easily verify that (I) implies (2.30). Notice that this choice of M yields $\|\xi_{t_0}\|^2 < M\|\xi_{t_0}\|^2 e^{-\lambda(t_1 - t_0)}$.

According to the principle of mathematical induction, we start showing that

$$V(t, \xi(t)) \leq rM\|\xi_{t_0}\|^2 e^{-\lambda(t - t_0)}, \quad \forall t \in [t_0, t_1), \tag{2.31}$$

holds by showing that the basis of mathematical induction

$$V(t, \xi(t)) \leq rM\|\xi_{t_0}\|^2 e^{-\lambda(t_1 - t_0)}, \quad \forall t \in [t_0, t_1), \tag{2.32}$$

holds. For the sake of contradiction, suppose that (2.32) does not hold. From (2.27) and (2.30), we infer that there exists $\bar{t} \in (t_0, t_1)$ such that

$$V(\bar{t}, \xi(\bar{t})) > rM\|\xi_{t_0}\|^2 e^{-\lambda(t_1 - t_0)}$$
$$> r\|\xi_{t_0}\|^2 \geq V(t_0 + s, \xi(t_0 + s)), \quad \forall s \in [-\check{d}, 0],$$

which implies that there exists $t^* \in (t_0, \bar{t})$ such that

$$V(t^*, \xi(t^*)) = rM\|\xi_{t_0}\|^2 e^{-\lambda(t_1 - t_0)},$$
$$V(t, \xi(t)) \leq V(t^*, \xi(t^*)), \quad \forall t \in [t_0 - \check{d}, t^*], \tag{2.33}$$

and there exists $t^{**} \in [t_0, t^*)$ such that

$$V(t^{**}, \xi(t^{**})) = r\|\xi_{t_0}\|^2,$$
$$V(t^{**}, \xi(t^{**})) \leq V(t, \xi(t)), \quad \forall t \in [t^{**}, t^*]. \tag{2.34}$$

Using (2.33) and (2.34), for any $s \in [-\check{d}, 0]$ we have

$$V(t+s, \xi(t+s)) \leq rM\|\xi_{t_0}\|^2 e^{-\lambda(t_1-t_0)} = Me^{-\lambda(t_1-t_0)}V(t^{**}, \xi(t^{**}))$$
$$\leq Me^{-\lambda(t_1-t_0)}V(t, \xi(t)), \quad \forall t \in [t^{**}, t^*]. \quad (2.35)$$

Let us now take s to be a function of time, that is, $s := s(t)$. From (2.28) and (2.35) with $s := s(t) = -d(t) \in [-\check{d}, 0]$ for all $t \in [t^{**}, t^*]$, we obtain

$$D^+ V(t, \xi(t)) \leq \left(r + \lambda_1 M e^{-\lambda(t_1-t_0)}\right) V(t, \xi(t)),$$

for all $t \in [t^{**}, t^*]$. Recall that $0 < t^* - t^{**} \leq t_1 - t_0 \leq \tau$. Having that said, it follows from (2.30), (2.33) and (2.34) that

$$V(t^*, \xi(t^*)) \leq V(t^{**}, \xi(t^{**})) e^{(r + \lambda_1 M e^{-\lambda(t_1-t_0)})(t^* - t^{**})}$$
$$= r\|\xi_{t_0}\|^2 e^{(r + \lambda_1 M e^{-\lambda(t_1-t_0)})(t^* - t^{**})}$$
$$\leq r\|\xi_{t_0}\|^2 e^{(r + \lambda_1 M e^{-\lambda\tau})\tau} \stackrel{\text{(I)}}{<} rM\|\xi_{t_0}\|^2 e^{-\lambda\tau}$$
$$\leq rM\|\xi_{t_0}\|^2 e^{-\lambda(t_1-t_0)} = V(t^*, \xi(t^*)),$$

which is a contradiction. Hence, (2.31) holds, i.e., (2.29) holds over $[t_0, t_1)$.

It is now left to show that (2.29) holds over $[t_{k-1}, t_k)$ for each $k \in \mathbb{N}$, $k \geq 2$. To that end, assume that (2.29) holds for each $k \in \{1, 2, \ldots, m\}$, where $m \in \mathbb{N}$, i.e.,

$$V(t, \xi(t)) \leq rM\|\xi_{t_0}\|^2 e^{-\lambda(t-t_0)}, \quad \forall t \in [t_0, t_k), \quad (2.36)$$

for every $k \in \{1, 2, \ldots, m\}$. Let us now show that (2.29) holds over $[t_m, t_{m+1})$ as well, i.e.,

$$V(t, \xi(t)) \leq rM\|\xi_{t_0}\|^2 e^{-\lambda(t-t_0)}, \quad \forall t \in [t_m, t_{m+1}). \quad (2.37)$$

For the sake of contradiction, suppose that (2.37) does not hold. Then, we can define

$$\bar{t} := \inf \left\{ t \in [t_m, t_{m+1}) \big| V(t, \xi(t)) > rM\|\xi_{t_0}\|^2 e^{-\lambda(t-t_0)} \right\}.$$

From (2.15b) with $\tilde{\nu} \equiv 0$ and (2.36), we know that

$$V(t_m^+, \xi(t_m^+)) = r\xi(t_m^+)^2 = rc^2\xi(t_m)^2 = \lambda_2 V(t_m^-, \xi(t_m^-))$$
$$\leq \lambda_2 rM\|\xi_{t_0}\|^2 e^{-\lambda(t_m-t_0)}$$
$$= \lambda_2 rM\|\xi_{t_0}\|^2 e^{\lambda(\bar{t}-t_m)} e^{-\lambda(\bar{t}-t_0)}$$
$$< \lambda_2 r e^{\lambda(t_{m+1}-t_m)} M\|\xi_{t_0}\|^2 e^{-\lambda(\bar{t}-t_0)}$$
$$< rM\|\xi_{t_0}\|^2 e^{-\lambda(\bar{t}-t_0)},$$

where $\lambda_2 \in (0, 1)$ is such that $\lambda_2 e^{\lambda(t_{m+1}-t_m)} \leq \lambda_2 e^{\lambda\tau} < 1$. One can easily verify that (II) implies $\lambda_2 e^{\lambda \tau} < 1$. Another fact to notice is that $\bar{t} \neq t_m$. Employing the continuity of $V(t, \xi(t))$ over the interval $[t_m, t_{m+1})$, we infer

$$V(\bar{t}, \xi(\bar{t})) = rM\|\xi_{t_0}\|^2 e^{-\lambda(\bar{t}-t_0)},$$
$$V(t, \xi(t)) \leq V(\bar{t}, \xi(\bar{t})), \qquad \forall t \in [t_m, \bar{t}]. \quad (2.38)$$

In addition, we know that there exists $t^* \in (t_m, \bar{t})$ such that

$$V(t^*, \xi(t^*)) = \lambda_2 r e^{\lambda(t_{m+1}-t_m)} M\|\xi_{t_0}\|^2 e^{-\lambda(\bar{t}-t_0)},$$
$$V(t^*, \xi(t^*)) \leq V(t, \xi(t)) \leq V(\bar{t}, \xi(\bar{t})), \qquad \forall t \in [t^*, \bar{t}]. \quad (2.39)$$

We proceed as follows, for any $t \in [t^*, \bar{t}]$ and any $s \in [-\check{d}, 0]$, then either $t + s \in [t_0 - \check{d}, t_m)$ or $t + s \in [t_m, \bar{t}]$. If $t + s \in [t_0 - \check{d}, t_m)$, then from (2.36) we have

$$V(t+s, \xi(t+s)) \leq rM\|\xi_{t_0}\|^2 e^{-\lambda(t+s-t_0)}$$
$$= rM\|\xi_{t_0}\|^2 e^{-\lambda(t-t_0)} e^{-\lambda s}$$
$$\leq rM\|\xi_{t_0}\|^2 e^{-\lambda(\bar{t}-t_0)} e^{\lambda(\bar{t}-t)} e^{\lambda \check{d}}$$
$$\leq r e^{\lambda \check{d}} e^{\lambda(t_{m+1}-t_m)} M\|\xi_{t_0}\|^2 e^{-\lambda(\bar{t}-t_0)}. \quad (2.40)$$

If $t + s \in [t_m, \bar{t}]$, then from (2.38) we have

$$V(t+s, \xi(t+s)) \leq rM\|\xi_{t_0}\|^2 e^{-\lambda(\bar{t}-t_0)}$$
$$\leq r e^{\lambda \check{d}} e^{\lambda(t_{m+1}-t_m)} M\|\xi_{t_0}\|^2 e^{-\lambda(\bar{t}-t_0)}. \quad (2.41)$$

Apparently, the upper bounds (2.40) and (2.41) are the same; hence, it does not matter whether $t + s \in [t_0 - \check{d}, t_m)$ or $t + s \in [t_m, \bar{t}]$. Therefore, from (2.39) and this upper bound, we have for any $s \in [-\check{d}, 0]$

$$V(t+s, \xi(t+s)) \leq \frac{e^{\lambda \check{d}}}{\lambda_2} V(t^*, \xi(t^*)) \leq \frac{e^{\lambda \check{d}}}{\lambda_2} V(t, \xi(t)), \quad (2.42)$$

for all $t \in [t^*, \bar{t}]$. Once more, let us take s to be a function of time, that is, $s := s(t) = -d(t) \in [-\check{d}, 0]$ for all $t \in [t^*, \bar{t}]$. Now, from (2.28) and (2.42), we have

$$D^+ V(t, \xi(t)) \leq \left(r + \frac{\lambda_1}{\lambda_2} e^{\lambda \check{d}}\right) V(t, \xi(t)), \qquad \forall t \in [t^*, \bar{t}].$$

Recall that $0 < \bar{t} - t^* \leq t_{m+1} - t_m \leq \tau$. Accordingly, we reach

$$V(\bar{t}, \xi(\bar{t})) \leq V(t^*, \xi(t^*)) e^{(r + \frac{\lambda_1}{\lambda_2} e^{\lambda \check{d}})(\bar{t}-t^*)}$$
$$= \lambda_2 r e^{\lambda(t_{m+1}-t_m)} M\|\xi_{t_0}\|^2 e^{-\lambda(\bar{t}-t_0)} e^{(r + \frac{\lambda_1}{\lambda_2} e^{\lambda \check{d}})(\bar{t}-t^*)}$$
$$\leq \lambda_2 e^{\lambda \tau} e^{(r + \frac{\lambda_1}{\lambda_2} e^{\lambda \check{d}})\tau} rM\|\xi_{t_0}\|^2 e^{-\lambda(\bar{t}-t_0)}$$
$$\overset{(II)}{<} rM\|\xi_{t_0}\|^2 e^{-\lambda(\bar{t}-t_0)} = V(\bar{t}, \xi(\bar{t})),$$

which is a contradiction; hence, (2.29) holds over $[t_m, t_{m+1})$. Employing mathematical induction, one immediately infers that (2.29) holds over $[t_{k-1}, t_k)$ for each $k \in \mathbb{N}$. From (2.27) and (2.29), it follows that

$$\|\xi(t)\| \leq \sqrt{M}\|\xi_{t_0}\|e^{-\frac{\lambda}{2}(t-t_0)}, \qquad \forall t \geq t_0.$$

$\square$

2.7.2 Proof of Theorem 2.1

The following two well-known results can be found in, for example, [168].

Lemma 2.2 (Young's Inequality). *Let $*$ denote convolution over an interval I, $f \in \mathcal{L}_p[I]$ and $g \in \mathcal{L}_q[I]$. The Young's inequality is $\|f * g\|_r \leq \|f\|_p \|g\|_q$ for $\frac{1}{r} = \frac{1}{p} + \frac{1}{q} - 1$ where $p, q, r > 0$.*

Theorem 2.3 (Riesz–Thorin Interpolation Theorem). *Let $F : \mathbb{R}^n \to \mathbb{R}^m$ be a linear operator and suppose that $p_0, p_1, q_0, q_1 \in [1, \infty]$ satisfy $p_0 < p_1$ and $q_0 < q_1$. For any $\theta \in [0, 1]$, define p_θ, q_θ by $1/p_\theta = (1-\theta)/p_0 + \theta/p_1$ and $1/q_\theta = (1 - \theta/q_0) + \theta/q_1$. Then, $\|F\|_{p_\theta \to q_\theta} \leq \|F\|_{p_0 \to q_0}^{1-\theta} \|F\|_{p_1 \to q_1}^{\theta}$, where $\|F\|_{p. \to q.}$ denotes the norm of the mapping F between the $\mathcal{L}_p$ and $\mathcal{L}_q$ space. In particular, if $\|F\|_{p_0 \to q_0} \leq M_0$ and $\|F\|_{p_1 \to q_1} \leq M_1$, then $\|F\|_{p_\theta \to q_\theta} \leq M_0^{1-\theta} M_1^{\theta}$.*

Proof. From the UGES assumption of the theorem, we infer that the fundamental matrix $\Phi(t, t_0)$ satisfies:

$$\|\Phi(t, t_0)\| \leq \sqrt{M} e^{-\frac{\lambda}{2}(t-t_0)}, \qquad \forall t \geq t_0,$$

uniformly in t_0. Refer to [4, Definition 3] for the exact definition of a fundamental matrix. Now, [4, Theorem 3.1] provides

$$\xi(t) = \Phi(t, t_0)\xi(t_0) + \int_{t_0}^{t} \Phi(t, s)\tilde{u}(s)ds$$

$$+ \int_{t_0}^{t} \Phi(t, s)a\xi_s(-d(t))ds + \sum_{t_0 < t_i \leq t} \Phi(t, t_i)\tilde{\nu}(t_i), \qquad \forall t \geq t_0, \quad (2.43)$$

where $\xi_s(-d(t)) = 0$ when $s - d(t) \geq t_0$. The above equality along with

$$\sum_{t_0 < t_i \leq t} e^{-\frac{\lambda}{2}(t-t_i)} \leq \sum_{i=1}^{\infty} e^{-\frac{\lambda}{2}\varepsilon i} = \frac{1}{e^{\frac{\lambda\varepsilon}{2}} - 1}$$

immediately yields

$$\|\xi(t)\| \leq \|\Phi(t,t_0)\|\|\xi_{t_0}\| + \int_{t_0}^{t} \|\Phi(t,s)\|\|\tilde{u}(s)\|ds$$

$$+ \int_{t_0}^{t} \|\Phi(t,s)\|\|a\|\|\xi_s(-d(t))\|ds + \sum_{t_0 < t_i \leq t} \|\Phi(t,t_i)\|\|\tilde{\nu}(t_i)\|$$

$$\leq \sqrt{M} e^{-\frac{\lambda}{2}(t-t_0)} \|\xi_{t_0}\| + \sqrt{M} \int_{t_0}^{t} e^{-\frac{\lambda}{2}(t-s)} \|\tilde{u}(s)\| ds$$

$$+ |a|\sqrt{M}\|\xi_{t_0}\| \int_{t_0}^{t_0+d(t)} e^{-\frac{\lambda}{2}(t-s)} ds + \tilde{K}_\nu \sqrt{M} \frac{1}{e^{\frac{\lambda\varepsilon}{2}}-1}, \quad \forall t \geq t_0, \quad (2.44)$$

uniformly in t_0.

Let us now estimate the contribution of the initial condition ξ_{t_0} toward $\|\xi[t_0,t]\|_p$ by setting $\tilde{u} \equiv 0$ and $\tilde{K}_\nu = 0$. In other words, we have

$$\|\xi(t)\| \leq \sqrt{M} e^{-\frac{\lambda}{2}(t-t_0)} \|\xi_{t_0}\| + |a|\sqrt{M}\|\xi_{t_0}\| \int_{t_0}^{t_0+\tilde{d}} e^{-\frac{\lambda}{2}(t-s)} ds, \quad \forall t \geq t_0. \quad (2.45)$$

In what follows, we use $(a+b)^p \leq 2^{p-1}a^p + 2^{p-1}b^p$ and $(a+b)^{\frac{1}{p}} \leq a^{\frac{1}{p}} + b^{\frac{1}{p}}$, where $a, b \geq 0$ and $p \in [1, \infty)$ (see, for example, [51, Lemmas 1 and 2]). Raising (2.45) to the $p^{\text{th}} \in [1, \infty)$ power, integrating over $[t_0, t]$ and taking the pth root yields

$$\|\xi[t_0,t]\|_p \leq \left(\sqrt{M} + |a|\sqrt{M}\frac{2}{\lambda}(e^{\frac{\tilde{d}\lambda}{2}}-1)\right) 2^{\frac{p-1}{p}} \left(\frac{2}{p\lambda}\right)^{\frac{1}{p}} \|\xi_{t_0}\|, \quad \forall t \geq t_0,$$

where we used

$$\left(\int_{t_0}^{\infty} \left(\int_{t_0}^{t_0+\tilde{d}} e^{-\frac{\lambda}{2}(t-s)} ds\right)^p dt\right)^{\frac{1}{p}} = \frac{2}{\lambda}(e^{\frac{\tilde{d}\lambda}{2}}-1)\left(\frac{2}{p\lambda}\right)^{\frac{1}{p}}. \quad (2.46)$$

When $p = \infty$, simply take the limit

$$\lim_{p \to \infty} 2^{\frac{p-1}{p}} \left(\frac{2}{p\lambda}\right)^{\frac{1}{p}} = 2.$$

Let us now estimate the contribution of the input $\tilde{u}(t)$ toward $\|\xi[t_0,t]\|_p$ by setting $\|\xi_{t_0}\| = 0$ and $\tilde{K}_\nu = 0$. In other words, we have

$$\|\xi(t)\| \leq \sqrt{M} \int_{t_0}^{t} e^{-\frac{\lambda}{2}(t-s)} \|\tilde{u}(s)\| ds, \quad \forall t \geq t_0. \quad (2.47)$$

Note that $\int_0^\infty e^{-\frac{\lambda}{2}s}\mathrm{d}s = \frac{2}{\lambda}$. Now, integrating the previous inequality over $[t_0, t]$ and using Lemma 2.2 with $p = q = r = 1$ yields the $\mathcal{L}_1$-norm estimate:

$$\|\xi[t_0, t]\|_1 \leq \frac{2}{\lambda}\sqrt{M}\|\tilde{u}[t_0, t]\|_1, \qquad \forall t \geq t_0. \tag{2.48}$$

Taking the max over $[t_0, t]$ in (2.47) and using Lemma 2.2 with $q = r = \infty$ and $p = 1$ yields the $\mathcal{L}_\infty$-norm estimate:

$$\|\xi[t_0, t]\|_\infty \leq \frac{2}{\lambda}\sqrt{M}\|\tilde{u}[t_0, t]\|_\infty, \qquad \forall t \geq t_0. \tag{2.49}$$

From (2.43), one infers that we are dealing with a linear operator, say F, that maps $\tilde{u}$ to e with bounds for the norms $\|F\|_1 \leq \|F\|_1^*$ and $\|F\|_\infty \leq \|F\|_\infty^*$, where $\|F\|_1^*$ and $\|F\|_\infty^*$ are given by (2.48) and (2.49), respectively. Because $\|F\|_1^* = \|F\|_\infty^*$, Theorem 2.3 gives that $\|F\|_p \leq \|F\|_1^* = \|F\|_\infty^*$ for all $p \in [1, \infty]$. This yields

$$\|\xi[t_0, t]\|_p \leq \frac{2}{\lambda}\sqrt{M}\|\tilde{u}[t_0, t]\|_p, \qquad \forall t \geq t_0, \tag{2.50}$$

for any $p \in [1, \infty]$.

Let us now estimate the contribution of the noise $\tilde{\nu}(t)$ towards $\|\xi[t_0, t]\|_p$ by setting $\|\xi_{t_0}\| = 0$ and $\tilde{u} \equiv 0$. In other words, we have

$$\|\xi(t)\| \leq \tilde{K}_\nu \sqrt{M}\frac{1}{e^{\frac{\lambda\varepsilon}{2}} - 1}, \qquad \forall t \geq t_0.$$

By identifying $b(t) \equiv b := \tilde{K}_\nu\sqrt{M}\frac{1}{e^{\frac{\lambda\varepsilon}{2}}-1}$, we immediately obtain

$$\|\xi[t_0, t]\|_p \leq \|b[t_0, t]\|_p, \qquad \forall t \geq t_0,$$

for all $p \in [1, \infty]$.

Finally, summing up the contributions of ξ_{t_0}, $\tilde{u}(t)$ and $\tilde{\nu}(t)$ produces

$$\|\xi[t_0, t]\|_p \leq 2\sqrt{M}\left(1 + |a|\frac{2}{\lambda}\left(e^{\frac{d\lambda}{2}} - 1\right)\right)\left(\frac{1}{p\lambda}\right)^{\frac{1}{p}}\|\xi_{t_0}\|$$
$$+ \frac{2}{\lambda}\sqrt{M}\|\tilde{u}[t_0, t]\|_p + \|b[t_0, t]\|_p, \qquad \forall t \geq t_0,$$

for any $p \in [1, \infty]$. $\square$

2.7.3 Proof of Theorem 2.2

Proof. Combining (i) of UGES protocols and (2.17), one obtains

$$\left\langle \frac{\partial W(i, e)}{\partial e}, g(t, x_t, e_t, \omega_t) \right\rangle \leq \frac{\overline{a}}{\underline{a}}LW(j, e(t - d(t))) + \|H(x_t, \omega_t)\| \tag{2.51}$$

for any $i, j \in \mathbb{N}$. Hence, the index i in $W(i, e)$ can be omitted in what follows. Now, we define $Z(t) := W(e(t))$ and reach

$$\frac{dZ(t)}{dt} \leq \frac{\overline{a}}{\underline{a}} L Z(t - d(t)) + \|H(x_t, \omega_t)\|, \qquad (2.52)$$

for almost all $t \notin \mathcal{T}$. For a justification of the transition from (2.51) to (2.52), refer to [131, Footnote 8]. Likewise, property (ii) of the UGES protocols yields

$$Z(t^+) \leq \rho Z(t) + \overline{a} \nu_j(t), \qquad (2.53)$$

for all $t \in \mathcal{T}$, where $\nu_j(t)$, $j \in \{1, \ldots, l\}$, is the jth NCS link noise given by (2.7) and upper bounded with K_ν. Notice that $|Z(t)| = |W(e(t))|$. Next, we use the comparison lemma for impulsive delayed systems [109, Lemma 2.2]. Basically, the fundamental matrix of (2.52)–(2.53) is upper bounded with the fundamental matrix of (2.15) with parameters $a := \frac{\overline{a}}{\underline{a}} L$ and $c := \rho$. Refer to [4, Definition 3] for the exact definition of a fundamental matrix. Of course, the corresponding transmission interval τ in (2.15b), and therefore in (2.53), has to allow for $M > 1$ and $\lambda > 0$ that satisfy (I), (II) and $\frac{2}{\lambda}\sqrt{M}\gamma_H < 1$ (as stated in Theorem 2.2). Essentially, (I) and (II) yield $\mathcal{L}_p$-stability with bias from H to W, while $\frac{2}{\lambda}\sqrt{M}\gamma_H < 1$ allows us to invoke the small-gain theorem. Following the proof of Theorem 2.1, one readily establishes $\mathcal{L}_p$-stability from H to W with bias, i.e.,

$$\|W[t_0, t]\|_p \leq K_W \|W_{t_0}\| + \gamma_W \|H[t_0, t]\|_p + \|b[t_0, t]\|_p, \qquad (2.54)$$

for any $t \geq t_0$ any $p \in [1, \infty]$, where

$$K_W := 2\sqrt{M}\left(1 + \frac{\overline{a}}{\underline{a}} L \frac{2}{\lambda}\left(e^{\frac{d\lambda}{2}} - 1\right)\right)\left(\frac{1}{p\lambda}\right)^{\frac{1}{p}},$$

$$\gamma_W := \frac{2}{\lambda}\sqrt{M}, \quad b := \frac{\overline{a} K_\nu \sqrt{M}}{e^{\frac{\lambda \varepsilon}{2}} - 1}.$$

Let us now infer $\mathcal{L}_p$-stability with bias from ω to (H, e) via the small-gain theorem. Inequality (2.16) implies

$$\|H[t, t_0]\|_p \leq K_H \|x_{t_0}\| + \gamma_H \|W[t, t_0]\|_p + \gamma_H \|\omega[t_0, t]\|_p, \qquad (2.55)$$

for all $t \geq t_0$. Combining the above with (2.54) and property (i) of the UGES protocols yields

$$\|e[t_0, t]\|_p \leq \frac{\overline{a} K_W / \underline{a}}{(1 - \gamma_W \gamma_H)} \|e_{t_0}\| + \frac{\gamma_W K_H / \underline{a}}{(1 - \gamma_W \gamma_H)} \|x_{t_0}\|$$
$$+ \frac{\gamma_W \gamma_H / \underline{a}}{(1 - \gamma_W \gamma_H)} \|\omega[t_0, t]\|_p + \frac{1/\underline{a}}{(1 - \gamma_W \gamma_H)} \|b[t_0, t]\|_p, \qquad (2.56)$$

$$\|H[t_0, t]\|_p \leq \frac{K_H}{1 - \gamma_W \gamma_H} \|x_{t_0}\| + \frac{\overline{a} K_H K_W}{1 - \gamma_W \gamma_H} \|e_{t_0}\|$$
$$+ \frac{\gamma_H}{1 - \gamma_W \gamma_H} \|\omega[t_0, t]\|_p + \frac{\gamma_H}{1 - \gamma_W \gamma_H} \|b[t_0, t]\|_p.$$

From the above two inequalities, $\mathcal{L}_p$-stability from ω to (H,e) with bias $\frac{\gamma_H + \frac{1}{a}}{1-\gamma_W\gamma_H}\frac{\overline{a}K_\nu\sqrt{M}}{e^{\frac{\lambda\epsilon}{2}}-1}$ and gain $\frac{\gamma_H(1+\frac{\gamma_W}{a})}{1-\gamma_W\gamma_H}$ is immediately obtained. □

2.7.4 Proof of Corollary 2.1

Proof. The $\mathcal{L}_p$-detectability of x from (W,ω,H) implies that there exist $K_d, \gamma_d \geq 0$ such that

$$\|x[t_0,t]\|_p \leq K_d\|x_{t_0}\| + \gamma_d\|H[t_0,t]\|_p + \gamma_d\|(W,\omega)[t_0,t]\|_p$$
$$\leq K_d\|x_{t_0}\| + \gamma_d\|H[t_0,t]\|_p + \gamma_d\|W[t_0,t]\|_p + \gamma_d\|\omega[t_0,t]\|_p \quad (2.57)$$

for all $t \geq t_0$. Plugging (2.55) into (2.57) leads to

$$\|x[t_0,t]\|_p \leq K_d\|x_{t_0}\| + \gamma_d K_H\|x_{t_0}\| + \gamma_d\gamma_H\|W[t,t_0]\|_p$$
$$+ \gamma_d\gamma_H\|\omega[t_0,t]\|_p + \gamma_d\|W[t_0,t]\|_p + \gamma_d\|\omega[t_0,t]\|_p$$
$$\leq K_d\|x_{t_0}\| + \gamma_d K_H\|x_{t_0}\| + (\overline{a}\gamma_d\gamma_H + \overline{a}\gamma_d)\|e[t,t_0]\|_p$$
$$+ (\gamma_d\gamma_H + \gamma_d)\|\omega[t_0,t]\|_p$$

for all $t \geq t_0$. Finally, we include (2.56) in (2.57) and add the obtained inequality to (2.56), which establishes $\mathcal{L}_p$-stability with bias from ω to (x,e). □

2.7.5 Proof of Proposition 2.1

Proof. For the case $p = \infty$, UGS of the interconnection Σ_n and Σ_e is immediately obtained using the definition of $\mathcal{L}_\infty$-norm. Therefore, the case $p \in [1,\infty)$ is more interesting. From the conditions of the proposition, we know that there exist $K \geq 0$ and $\gamma \geq 0$ such that

$$\|(x,e)[t_0,t]\|_p \leq K\|(x_{t_0}, e_{t_0})\| + \gamma\|\omega[t_0,t]\|_p, \; \forall t \geq t_0.$$

Recall that $\omega \equiv \mathbf{0}_{n_\omega}$ when one is interested in asymptotic stability. By raising both sides of the above inequality to the pth power, we obtain:

$$\int_{t_0}^{t}\|(x,e)(s)\|^p ds \leq K_1\|(x_{t_0}, e_{t_0})\|^p, \quad \forall t \geq t_0, \quad (2.58)$$

where $K_1 := K^p$.

First, we need to establish UGS of the interconnection Σ_n and Σ_e when $\omega \equiv \mathbf{0}_{n_\omega}$. Before we continue, note that the jumps in (2.13) and (2.14) are such that $\|(x,e)(t^+)\| \leq \|(x,e)(t)\|$ for each $t \in \mathcal{T}$. Apparently, jumps are not destabilizing and can be disregarded in what follows. Along the lines of the proof for [165, Theorem 1], we pick any $\epsilon > 0$. Let $\mathcal{K}$ be the set $\{(x,e)|\frac{\epsilon}{2} \leq \|(x,e)\| \leq \epsilon\}$, $\mathcal{K}_1$ be the set $\{(x,e)|\|(x,e)\| \leq \epsilon\}$, and take

$$a := \sup_{\substack{t\in\mathbb{R}\\(x,e)(t)\in\mathcal{K}\\(x,e)_t\in PC([-\overline{d},0],\mathcal{K}_1)}}\|\big(f(t,x_t,e,\mathbf{0}_{n_\omega}), g(t,x_t,e_t,\tilde{\mathbf{0}}_{n_\omega})\big)\|,$$

where $f(t, x_t, e, \mathbf{0}_{n_\omega})$ and $g(t, x_t, e_t, \tilde{\mathbf{0}}_{n_\omega})$ are given by (2.10) and (2.12), respectively. This supremum exists because the underlying dynamics are Lipschitz uniformly in t. Next, choose $0 < \delta < \frac{\epsilon}{2}$ such that $r < \delta$ implies $K_1 \|(x_{t_0}, e_{t_0})\|^p < s_0$, where $s_0 := \frac{\epsilon(\epsilon/2)^p}{2a}$. Let $\|(x_{t_0}, e_{t_0})\|_d < \delta$. Then $\|(x, e)(t)\| < \epsilon$ for all $t \geq t_0$. Indeed, suppose that there exists some $t > t_0$ such that $\|(x, e)(t)\| \geq \epsilon$. Then, there is an interval $[t_1, t_2]$ such that $\|(x, e)(t_1)\| = \frac{\epsilon}{2}$, $\|(x, e)(t_2)\| = \epsilon$, $(x, e)(t) \in \mathcal{K}$, $(x, e)_t \in PC([-\bar{d}, 0], \mathcal{K}_1)$, for all $t \in [t_1, t_2]$. Hence,

$$\int_{t_0}^{\infty} \|(x, e)(s)\|^p \mathrm{d}s \geq \int_{t_1}^{t_2} \|(x, e)(s)\|^p \mathrm{d}s \geq (t_2 - t_1)\left(\frac{\epsilon}{2}\right)^p.$$

On the other hand,

$$\frac{\epsilon}{2} \leq \|(x, e)(t_2) - (x, e)(t_1)\| \leq \int_{t_1}^{t_2} \|(f(t, x_t, e, \mathbf{0}_{n_\omega}), g(t, x_t, e_t, \tilde{\mathbf{0}}_{n_\omega}))\| \mathrm{d}s$$
$$\leq a(t_2 - t_1),$$

and, combining the above with (2.58), we conclude that

$$s_0 \leq \int_{t_0}^{\infty} \|(x, e)(s)\|^p \mathrm{d}s \leq K_1 \|(x_{t_0}, e_{t_0})\|^p,$$

which is a contradiction.

Second, let us show asymptotic convergence of $\|(x, e)(t)\|$ to zero. Using (2.58), we infer that $(x, e)(t) \in \mathcal{L}_p$. Owing to the Lipschitz dynamics of the corresponding system, we conclude that $(\dot{x}, \dot{e})(t) \in \mathcal{L}_p$ as well. Now, one readily establishes asymptotic convergence of $\|x\|$ to zero using [171, Facts 1–4]. Consequently, asymptotic convergence of $\|e\|$ to zero follows from (2.44) by observing that $\xi(t)$ and $\tilde{u}(t)$ in (2.44) correspond to $W(t)$ and $H(t)$, respectively, and using (i) of Definition 2.6. □

3

Input-Output Triggering

CONTENTS

3.1 Motivation, Applications and Related Works 52
    3.1.1 Motivational Example: Autonomous Cruise Control 52
    3.1.2 Applications and Literature Review 54
3.2 Impulsive Switched Systems and Related Stability Notions 57
3.3 Problem Statement: Self-Triggering from Input and Output Measurements 60
3.4 Input-Output Triggered Mechanism 62
    3.4.1 Why $\mathcal{L}_p$-gains over a Finite Horizon? 63
    3.4.2 Proposed Approach 64
    3.4.3 Design of Input-Output Triggering 65
        3.4.3.1 Cases 3.1 and 3.2 66
        3.4.3.2 Case 3.3 68
    3.4.4 Implementation of Input-Output Triggering 68
3.5 Example: Autonomous Cruise Control 70
3.6 Conclusions and Perspectives 73
3.7 Proofs of Main Results 76
    3.7.1 Properties of Matrix Functions 76
    3.7.2 Proof of Theorem 3.1 77
    3.7.3 Proof of Theorem 3.2 78
    3.7.4 Proof of Results in Section 3.4.3 80
        3.7.4.1 $\mathcal{L}_p$ property over an arbitrary finite interval with constant δ 80
        3.7.4.2 Extending bounds to (an arbitrarily long) finite horizon 81
        3.7.4.3 Proof of Theorem 3.3 82
        3.7.4.4 Proof of Theorem 3.4 82

The previous chapter was characterized by constant MATIs that, at best, result in periodic transmissions of up-to-date information and periodic executions of control laws when one aims for utmost feedback intermittency. The goal of the present chapter is to relax this periodicity (i.e., MATI constancy) using the self-triggered transmission paradigm. The MATIs in this chapter are designed for the worst-case scenario and preclude potential prolongation of transmission intervals utilizing the information arriving from the plant (e.g., plant output and external input). It is worth mentioning that this chapter does not consider delays or scheduling protocols. The proposed approach is successfully applied to an Autonomous Cruise Control (ACC) scenario.

3.1 Motivation, Applications and Related Works

In this chapter, we devise *input-output triggering* employing $\mathcal{L}_p$-gains over a finite horizon in the small-gain theorem. By resorting to convex programming, a method to compute $\mathcal{L}_p$-gains over a finite horizon is devised. Under the term *input-output triggering*, we refer to self-triggering based on the currently available values of the plant's external input and output. The triggering event, which is precluded in our approach, is the violation of the small-gain condition. Depending on the noise in the environment, the developed mechanism yields stable, asymptotically stable, and $\mathcal{L}_p$-stable (with bias) closed-loop systems. Control loops are modeled as interconnections of impulsive switched systems for which several $\mathcal{L}_p$-stability results are presented.

3.1.1 Motivational Example: Autonomous Cruise Control

Using laser-based or radar-based sensors, ACC technology allows a vehicle to slow down when approaching another vehicle and accelerate to the desired speed when traffic allows. Besides reducing driver fatigue, improving comfort and fuel economy, ACC is also intended to keep cars from crashing [85]. The sampling periods of ACC loops are typically fixed and designed for the worst-case scenario (e.g., fast and heavy traffic). Furthermore, these fixed sampling periods are often determined experimentally and are based on the traditional rules of thumb (e.g., 20 times the time constant of the dominant pole). Intuitively, the sampling periods of ACC loops should not remain constant as the desired speed, distance between the cars, the environment (urban on non-urban), and paths (straight or turns) change. The present chapter quantifies this intuition.

Consider the *trajectory tracking* controller in [40] as an example of a simple ACC. In [40], a velocity-controlled unicycle robot R_1 given by

$$\dot{x}_{R1} = v_{R1}\cos\theta_{R1}, \; \dot{y}_{R1} = v_{R1}\sin\theta_{R1}, \; \dot{\theta}_{R1} = \omega_{R1} \tag{3.1}$$

tracks a trajectory generated by a virtual velocity-controlled unicycle robot R_2 with states x_{R2}, y_{R2} and θ_{R2}, and linear and angular velocities v_{R2} and ω_{R2}, respectively. See Figure 3.1 for an illustration. Besides trajectory tracking, this controller can be employed in leader-follower, target-pursuit, obstacle avoidance and waypoint-following problems. The tracking error x_p in the co-

Input-Output Triggering

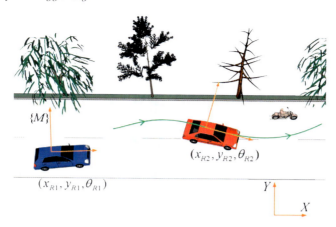

FIGURE 3.1
An illustration of the trajectory tracking problem considered in this chapter.

ordinate frame $\{M\}$ of robot R_1 is

$$x_p = \begin{bmatrix} x_{p1} \\ x_{p2} \\ x_{p3} \end{bmatrix} = \begin{bmatrix} \cos\theta_{R1} & \sin\theta_{R1} & 0 \\ -\sin\theta_{R1} & \cos\theta_{R1} & 0 \\ 0 & 0 & 1 \end{bmatrix} \begin{bmatrix} x_{R2} - x_{R1} \\ y_{R2} - y_{R1} \\ \theta_{R2} - \theta_{R1} \end{bmatrix}. \quad (3.2)$$

After differentiating (3.2), we obtain:

$$\dot{x}_p = \begin{bmatrix} \omega_{R1} x_{p2} - v_{R1} + v_{R2} \cos x_{p3} \\ -\omega_{R1} x_{p1} + v_{R2} \sin x_{p3} \\ \omega_{R2} - \omega_{R1} \end{bmatrix}. \quad (3.3)$$

System (3.3) can be interpreted as a plant with state x_p and external inputs v_{R2} and ω_{R2}. Take the output of the plant to be $y = x_p$ and introduce $\omega_p := [v_{R2} \ \omega_{R2}]^\top$. The plant is controlled through control signals v_{R1} and ω_{R1}. In order to compute v_{R1} and ω_{R1}, and track an unknown trajectory (a trajectory is given by $v_{R2}(t)$, $\omega_{R2}(t)$ and initial conditions of x_{R2}, y_{R2} and θ_{R2}), robot R_1 needs to know the state of the plant x_p and the inputs to R_2, i.e., v_{R2} and ω_{R2}. Following [40], choose the following control law:

$$v_{R1} = v_{R2} \cos x_{p3} + k_1 x_{p1},$$
$$\omega_{R1} = \omega_{R2} + k_2 v_{R2} \frac{\sin x_{p3}}{x_{p3}} x_{p2} + k_3 x_{p3}, \quad (3.4)$$

where k_1, k_2 and k_3 are positive control gains. Let us introduce $u := [v_{R1} \ \omega_{R1}]^\top$. Proposition 3.1 in [40] shows that the control law (3.4) makes the origin $x_p = [0 \ 0 \ 0]^\top$ of the plant (3.3) globally asymptotically stable provided that $v_{R2}(t)$, $\omega_{R2}(t)$ and their derivatives are bounded for all times $t \geq 0$ and $\lim_{t \to \infty} v_{R2}(t) \neq 0$ or $\lim_{t \to \infty} \omega_{R2}(t) \neq 0$.

The above asymptotic stability result is obtained assuming instantaneous and continuous information. In real-life applications, continuous access to the values of y and ω_p is rarely achievable. In other words, the control signal u is typically computed using intermittent measurements corrupted by noise. The measurements of the outputs and external inputs of the plant are denoted $\hat{y}$ and $\hat{\omega}_p$, respectively. In general, as new up-to-date values of $\hat{y}$ and $\hat{\omega}_p$ arrive, the control signal may change abruptly. Afterward, the newly computed values u are sent to actuators. These values might be noisy and intermittently updated as well. Hence, the plant is not controlled by u but instead by $\hat{u}$. An illustration of such a control system is provided in Figure 3.2.

A goal of this chapter is to take advantage of the available information from the plant, i.e., of $\hat{y}$ and $\hat{\omega}_p$, and design sampling/control update instants $\mathcal{T} = \{t_1, t_2, \ldots\}$ such that stability of the control system is preserved. As our intuition suggests, different $\hat{y}$ and $\hat{\omega}_p$ may yield different time instants in $\mathcal{T}$. In fact, the intersampling intervals $\tau_1 = t_2 - t_1$, $\tau_2 = t_3 - t_2, \ldots$, for R_1 are determined based on the distance from the desired trajectory (i.e., $\hat{y}$) and the nature of the trajectory (i.e., $\hat{\omega}_p$). Driven by the desire to obtain intersampling instants τ_i's as large as possible, we adopt impulsive switched system modeling in this chapter. Figure 3.3 contrasts different methods for computing τ_i's. The solid blue line in Figure 3.3 represents τ_i's computed via the methodology devised in this chapter. Apparently, the use of finite horizon $\mathcal{L}_p$-gains (this notion somewhat corresponds to the notion of individual $\mathcal{L}_p$-gains considered in [207]) produces larger τ_i's in comparison with the use of unified gains. Unified gains are simply the maximum of all individual gains of a switched system (which corresponds to the worst-case scenario). As discussed in [207], unified gains are a valid (although quite conservative) choice for the $\mathcal{L}_p$-gain of a switched system. However, even such conservative $\mathcal{L}_p$-gains of interconnected switched systems, when used in the small-gain theorem, do not suffice to conclude stability of the closed-loop system (see [207] or condition (ii) of Theorem 3.1). Altogether, one should thoughtfully use the finite horizon $\mathcal{L}_p$-gains of interconnected switched systems in order to decrease conservativeness, i.e., maximize τ_i's, when applying the small-gain theorem.

3.1.2 Applications and Literature Review

Event-triggered and *self-triggered* realizations of intermittent feedback are proposed in [169, 5, 6, 103, 200]. In these event-driven approaches, one defines a desired performance, and sampling (i.e., transmission of up-to-date information) is triggered when an event representing the unwanted performance occurs. The work in [103] applies event-triggering to control, estimation and optimization tasks. The work in [200] utilizes the dissipative formalism of nonlinear systems, and employs passivity properties of feedback interconnected systems in order to reach an event-triggered control strategy for stabiliza-

Input-Output Triggering

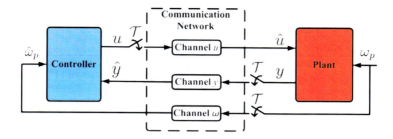

FIGURE 3.2
A diagram of a control system with the plant and controller interacting over a communication network with intermittent information updates. The three switches indicate that the information between the plant and controller are exchanged at discrete time instants belonging to a set $\mathcal{T}$.

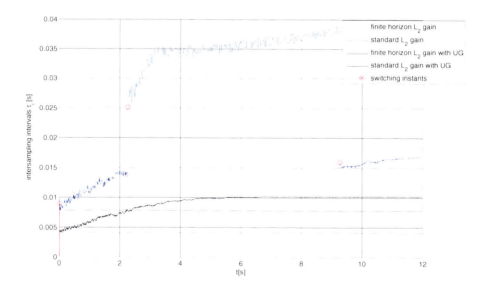

FIGURE 3.3
A comparison of the intersampling intervals τ_i's obtained for different notions of $\mathcal{L}_p$-gains. The abbreviation UG stands for "Unified Gain". Red stems indicate time instants when changes in $\hat{\omega}_p$ occur. The solid blue line indicates τ_i's generated via the methodology devised in this chapter. Definitions of the said notions appear in Sections 3.2 and 3.4.

tion of passive and output passive systems. In [203], the authors propose an event-triggered output feedback control approach for time-invariant input-feedforward output-feedback passive plants and controllers (in addition to some other technical conditions imposed on plants and controllers). A particular limitation of the results in [203] is that the number of inputs and outputs of the plant and controller must be equal (due to this, this approach cannot be applied to the ACC example in Section 3.5). The event-triggered approach of [149] converts a trajectory tracking control problem into a stabilization problem for an autonomous system and then applies an invariance principle for hybrid systems in order to solve the tracking problem. It is worth mentioning that [149] is tailored for a trajectory tracking problem of unicycles, which is quite similar to our case study in Section 3.5. However, while the results of [149] hold for a certain class of trajectories (i.e., trajectories generated with a non-constant linear velocity and angular velocity which does not change the sign or converge to zero), we do not impose such requirements on the trajectories. In addition, [149] assumes that the trajectory is known a priori by the controller while we make no such assumption. Lastly, notice that the self-triggered counterparts of the event-triggered approaches from [203] and [149] are yet to be devised.

In self-triggered approaches, the current sample is used to determine the next sampling instant, i.e., to predict the occurrence of the triggering event. In comparison with event-triggering, where sensor readings are constantly obtained and analyzed in order to detect events (even though the control signals are updated only upon event detection), self-triggering decreases requirements posed on sensors and processors in embedded systems. The pioneering work on self-triggering, intended to maintain $\mathcal{L}_2$-gains of linear control systems below a certain value, is found in [191] and [192]. A comprehensive comparison of our work with [191] and [192] is provided in [184]. The authors in [6] extend the event-triggering presented in [169] and develop *state-triggering*: self-triggering based on the value of the system state in the last feedback transmission. The work in [170] utilizes Lyapunov theory and develops event-triggered trajectory tracking for nonlinear systems affine in controls.

The approach in Chapter 2 is developed for general nonlinear controllers and plants that render closed-loop systems stable, asymptotically stable or exponentially stable in the absence of a communication network. The generality of nonlinear models considered in Chapter 2 allows for analysis of time-varying closed-loop systems with external inputs/disturbances, output feedback and dynamic controllers. In addition, the methodology from Chapter 2 yields stable, asymptotically stable, exponentially stable or $\mathcal{L}_p$-stable control systems under the intermittent information and settings of Chapter 3 as illustrated in the remainder of this chapter. The principal requirement in Chapter 2 is $\mathcal{L}_p$-stability of the closed-loop system. In other words, if a certain controller does not yield the closed-loop system $\mathcal{L}_p$-stable, one can seek another controller. Hence, unlike related works, our requirements are on the closed-loop systems and not on the plant and controller per se. On the other hand, the work in [47] does not consider external inputs, and the results for nonlinear systems are provided for a class of exponentially stable closed-loop systems in the absence of communication networks. The authors in [104] do not consider external inputs and exponential stability of systems with specific nonlinearities is analyzed. The work in [170] investigates state feedback for nonlinear systems affine in controls and static controllers. A comparison of our approach and the approach from [170] can be found in [183]. The Input-to-State Stability (ISS) approaches [169, 5, 6, 103] assume state feedback, static controllers, and do not consider external inputs. In addition, the results of [5] and [6] are applicable to state-dependent homogeneous systems and polynomial systems. The work in [200] analyzes passive plants, proportional controllers and does not take into account external inputs.

The main limitation of the approach in Chapter 2 is periodicity of the transmission instants inherited from the standard definition of $\mathcal{L}_p$-gains. Recall that the standard $\mathcal{L}_p$-gain is not a function of time (i.e., there is no prediction of when some event might happen) or state. To circumvent this limitation (while retaining most of the generality of Chapter 2), we devise an input-output triggered approach employing $\mathcal{L}_p$-gains over a finite horizon in the small-gain theorem. Under the term *input-output triggering*, we refer to self-triggering based on the values of the plant's external input and output in the last feedback transmission. The triggering event, which is precluded in our approach, is the violation of the small-gain condition. It is worth pointing out that our approach does not require construction of storage or Lyapunov functions which can be a difficult task for a given problem.

The remainder of this chapter is organized as follows. Section 3.2 presents the utilized notation and definitions. Section 3.3 formulates the problem of input-output triggered intermittent feedback under various assumptions. The methodology brought together to solve the problem of interest is presented in Section 3.4. The proposed input-output triggered sampling policy is verified on a trajectory tracking problem in Section 3.5. Conclusions and future challenges are in Section 3.6. Proofs and technical results are included in Section 3.7.

3.2 Impulsive Switched Systems and Related Stability Notions

Given a (finite or infinite) sequence of time instances $t_1, t_2, \ldots, t_i, \ldots$ with $t_{i+1} > t_i$, defining the set $\mathcal{T}$, and an initial time $t_0 < t_1$, this chapter considers switched systems with impulses written as

$$\Sigma^\delta \begin{cases} \chi(t^+) = h_\chi^\delta(t, \chi(t)) & t \in \mathcal{T} \\ \left. \begin{array}{l} \dot\chi = f_\chi^\delta(t, \chi, \omega) \\ y = \ell_\chi^\delta(t, \chi, \omega) \end{array} \right\} & \text{otherwise}, \end{cases} \quad (3.5)$$

where $\chi \in \mathbb{R}^{n_\chi}$ is the state, $(\omega, \delta) \in \mathbb{R}^{n_\omega + n_\omega}$ is the input, and $y \in \mathbb{R}^{n_y}$ is the output. The input δ is given by a piecewise constant, right-continuous function of time $\delta : [t_0, \infty) \to \mathcal{P}$, which we refer to as the *switching signal*, with $\mathcal{P}$ being an index set (not necessarily a finite set). The functions f_χ^δ and h_χ^δ are regular enough to guarantee the existence of solutions, which, given the initial state χ_0, initial time t_0, and a switching signal $\delta : [t_0, \infty) \to \mathcal{P}$, are given by right-continuous functions $t \mapsto \chi(t)$. Jumps of the state χ occur at each $t \in \mathcal{T}$. The value of the state after a jump is given by $\chi(t^+) = \lim_{t' \searrow t} \chi(t')$ for each $t \in \mathcal{T}$. The switching signal δ changes at time instances t_i^δ, defining the set $\mathcal{T}^\delta$, which is a subset of $\mathcal{T}$. For notational convenience, we define $t_0^\delta := t_0$, which is not a switching time, the intersampling intervals $\tau_i = t_{i+1} - t_i$ for each t_{i+1}, t_i in $\mathcal{T} \cup \{t_0\} =: \mathcal{T}_0$, and the interswitching intervals $\tau_i^\delta := t_{i+1}^\delta - t_i^\delta$ for each $t_{i+1}^\delta, t_i^\delta$ in $\mathcal{T}^\delta \cup \{t_0^\delta\} =: \mathcal{T}_0^\delta$.

The following stability notions for impulsive switched systems Σ^δ as in (3.5) are employed in this chapter.

Definition 3.1 ($\mathcal{L}_p$-stability with bias b)**.** Let $p \in [1, \infty]$. Given a switching signal $t \mapsto \delta(t)$, the impulsive switched system Σ^δ is $\mathcal{L}_p$-stable with bias $b(t) \equiv b \geq 0$ from ω to y with (linear) gain $\gamma \geq 0$ if there exists $K \geq 0$ such that, for each $t_0 \in \mathbb{R}$ and each $\chi_0 \in \mathbb{R}^{n_\chi}$, each solution to Σ^δ from χ_0 at $t = t_0$ we have that $\|y[t_0, t]\|_p \leq K\|\chi_0\| + \gamma\|\omega[t_0, t]\|_p + \|b[t_0, t]\|_p$ for each $t \geq t_0$.

Definition 3.2 ($\mathcal{L}_p$-stability with bias b over a finite horizon τ)**.** Let $p \in [1, \infty]$. Given a switching signal $t \mapsto \delta(t)$ and $\tau \geq 0$, the impulsive switched system Σ^δ is $\mathcal{L}_p$-stable over a finite horizon of length τ with bias $b(t) \equiv b \geq 0$ from ω to y with (linear) constant gain $\widetilde{\gamma}(\tau) \geq 0$ if there exists a constant[1] $\widetilde{K}(\tau) \geq 0$ such that, for each $t_0 \in \mathbb{R}$ and each $\chi_0 \in \mathbb{R}^{n_\chi}$, each solution to Σ^δ from χ_0 at $t = t_0$ satisfies $\|y[t_0, t]\|_p \leq \widetilde{K}(\tau)\|\chi_0\| + \widetilde{\gamma}(\tau)\|\omega[t_0, t]\|_p + \|b[t_0, t]\|_p$ for each $t \in [t_0, t_0 + \tau)$.

Definition 3.3 (detectability)**.** Let $p \in [1, \infty]$. Given a switching signal $t \mapsto \delta(t)$, the state χ of Σ^δ is $\mathcal{L}_p$-detectable from (y, ω) to χ with (linear) gain $\gamma \geq 0$ if there exists $K \geq 0$ such that, for each $t_0 \in \mathbb{R}$ and each $\chi_0 \in \mathbb{R}^{n_\chi}$, each solution to Σ^δ from χ_0 at $t = t_0$ satisfies $\|\chi[t_0, t]\|_p \leq K\|\chi_0\| + \gamma\|y[t_0, t]\|_p + \gamma\|\omega[t_0, t]\|_p$ for each $t \geq t_0$.

Proposition 3.1. *Given a switching signal $t \mapsto \delta(t)$, if Σ^δ is $\mathcal{L}_p$-stable with*

---

[1] The parenthesis in $\widetilde{K}$ and $\widetilde{\gamma}$ denote explicitly the dependency of these constants on the already chosen τ.

Input-Output Triggering 59

bias $b(t) \equiv b \geq 0$ *from ω to y with gain $\gamma \geq 0$ and $\mathcal{L}_p$-detectable from (y, ω) to χ with gain $\gamma' \geq 0$, then Σ^δ is $\mathcal{L}_p$-stable with bias $\gamma' b$ from ω to state χ for the given switching signal.*

Proof. From the $\mathcal{L}_p$-stability with bias assumption, Definition 3.1 implies that there exists $K \geq 0$ such that for each t_0 and each χ_0 we have

$$\|y[t_0, t]\|_p \leq K\|\chi_0\| + \gamma\|\omega[t_0, t]\|_p + \|b[t_0, t]\|_p \quad \forall t \geq t_0,$$

while $\mathcal{L}_p$-detectability from (y, ω) to χ with gain γ' implies that there exists $K' \geq 0$ such that

$$\|\chi[t_0, t]\|_p \leq K'\|\chi_0\| + \gamma'\|y[t_0, t]\|_p + \gamma'\|\omega[t_0, t]\|_p$$

for all $t \geq t_0$. Then, we obtain

$$\|\chi[t_0, t]\|_p \leq (K\gamma' + K')\|\chi_0\| + (\gamma\gamma' + \gamma')\|\omega[t_0, t]\|_p + \gamma'\|b[t_0, t]\|_p \quad (3.6)$$

for all $t \geq t_0$. This proves the claim since (3.6) corresponds to $\mathcal{L}_p$-stability with bias $\gamma' b$, gain $\gamma\gamma' + \gamma'$, and constant $K\gamma' + K'$. □

For a given switching signal δ, the following result provides a set of sufficient conditions for $\mathcal{L}_p$-stability of a system Σ^δ with $\mathcal{L}_p$-stable (over a finite horizon) subsystems.

Theorem 3.1. *Given a switching signal $t \mapsto \delta(t)$, consider the impulsive switched system Σ^δ given by (3.5). Let $\overline{K} \geq 0$ and $p \in [1, \infty)$. Suppose the following properties hold:*

(i) For each $t_i^\delta \in \mathcal{T}_0^\delta$, there exist constants $\widetilde{K}(\tau_i^\delta)$ and $\widetilde{\gamma}(\tau_i^\delta)$ such that[2]

$$\|y[t_i^\delta, t']\|_p \leq \widetilde{K}(\tau_i^\delta)\|\chi(t_i^{\delta+})\| + \widetilde{\gamma}(\tau_i^\delta)\|\omega[t_i^\delta, t']\|_p \quad (3.7)$$

for all $t' \in [t_i^\delta, t_{i+1}^\delta]$ if t_i^δ is not the largest switching time in $\mathcal{T}^\delta$ (in which case $\tau_i^\delta = t_{i+1}^\delta - t_i^\delta$) or for all $t' \in [t_i^\delta, \infty)$ if t_i^δ is the largest switching time in $\mathcal{T}^\delta$ (in which case it corresponds to $\mathcal{L}_p$-stability and one can write $\widetilde{K}$ and $\widetilde{\gamma}$), and such that

$$K_M := \sup_i \widetilde{K}(\tau_i^\delta), \quad (3.8)$$

$$\gamma_M := \sup_i \widetilde{\gamma}(\tau_i^\delta), \quad (3.9)$$

exist.

(ii) The condition

$$\sum_i \|\chi(t_i^{\delta+})\| \leq \overline{K}\|\chi(t_0)\|, \quad (3.10)$$

holds.

---
[2] For $i = 0$, we have $\chi(t_0^{\delta+}) = \chi(t_0^\delta)$, which is equal to $\chi(t_0)$.

Then, Σ^δ is $\mathcal{L}_p$-stable from ω to y with constant $K_M \overline{K}$ and gain γ_M for the given δ. For $p = \infty$, the same result holds with the constant $K_M \overline{K}$ and gain γ_M when (3.10) is replaced with $\sup_i \|\chi(t_i^{\delta+})\| \le \overline{K}\|\chi(t_0)\|$.

See Subsection 3.7.2 for a proof.

Remark 3.1. *The sampling policy designed herein (along with Assumption 3.2 provided below) ensures that hypothesis (i) of Theorem 3.1 always holds. Note that it is not straightforward to verify (3.10) beforehand. In fact, the control policy needs to make decisions "on the fly", based on the available information regarding previous switching instants, in order to enforce (3.10) (refer to Section 3.5 for further details). For example, condition (3.10) can be enforced by requiring $\|\chi(t_{i+1}^{\delta+})\| \le \lambda \|\chi(t_i^{\delta+})\|$, where $\lambda \in [0,1)$, which is similar to the property exploited in the design of uniformly globally exponentially stable protocols in Chapter 2. Notice that decision making "on the fly" by exploiting previously received information is a salient feature of self-triggering.*

Building from ideas in [168], the next result proposes an expression of the $\mathcal{L}_p$-gain over a finite horizon for a generic nonlinear system $\dot{\chi} = \widetilde{g}(t, \chi, v)$ with state χ and input v.

Theorem 3.2. *Given $\tau \ge 0$ and $t_0 \in \mathbb{R}$, suppose that there exist $A \in \mathcal{A}_{n_\chi}^+$ with $\|A\| < \infty$, a continuous function $\widetilde{y} : \mathbb{R} \times \mathbb{R}^{n_\chi} \times \mathbb{R}^{n_v} \to \mathbb{R}_+^{n_\chi}$ such that*

$$\overline{\dot{\chi}} = \overline{\widetilde{g}(t, \chi, v)} \preceq A\overline{\chi} + \widetilde{y}(t, \chi, v), \quad \forall (t, \chi, v) \in [t_0, t_0 + \tau] \times \mathbb{R}^{n_\chi} \times \mathbb{R}^{n_v}. \quad (3.11)$$

Then, for each solution to $\dot{\chi} = \widetilde{g}(t, \chi, v)$ we have

$$\|\chi[t_0, t_0 + \tau]\|_p \le \widetilde{K}(\tau)\|\chi(t_0)\| + \widetilde{\gamma}(\tau)\|\widetilde{y}[t_0, t_0 + \tau]\|_p, \quad (3.12)$$

for all $t \in [t_0, t_0 + \tau)$, where

$$\widetilde{K}(\tau) = \left(\frac{\exp(\|A\|p\tau) - 1}{p\|A\|}\right)^{\frac{1}{p}}, \quad (3.13)$$

$$\widetilde{\gamma}(\tau) = \frac{\exp(\|A\|\tau) - 1}{\|A\|}. \quad (3.14)$$

Theorem 3.2 proposes conditions under which $\dot{\chi} = \widetilde{g}(t, \chi, v)$ is $\mathcal{L}_p$-stable from $\widetilde{y}$ to χ over the finite horizon τ for any $p \in [1, \infty]$. Its proof is in Subsection 3.7.3.

3.3 Problem Statement: Self-Triggering from Input and Output Measurements

Consider a nonlinear feedback control system consisting of a plant

$$\dot{x}_p = f_p(t, x_p, u, \omega_p),$$
$$y = g_p(t, x_p), \quad (3.15)$$

Input-Output Triggering 61

and a controller

$$\begin{aligned}\dot{x}_c &= f_c(t, x_c, u_c, \omega_c), \\ y_c &= g_c(t, x_c),\end{aligned} \qquad (3.16)$$

interconnected via the assignment

$$u = y_c, \qquad u_c = y, \qquad \omega_c = \omega_p, \qquad (3.17)$$

where $x_p \in \mathbb{R}^{n_p}$ and $x_c \in \mathbb{R}^{n_c}$ are the states, $y \in \mathbb{R}^{n_y}$ and $y_c \in \mathbb{R}^{n_u}$ are the outputs, and $(u, \omega_p) \in \mathbb{R}^{n_u} \times \mathbb{R}^{n_\omega}$ and $(u_c, \omega_c) \in \mathbb{R}^{n_y} \times \mathbb{R}^{n_\omega}$ are the inputs of the plant and controller, respectively, where ω_p is an exogenous input to the plant. Following the assignment (3.17), we model the connections (or links) between the plant and the controller as communication networks over which intermittent exchange of information due to sampling takes place. Figure 3.2 depicts this setting, where the value of u computed by the controller that arrives at the plant is denoted $\hat{u}$. Similarly, the values of y and ω_p that the controller actually receives are denoted $\hat{y}$ and $\hat{\omega}_p$, respectively. In this setting, the quantity $\hat{u}$ is the input fed to the plant (3.15) while the quantities $\hat{y}$ and $\hat{\omega}_p$ are the measurement of y and ω_p received by the controller (3.16).

To study the properties of the feedback control system in Figure 3.2, define

$$e = \begin{bmatrix} e_y \\ e_u \end{bmatrix} := \begin{bmatrix} \hat{y} - y \\ \hat{u} - u \end{bmatrix} \qquad (3.18)$$

and

$$e_\omega := \hat{\omega}_p - \omega_p. \qquad (3.19)$$

To model intermittent transmission (or sampling) of the values of y and u, the quantities $\hat{y}$ and $\hat{u}$ are updated at time instances $t_1, t_2, \ldots, t_i, \ldots$ in $\mathcal{T}$, i.e.,[3]

$$\left.\begin{aligned}\hat{y}(t_i^+) &= y(t_i) + h_y(t_i) \\ \hat{u}(t_i^+) &= u(t_i) + h_u(t_i)\end{aligned}\right\} \quad t_i \in \mathcal{T}, \qquad (3.20)$$

where $h_y : \mathbb{R} \to \mathbb{R}^{n_y}$ and $h_u : \mathbb{R} \to \mathbb{R}^{n_u}$. Similarly, the quantity $\hat{\omega}_p$ may change discretely reflecting changes in the plant exogenous input or in the level of plant disturbances. The time instants at which jumps of $\hat{\omega}_p$ occur are denoted t_i^δ and belong to the set $\mathcal{T}^\delta$, which is a subset of $\mathcal{T}$. We assume that the received values of y, u, and ω_p given by $\hat{y}$, $\hat{u}$, and $\hat{\omega}_p$, respectively, remain constant in between updates, i.e., for each $t \in [t_0, \infty) \setminus \mathcal{T}$,

$$\dot{\hat{y}} = 0, \qquad \dot{\hat{u}} = 0, \qquad \dot{\hat{\omega}}_p = 0, \qquad (3.21)$$

which is known as the Zero-Order-Hold (ZOH) strategy [80].

The following standing assumption summarizes the properties imposed on the feedback control system in Figure 3.2 throughout this chapter.

---

[3] The formulation of the update law in (3.20) implies that the jump times at the controller and plant end coincide.

Assumption 3.1 (standing assumption). *The jump times at the controller and plant end coincide. The set of sampling instants $\mathcal{T}^\delta := \{t_1^\delta, t_2^\delta, \ldots, t_i^\delta, \ldots\}$ at which $\hat{\omega}_p$ changes its value satisfies $\mathcal{T}^\delta \subset \mathcal{T}$, where $\mathcal{T} := \{t_1, t_2, \ldots, t_i, \ldots\}$, $t_{i+1} > t_i$ for each t_{i+1}, t_i in $\mathcal{T}$.*

We are now ready to state the problem studied in this chapter.

Problem 3.1. *Determine the set of sampling instants $\mathcal{T}$ and $\mathcal{T}^\delta$ to update $(\hat{y}, \hat{u})$ and $\hat{\omega}_p$, respectively, such that the closed-loop system (3.15)–(3.16) is stable in the $\mathcal{L}_p$ sense.*

The following specific scenarios are investigated:

Case 3.1. *The signals $\hat{u}$, $\hat{y}$, and $\hat{\omega}_p$ are not corrupted by noise, and ω_p is constant between consecutive t_i^δ's.*

Case 3.2. *The signals $\hat{u}$ and $\hat{y}$ are not corrupted by noise while $\hat{\omega}_p$ is corrupted by noise. In addition, ω_p is arbitrary between consecutive t_i^δ's.*

Case 3.3. *The signals $\hat{u}$, $\hat{y}$, and $\hat{\omega}_p$ are corrupted by noise. In addition, ω_p is arbitrary between two consecutive t_i^δ's.*

3.4 Input-Output Triggered Mechanism

Following Chapter 2, our solution to Problem 3.1 determines the sampling instants $\mathcal{T}$ using input-output information of the system resulting from the interconnection between (3.15) and (3.16), namely

$$\left.\begin{aligned} x(t^+) &= x(t) \\ e(t^+) &= h(t) \end{aligned}\right\} \quad t \in \mathcal{T} \tag{3.22a}$$

$$\left.\begin{aligned} \dot{x} &= f(t, x, e, \hat{\omega}_p, e_\omega) \\ \dot{e} &= g(t, x, e, \hat{\omega}_p, e_\omega) \end{aligned}\right\} \quad \text{otherwise}, \tag{3.22b}$$

Input-Output Triggering

where $x := (x_p, x_c)$, and functions f, g and h are given by

$$f(t, x, e, \hat{\omega}_p, e_\omega) := \begin{bmatrix} f_p(t, x_p, g_c(t, x_c) + e_u, \hat{\omega}_p - e_\omega) \\ f_c(t, x_c, g_p(t, x_p) + e_y, \hat{\omega}_p) \end{bmatrix}, \quad (3.23)$$

$$h(t_i) := \begin{bmatrix} h_y(t_i) \\ h_u(t_i) \end{bmatrix}, \quad (3.24)$$

$$g(t, x, e, \hat{\omega}_p, e_\omega) :=$$
$$\begin{bmatrix} \underbrace{\hat{f}_p(t, x_p, x_c, g_p(t, x_p) + e_y, g_c(t, x_c) + e_u, \hat{\omega}_p - e_\omega)}_{\equiv 0 \text{ for zero-order-hold estimation strategy}} - \\ \hat{f}_c(t, x_p, x_c, g_p(t, x_p) + e_y, g_c(t, x_c) + e_u, \hat{\omega}_p) - \\ -\frac{\partial g_p}{\partial t}(t, x_p) - \frac{\partial g_p}{\partial x_p}(t, x_p) f_p(t, x_p, g_c(t, x_c) + e_u, \hat{\omega}_p - e_\omega) \\ -\frac{\partial g_c}{\partial t}(t, x_c) - \frac{\partial g_c}{\partial x_c}(t, x_c) f_c(t, x_c, g_p(t, x_p) + e_y, \hat{\omega}_p) \end{bmatrix}$$
$$(3.25)$$

where h_y and h_y are introduced in (3.20).

By identifying the switching signal δ in (3.5) with $\hat{\omega}_p$[4], we write (3.22) in the form of an impulsive switched system Σ^δ as in (3.5) as follows:

$$\left. \begin{array}{l} x(t^+) = x(t) \\ e(t^+) = h(t) \end{array} \right\} t \in \mathcal{T} \quad (3.26a)$$

$$\left. \begin{array}{l} \dot{x} = f^\delta(t, x, e, e_\omega) \\ \dot{e} = g^\delta(t, x, e, e_\omega) \end{array} \right\} \text{otherwise} . \quad (3.26b)$$

In this chapter, we are interested in changes in $\mathcal{L}_p$-gains (over a finite horizon) of (3.26) for different values of the switching signal $\delta = \hat{\omega}_p$. Note that $f^\delta(t, x, e, e_\omega)$ and $g^\delta(t, x, e, e_\omega)$ are alternative (but equivalent) labels for $f(t, x, e, \hat{\omega}_p, e_\omega)$ and $g(t, x, e, \hat{\omega}_p, e_\omega)$. (For convenience, we use the former when the switching component of our model is explicitly utilized.)

3.4.1 Why $\mathcal{L}_p$-gains over a Finite Horizon?

$\mathcal{L}_p$-gains over a finite horizon allow prediction of the triggering event in this chapter. In addition, as suggested by the example in Section 3.1.1, they produce less conservative intertransmission intervals τ_i's than classical $\mathcal{L}_p$-gains when used in the small-gain theorem. This is due to the fact that $\mathcal{L}_p$-gains over a finite horizon are monotonically nondecreasing in τ. To show this fact, we use the following characterization for $p \in [1, \infty)$ taken from [78], [79] and [81]:

$$[\tilde{\gamma}(\tau)]^p := \sup_{\omega \in \mathcal{L}_p[t_0, t_0+\tau]} \left\{ \frac{\int_{t_0}^{t_0+\tau} \|y(t)\|^p dt}{\int_{t_0}^{t_0+\tau} \|\omega(t)\|^p dt} \right\}, \quad (3.27)$$

---
[4]This assignment requires the space of δ and $\hat{\omega}_p$ to match, i.e., $\mathcal{P} \subseteq \mathbb{R}^{n_\omega}$.

where $\|x(t_0)\| = 0$, $\|\omega[t_0, t_0+\tau]\|_p \neq 0$, bias $b = 0$ and δ is fixed to be constant to generate an output y and solution x of Σ^δ. The case $p = \infty$ is similar.

Proposition 3.2. *The function $\tau \mapsto \widetilde{\gamma}(\tau)$ is monotonically nondecreasing.*

Proof. Take $\tau > 0$ and choose any τ' such that $\tau' > \tau$. According to (3.27), for the horizon $[t_0, t_0 + \tau']$ we can write

$$[\widetilde{\gamma}(\tau')]^p = \sup_{\omega \in \mathcal{L}_p[t_0, t_0+\tau']} \left\{ \frac{\int_{t_0}^{t_0+\tau} \|y(t)\|^p dt + \int_{t_0+\tau}^{t_0+\tau'} \|y(t)\|^p dt}{\int_{t_0}^{t_0+\tau} \|\omega(t)\|^p dt + \int_{t_0+\tau}^{t_0+\tau'} \|\omega(t)\|^p dt} \right\}.$$

Now, choose $\omega \in \mathcal{L}_p[t_0, t_0 + \tau']$ such that $\omega(t) = 0$ for $t \in (t_0 + \tau, t_0 + \tau']$. This yields

$$[\widetilde{\gamma}(\tau')]^p = \sup_{\omega \in \mathcal{L}_p[t_0, t_0+\tau']} \left\{ \frac{\int_{t_0}^{t_0+\tau} \|y(t)\|^p dt + \int_{t_0+\tau}^{t_0+\tau'} \|y(t)\|^p dt}{\int_{t_0}^{t_0+\tau} \|\omega(t)\|^p dt} \right\} \geq$$

$$\geq \sup_{\omega \in \mathcal{L}_p[t_0, t_0+\tau]} \left\{ \frac{\int_{t_0}^{t_0+\tau} \|y(t)\|^p dt}{\int_{t_0}^{t_0+\tau} \|\omega(t)\|^p dt} \right\} = [\widetilde{\gamma}(\tau)]^p.$$

Taking the pth root of the above inequality shows the claim. $\square$

Since a standard (i.e., infinite horizon or classical) $\mathcal{L}_p$-gain γ can be defined as

$$\gamma := \sup_{\tau \geq 0} \widetilde{\gamma}(\tau), \tag{3.28}$$

we conclude that $\widetilde{\gamma}(\tau) \leq \gamma$ for all $\tau \geq 0$. Lastly, notice that some systems are $\mathcal{L}_p$-stable only over a finite horizon.

3.4.2 Proposed Approach

The approach proposed to provide a solution to Problem 3.1 is as follows. Suppose that t_i^δ and t_{i+1}^δ are two consecutive switching instants of the switching signal $t \mapsto \delta(t)$. Then, the switching signal $t \mapsto \delta(t)$ remains constant over $[t_i^\delta, t_{i+1}^\delta)$, i.e., $\delta(t) = r$ for all $t \in [t_i^\delta, t_{i+1}^\delta)$ for some $r \in \mathcal{P}$. To determine if a sample should be taken within $(t_i^\delta, t_{i+1}^\delta)$, i.e., to determine τ_i with $\tau_i \leq t_{i+1}^\delta - t_i^\delta$, suppose that for each $t \in [t_i^\delta, t_i^\delta + \tau_i)$, and for some $p \in [1, \infty]$, the solution $t \mapsto (x(t), e(t))$ to (3.26) resulting from an input $t \mapsto e_\omega(t)$ satisfies

$$\dot{\overline{e}}(t) \preceq A^r \overline{e}(t) + \widetilde{y}^r(t, x(t), e_\omega(t)), \tag{3.29}$$

$$\|\widetilde{y}^r[t_i^\delta, t]\|_p \leq K_n^r \|x(t_i^\delta)\| + \gamma_n^r \|(e, e_\omega)[t_i^\delta, t]\|_p, \tag{3.30}$$

where $A^r \in \mathcal{A}_{n_e}^+$ with $\|A^r\| < \infty$, $\widetilde{y}^r : \mathbb{R} \times \mathbb{R}^{n_x} \times \mathbb{R}^{n_\omega} \to \mathbb{R}_+^{n_e}$ is continuous, and K_n^r and γ_n^r are positive constants. From (3.29), it follows that, for each $t \in [t_i^\delta, t_i^\delta + \tau_i)$, we have

$$\|e[t_i^\delta, t]\|_p \leq \widetilde{K}_e^r(\tau_i)\|e(t_i^\delta)\| + \widetilde{\gamma}_e^r(\tau_i)\|\widetilde{y}^r[t_i^\delta, t]\|_p,$$

Input-Output Triggering 65

with $\tau \mapsto \widetilde{K}_e^r(\tau)$ and $\tau \mapsto \widetilde{\gamma}_e^r(\tau)$ given as

$$\widetilde{K}_e^r(\tau) = \left(\frac{\exp(\|A^r\|p\tau)-1}{p\|A^r\|}\right)^{\frac{1}{p}}, \quad \widetilde{\gamma}_e^r(\tau) = \frac{\exp(\|A^r\|\tau)-1}{\|A^r\|} \quad (3.31)$$

(see (3.13) and (3.14), respectively). Then, interpreting system (3.26) as the interconnection shown in Figure 3.4 between the *nominal system* Σ_n^δ given by

$$x(t^+) = x(t) \qquad \} \; t \in \mathcal{T} \qquad (3.32a)$$
$$\dot{x} = f^\delta(t, x, e, e_\omega) \qquad \} \; \text{otherwise}, \qquad (3.32b)$$

with input (e, e_ω) and output $\tilde{y}^r$, and the *error system* Σ_e^δ given by

$$e(t^+) = h(t) \qquad \} \; t \in \mathcal{T} \qquad (3.33a)$$
$$\dot{e} = g^\delta(t, x, e, e_\omega) \qquad \} \; \text{otherwise}, \qquad (3.33b)$$

with input $\tilde{y}^r$ and output e, and using a small-gain argument, we propose to choose $\tau > 0$ such that $\widetilde{\gamma}_e^r(\tau) \leq \kappa/\gamma_n^r$, where $\kappa \in (0,1)$. In this way, the small-gain condition $\gamma_n^r \widetilde{\gamma}_e^r(\tau) < 1$ is satisfied and a stabilizing sampling policy is given by $\tau^r \in (0, \tau^{r,*}]$, where[5]

$$\tau^{r,*} = \frac{1}{\|A^r\|} \ln\left(\kappa \frac{\|A^r\|}{\gamma_n^r} + 1\right). \quad (3.35)$$

This policy yields the closed-loop system (3.26) $\mathcal{L}_p$-stable over a horizon τ when, over this horizon, the value of the switching signal is equal to r. We emphasize that the policy (3.35) is utilized in all subsequent results of this chapter. However, different settings (i.e., Cases 3.1, 3.2 and 3.3) lead to different stability properties of closed-loop systems (refer to the following subsection).

3.4.3 Design of Input-Output Triggering

In this subsection, we provide a solution to Problem 3.1 by designing intersampling intervals using input-output information. The following assumption is imposed in the results to follow.

---

[5] Using the expression for $\widetilde{\gamma}_e^r$ in (3.31), we obtain

$$\frac{\exp(\|A^r\|\tau)-1}{\|A^r\|} \leq \frac{\kappa}{\gamma_n^r}. \quad (3.34)$$

Then, solving the above inequality for τ and taking the largest possible value, denoted $\tau^{r,*}$, yields

$$\tau^{r,*} = \frac{1}{\|A^r\|} \ln\left(\kappa \frac{\|A^r\|}{\gamma_n^r} + 1\right).$$

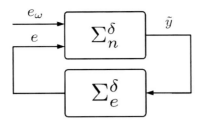

FIGURE 3.4
Interconnection of the nominal switched system Σ_n^δ and the output error impulsive switched system Σ_e^δ.

Assumption 3.2. *For each $r \in \mathcal{P}$, there exists a function $\tilde{y}^r$ such that the following hold:*

(i) *There exists A^r such that*

$$\overline{g^r(t,x,e,e_\omega)} \preceq A^r \overline{e} + \tilde{y}^r(t,x,e_\omega), \; \forall (t,x,e,e_\omega) \in [t_0,\infty) \times \mathbb{R}^{n_x} \times \mathbb{R}^{n_e} \times \mathbb{R}^{n_\omega}.$$

Furthermore, there exists $\overline{\eta}$ such that $\sup_{r \in \mathcal{P}} \|A^r\| \leq \overline{\eta}$.

(ii) *The system Σ_n^r is $\mathcal{L}_p$-stable from (e_ω, e) to $\tilde{y}^r$ with a constant K_n^r and gain γ_n^r. Furthermore, there exist constants $\overline{K_n}$ and $\overline{\gamma_n}$ such that $\sup_{r \in \mathcal{P}} K_n^r \leq \overline{K_n}$ and $\sup_{r \in \mathcal{P}} \gamma_n^r \leq \overline{\gamma_n}$.*

(iii) *The state x of Σ_n^r is $\mathcal{L}_p$-detectable from $(\tilde{y}, e_\omega, e)$ with a constant K_d^r and gain γ_d^r. Furthermore, there exist constants K and γ such that $\sup_{r \in \mathcal{P}} K_d^r \leq K$ and $\sup_{r \in \mathcal{P}} \gamma_d^r \leq \gamma$.*

3.4.3.1 Cases 3.1 and 3.2

In each of these cases, $\hat{u}$ and $\hat{y}$ are transmitted without distortions (and without transmission delays). Then, the plant and the controller receive precise values of u and y at transmission instants; namely, h_y and h_u in (3.20) are equal to zero. Therefore, the error state e is reset to zero at every sampling event, in which case, the impulsive switched system (3.26) becomes

$$\left. \begin{array}{l} x(t^+) = x(t) \\ e(t^+) = 0 \end{array} \right\} \; t \in \mathcal{T} \quad (3.36\text{a})$$

$$\left. \begin{array}{l} \dot{x} = f^\delta(t,x,e,e_\omega) \\ \dot{e} = g^\delta(t,x,e,e_\omega) \end{array} \right\} \; \text{otherwise} . \quad (3.36\text{b})$$

In particular, for Case 3.1, the signal e_ω is also reset to zero at transmission instants, and since ω_p is constant between two consecutive transmission instants, we have that $e_\omega \equiv 0$ if $\hat{\omega}_p(0) = \omega_p(0)$. The following $\mathcal{L}_p$-stability properties from e_ω to (x,e) are guaranteed by the proposed policies.

Input-Output Triggering

Theorem 3.3. *Given $p \in [1, \infty]$, suppose that Assumption 3.2 holds. Let the sampling instants in $\mathcal{T}$ be given by (3.35), computed for given values of r, that are constant on each intersampling interval, and define the switching signal $\delta : [t_0, \infty) \to \mathcal{P}$ with $\mathcal{T}^\delta \subset \mathcal{T}$. Suppose there exists $\overline{K}$ such that, for the given switching signal δ, for each $(x(t_0), e(t_0), t_0)$ and each $t \mapsto e_\omega(t)$, each solution to (3.36) is such that its x component satisfies*

$$\sum_{t_i^\delta \in \mathcal{T}_0^\delta} \|x(t_i^{\delta+})\| \leq \overline{K}\|x(t_0)\|. \tag{3.37}$$

Then, there exists $\tau^*_{\min} > 0$ such that $\tau^{r,*} \geq \tau^*_{\min}$ for all $r \in \mathcal{P}$ and, for each (x_0, t_0), each solution to (3.36) satisfies

$$\|(x,e)[t_0,t]\|_p \leq \overline{K}\widehat{K}\|(x,e)(t_0)\| + \widehat{\gamma}\|e_\omega[t_0,t]\|_p, \qquad \forall t \geq t_0, \tag{3.38}$$

for the switching signal δ. (Constants $\widehat{K}$ and $\widehat{\gamma}$ are defined in Subsection 3.7.4.)

Remark 3.2. *Property (3.38) corresponds to $\mathcal{L}_p$-stability from e_ω to (x,e). The existence of $\tau^*_{\min}$ follows from the upper bound on $\|A^r\|$ and on γ_n^r in items (ii) and (iii) of Assumption 3.2, respectively. From (3.35) we obtain*

$$\tau^{r,*} = \frac{1}{\|A^r\|} \ln\left(\kappa \frac{\|A^r\|}{\gamma_n^r} + 1\right) \geq \frac{1}{\overline{\eta}} \ln\left(\kappa \frac{\overline{\eta}}{\gamma_n^r} + 1\right) > 0.$$

Note that since, without loss of generality, there exists $\underline{\gamma_n} > 0$ such that $\inf_{r \in \mathcal{P}} \gamma_n^r \geq \underline{\gamma_n}$, then we have that $\tau^{r,*}$ is upper bounded by $\tau^*_{\max} = \lim_{\|A^r\| \searrow 0} \frac{1}{\|A^r\|} \ln\left(\kappa \frac{\|A^r\|}{\underline{\gamma_n}} + 1\right) = \frac{\kappa}{\underline{\gamma_n}}$.

Remark 3.3. *For Case 3.1, inequality (3.38) becomes $\|(x,e)[t_0,t]\|_p \leq \overline{K}\widehat{K}\|(x,e)(t_0)\|$. In addition, notice that $\|(x,e)(t_i^+)\| \leq \|(x,e)(t_i)\|$, $t_i \in \mathcal{T}$. Consequently, one can infer stability, asymptotic and exponential stability after imposing additional structure on $f^\delta(t,x,e,e_\omega)$ and $g^\delta(t,x,e,e_\omega)$ in (3.36) (refer to [131, 168, Section II], [165], [172] and Chapter 6).*

Remark 3.4. *Let us consider the case where u or y (or both) is the output of a state observer. In other words, the plant or controller (or both) is fed with an estimate provided by an observer. Consequently, in (3.36a) we have $e(t^+) = h(t)$ for all $t \in \mathcal{T}$, where $h(t)$ is the observer error. For $p \in [1, \infty)$, if the observer error satisfies the condition that there exists $\overline{K}_h \geq 0$ such that $\sum_{t_i^\delta \in \mathcal{T}_0^\delta} \|h(t_i^\delta)\| \leq \overline{K}\|e(t_0)\|$ (e.g., exponentially converging observers), then Theorem 3.3 holds with*

$$\|(x,e)[t_0,t]\|_p \leq K_1 \widehat{K}\|(x,e)(t_0)\| + \widehat{\gamma}\|e_\omega[t_i,t]\|_p, \qquad \forall t \geq t_0,$$

where $K_1 = \sqrt{n_x + n_e} \max\{\overline{K}, \overline{K}_h\}$. For $p = \infty$, the observer error has to be bounded, i.e., there exists $\overline{K}_h \geq 0$ such that $\|h(t)\| \leq \overline{K}_h\|e(t_0)\|$ for every $t \geq t_0$, in order for Theorem 3.3 to hold.

3.4.3.2 Case 3.3

This is the most general case, for which (3.26) is rewritten as

$$\left.\begin{array}{l} x(t^+) = x(t) \\ e(t^+) = h(t) \end{array}\right\} \quad t \in \mathcal{T} \tag{3.39a}$$

$$\left.\begin{array}{l} \dot{x} = f^\delta(t, x, e, e_\omega) \\ \dot{e} = g^\delta(t, x, e, e_\omega) \end{array}\right\} \quad \text{otherwise,} \tag{3.39b}$$

where $t \mapsto h(t)$ models measurement noise, quantization error, and related perturbations. For this system, we have the following result.

Theorem 3.4. *Given $p \in [1, \infty]$, suppose that Assumption 3.2 holds and that there exists $\overline{K}_h \geq 0$ such that $\|h(t)\| \leq \overline{K}_h$ for all $t \geq t_0$. Let the sampling instants in $\mathcal{T}$ be given by (3.35), computed for given values of r that, at least, are constant on each interval and define the switching signal $\delta : [t_0, \infty) \to \mathcal{P}$ with $\mathcal{T}^\delta \subset \mathcal{T}$. Suppose there exists $\overline{K}$ such that, for the given switching signal δ, for each $(x(t_0), e(t_0), t_0)$ and each $t \mapsto e_\omega(t)$, each solution to (3.39) is such that its x component satisfies (3.37). Then, there exists $\tau^*_{\min} > 0$ such that $\tau^{r,*} \geq \tau^*_{\min}$ for all $r \in \mathcal{P}$ and, for each $(x(t_0), e(t_0), t_0)$, each solution to (3.39) satisfies*

$$\|(x, e)[t_0, t]\|_p \leq \overline{K}\widehat{K}\|(x, e)(t_0)\| + \widehat{\gamma}\|e_\omega[t_0, t]\|_p + \|\widehat{b}[t_0, t]\|, \quad \forall t \geq t_0 \tag{3.40}$$

for the switching signal δ. (Expressions for $\widehat{K}$, $\widehat{\gamma}$ and $\widehat{b}(t)$ are provided in Subsection 3.7.4.)

Property (3.40) corresponds to $\mathcal{L}_p$-stability from e_ω to (x, e) with bias $\widehat{b}(t) \equiv \widehat{b} \geq 0$. See Subsection 3.7.4 for a proof. Remarks 2.5 and 2.6 are applicable to Theorem 3.4 as well.

Remark 3.5. *In order to account for possible delays introduced by the communication networks in Figure 3.2, one can use scattering transformation for the small-gain theorem [76]. Provided that Σ_n^δ and Σ_e^δ are input-feedforward output-feedback passive systems satisfying certain conic relations, the work in [76] makes stability properties of (3.26) independent of constant time delays and Theorem 3.4 is applicable again. In light of [201] and [203], these constant time delays are allowed to be larger than the intersampling intervals $\tau^{r,*}$'s.*

3.4.4 Implementation of Input-Output Triggering

Note that condition (3.29) is only needed over a horizon $[t_i^\delta, t_i^\delta + \tau_i)$, where τ_i is yet to be determined. Asking for this condition to hold for every (x, e_ω) would be too restrictive or lead to conservative sampling times. Since $x(t_i^\delta)$ and $e(t_i^\delta)$ are known after every sampling event, then, for the current constant value of

Input-Output Triggering

$t \mapsto \delta(t)$ and the input $t \mapsto e_\omega(t)$ over $[t_i^\delta, t_i^\delta + \tau_i)$, the required property is guaranteed when

$$\overline{g^\delta(t, x, e, e_\omega)} \preceq A^\delta \bar{e} + \tilde{y}^\delta(t, x, e_\omega(t))$$

holds for each

$$t \in [t_i^\delta, t_i^\delta + \tau_i], \qquad (x, e) \in \mathcal{S}_i := \text{Reach}_{t_i^\delta, t-t_i^\delta}((x(t_i^\delta), e(t_i^\delta))),$$

where $\text{Reach}_{t_i^\delta, t-t_i^\delta}((x(t_i^\delta), e(t_i^\delta)))$ is the *reachable set* of

$$\dot{x} = f^\delta(t, x, e, e_\omega), \qquad \dot{e} = g^\delta(t, x, e, e_\omega) \qquad (3.41)$$

from $(x(t_i^\delta), e(t_i^\delta))$ at t_i^δ after $t - t_i^\delta$ units of time, namely

$$\text{Reach}_{\underline{t}, \tau}(z_0) := \{z(t') \ : \ z \text{ is a solution to (3.41) from } z_0 \text{ at } \underline{t}, t' \in [\underline{t}, \underline{t} + \tau]\}.$$

Then, exploiting the reachability ideas outlined above, the following algorithm can be used at each time instant t_i, $i = 0, 1, 2, \ldots$:

Step 1. Obtain measurements $\hat{y}(t_i)$ and $\hat{\omega}_p(t_i)$.

Step 2. Extract state estimate $\hat{x}(t_i)$ from the measurements.

Step 3. Update the control law (3.16) with $\hat{y}(t_i)$ and, if (3.37) is not compromised, with $\hat{\omega}_p(t_i)$.

Step 4. Actuate the plant with $\hat{u}(t_i)$.

Step 5. Estimate $\mathcal{S}_i$ from (3.15)–(3.16) using reachability analysis.

Step 6. Compute (3.35) and pick τ_i.

With the objective of obtaining values of $\tilde{\gamma}_e(\tau)$ in (3.14) as small as possible, we propose to minimize $\|A\|$. This leads to less conservative τ_i's in a solution to Problem 3.1. Following the statement of Theorem 3.2, we consider the following optimization problem: given $r \in \mathcal{P}$, $\underline{t}, \tau > 0$, and $e_\omega : [\underline{t}, \underline{t}+\tau] \to \mathbb{R}^{n_\omega}$,

$$\begin{aligned}
\text{minimize} \quad & \|A\| & (3.42a) \\
\text{subject to:} \quad & A \in \mathcal{A}_{n_e}^+ & (3.42b) \\
& \overline{g^r(t, x, e, e_\omega)} \preceq A\bar{e} + \tilde{y}^r(t, x, e_\omega) & (3.42c)
\end{aligned}$$

for all $t \in [\underline{t}, \underline{t} + \tau]$, $(x, e) \in \mathcal{S} := \text{Reach}_{\underline{t}, \tau}((x(\underline{t}), e(\underline{t})))$.

Proposition 3.3. *The optimization problem (3.42) is convex.*

Proof. It is well known that $\|A\|$ is a convex function of A (see [30], Chapter 3). Now, let us prove that constraints (3.42b) and (3.42c) yield a convex set. First, a convex combination of two matrices in $\mathcal{A}_{n_e}^+$ is again in $\mathcal{A}_{n_e}^+$. This is due to the fact that symmetric matrices with nonnegative elements remain symmetric with nonnegative elements when multiplied with nonnegative scalars and when added together. Let us now show that inequality (3.42c) yields a convex set in A. For any $t' \in [\underline{t}, \underline{t} + \tau]$, pick any $(x', e') \in \mathcal{S}$ and $e'_\omega \in \mathbb{R}^{n_\omega}$. Now, let us introduce substitutions $E(\bar{e}) = g^r(t', x', e', e'_\omega)$ and $F = \tilde{y}^r(t', x', e'_\omega)$. Our goal is to show that if

$$E(\bar{e}') \preceq A_1 \bar{e}' + F, \qquad (3.43)$$
$$E(\bar{e}') \preceq A_2 \bar{e}' + F, \qquad (3.44)$$

then

$$E(\bar{e}') \preceq [(1-\alpha)A_1 + \alpha A_2]\bar{e}' + F \qquad (3.45)$$

where $\alpha \in [0,1]$. Using (3.43) and (3.44), we obtain

$$(1-\alpha)A_1 \bar{e}' + \alpha A_2 \bar{e}' \succeq (1-\alpha)(E(\bar{e}') - F) + \alpha(E(\bar{e}') - F) = E(\bar{e}') - F,$$

which is equivalent to (3.45). Since t', (x', e') and e'_ω were picked arbitrarily from $[\underline{t}, \underline{t}+\tau]$, $\mathcal{S}$ and $\mathbb{R}^{n_\omega}$, respectively, therefore (3.45) holds for all $t \in [\underline{t}, \underline{t}+\tau]$ and all $(x,e) \in \mathcal{S}$. The fact that the intersection of a family of convex sets is a convex set concludes the proof. $\square$

3.5 Example: Autonomous Cruise Control

In this section, we apply the input-output triggered update policy (3.35) to the *trajectory tracking* controller presented in Section 3.1.1 and exploit the ideas from Section 3.4.4. Since the controller (3.4) is not a dynamic controller, we have $f_c \equiv 0$. Next, we take $x_c = (v_{R1}, \omega_{R1})$ and

$$u = g_c(t, x_c) = x_c. \qquad (3.46)$$

Recall that the states of the plant (3.3) are measured directly, i.e.,

$$y = g_p(t, x_p) = x_p, \qquad (3.47)$$

and assume that the communication network for transmitting the control input (v_{R1}, ω_{R1}) to the actuators of R_1 can be neglected due to on-board controllers. Because of the absence of the communication network for transmitting u and $f_c \equiv 0$, we have that $e_u \equiv 0$. Consequently, we can exclude x_c from x

Input-Output Triggering 71

and take $x = x_p$. Notice that f_p is given by (3.3). Recall that the external input is $\omega_p = (v_{R2}, \omega_{R2})$. Now we have

$$e = \hat{x} - x = [e_1 \ e_2 \ e_3]^\top, \tag{3.48}$$

and

$$e_\omega = \hat{\omega}_p - \omega_p = [e_{\omega,1} \ e_{\omega,2}]^\top. \tag{3.49}$$

After substituting (3.3), (3.4), (3.46), (3.47), (3.48) and (3.49) into (3.23) and (3.25), we obtain (compare with expression (3.26))

$$\dot{e} = -\dot{x} = \underbrace{\begin{bmatrix} -Q - P(x_2 + e_2)x_2 - k_3(x_3 + e_3)x_2 + R + k_1(x_1 + e_1) - S \\ \hat{\omega}_{R2}x_1 + P(x_2 + e_2)x_1 + k_3(x_3 + e_3)x_1 - T \\ e_{\omega,2} + P(x_2 + e_2) + k_3(x_3 + e_3) \end{bmatrix}}_{= g^\delta(t, x, e, e_\omega) = -f^\delta(t, x, e, e_\omega)} \tag{3.50}$$

and (compare with expression (3.11))

$$\bar{\dot{e}} \preceq \underbrace{\begin{bmatrix} k_1 & k_2|\hat{v}_{R2}|M_2 & \max\{|\hat{v}_{R2}|, k_3 M_2\} \\ k_2|\hat{v}_{R2}|M_2 & k_2|\hat{v}_{R2}|M_1 & \max\{k_2|\hat{v}_{R2}|, k_3 M_1\} \\ \max\{|\hat{v}_{R2}|, k_3 M_2\} & \max\{k_2|\hat{v}_{R2}|, k_3 M_1\} & k_3 \end{bmatrix}}_{A \ = \ \text{initial point for the convex program (3.42)}} \bar{e} +$$

$$+ \underbrace{\begin{bmatrix} k_2|\hat{v}_{R2}|x_2^2 + |k_1 x_1 + e_{\omega,1}\cos x_3 - \hat{\omega}_{R2}x_2 - k_3 x_2 x_3| \\ k_2|\hat{v}_{R2}|x_1 x_2| + |\hat{\omega}_{R2}x_1 + k_3 x_1 x_3 - T| \\ k_2|\hat{v}_{R2}|x_2| + |e_{\omega,2} + k_3 x_3| \end{bmatrix}}_{\tilde{y}(t, x, \hat{\omega}_p, e_\omega)} \tag{3.51}$$

where

$$P = k_2 \hat{v}_{R2} \frac{\sin(x_3 + e_3)}{x_3 + e_3},$$
$$Q = \hat{\omega}_{R2} x_2,$$
$$R = \hat{v}_{R2} \cos(x_3 + e_3),$$
$$S = (\hat{v}_{R2} - e_{\omega,1}) \cos x_3,$$
$$T = (\hat{v}_{R2} - e_{\omega,1}) \sin x_3,$$

and $|x_1| \leq M_1$, $|x_2| \leq M_2$. Constants M_1 and M_2 are obtained from the sets $\mathcal{S}_i$'s (see the next paragraph for more details about computing $\mathcal{S}_i$'s). We choose $k_1 = 1.5$, $k_2 = 1.2$ and $k_3 = 1.1$. In addition, in order to make this example more realistic, ω_p takes values in $[-3, 3] \times [-3, 3]$. Consequently, reachability sets $\mathcal{S}_i$ are compact which, in turn, implies finite M_1 and M_2. Since scenarios including Case 3.3 are more realistic, this section includes numerical results for such a scenario. When emulating noisy environments, we use $e_\omega \in U([-0.3, 0.3] \times [-0.3, 0.3])$ and $h(t) \in U([-0.15, 0.15] \times [-0.15, 0.15] \times [-0.15, 0.15])$ where $U(\mathcal{B})$ denotes the uniform distribution over a set $\mathcal{B}$.

Before verifying that the hypotheses of Theorem 3.4 hold and presenting numerical results, let us provide details behind Steps 1–6. Since the bounds on ω_p are known, we confine measurements $\hat{\omega}_p$ to the same set, i.e., $r \in \mathcal{P} = [-3, 3] \times [-3, 3]$. In other words, if we obtain $\hat{\omega}_p(t_i) \notin \mathcal{P}$ (for instance, because of measurement noise), we use the closest value in $\mathcal{P}$ (with respect to the Euclidean distance) for $\hat{\omega}_p(t_i)$. Now, after receiving $\hat{\omega}_p(t_i)$ and $\hat{y}(t_i)$, we update control signal $u(t_i)$. Next, we need to determine when to sample again. Utilizing $\hat{y}(t_i)$, we obtain $\hat{x}(t_i) \in \hat{y}(t_i) \pm [-0.15, 0.15] \times [-0.15, 0.15] \times [-0.15, 0.15]$. Starting from these $\hat{x}(t_i)$, and due to the fact that $u(t_i)$ is the linear and angular velocity of $R1$ and remains constant until t_{i+1}, we readily compute reachable states $\hat{x}(t_i + \tau_i)$ for any $\tau_i \geq 0$ using $u(t_i)$ and the bounds on ω_p. Because of (3.48), reachable output errors $e(t_i + \tau_i)$ are immediately computed and sets $\mathcal{S}_i$ are obtained. Inspecting the form of A in (3.51), we infer that the first two components of x are important for (3.51) to hold on some $\mathcal{S}_i$. Due to the properties of $\preceq$, we choose the maximum values of $|x_1|$ and $|x_2|$ in $\mathcal{S}_i$ denoted M_1 and M_2, respectively.

Let us verify that the hypotheses of Theorem 3.4 hold. Due to (3.51), we infer that item (i) of Assumption 3.2 is fulfilled provided that M_1 and M_2 are finite. For this reason, we restrict the analysis to a set for the state x given by bounded $|x_1|$ and $|x_2|$. Next, combining the approach of [78] and the power iterations method [77] (as this chapter is void of delays), we estimate $\mathcal{L}_2$-gains Σ_n^δ over $\mathcal{P}$ and obtain $\overline{\gamma_n} = 96$. The upper bound $\overline{K_n}$ is obtained similarly. Hence, item (ii) of Assumption 3.2 holds. Item (iii) of Assumption 3.2 is inferred from (3.50) and (3.51) as follows. It can be shown that

$$\|x\|^2 \leq k(\|\tilde{y}\|^2 + \|(e, e_\omega)\|^2)$$

for any $k \geq 2$. Integrating both sides of the last inequality over $[t_0, t]$ for any $t \geq t_0$ and taking the square root, yields

$$\|x[t_0, t]\|_2 \leq \sqrt{k}\|\tilde{y}[t_0, t]\|_2 + \sqrt{k}\|(e, e_\omega)[t_0, t]\|_2.$$

In other words, the state x of the system Σ_n^r is $\mathcal{L}_2$-detectable from $(e, e_\omega, \tilde{y})$ with the upper bounds $K = 0$ and $\gamma = \sqrt{k}$. Finally, the condition (3.37) of Theorem 3.4 is easily verified for an arbitrary user-selected $\overline{K} \geq 0$, where $\overline{K}$ captures the desired impact of the initial conditions on the tracking performance according to (3.40), as the simulation progresses for a given $\delta : [t_0, \infty) \to \mathcal{P}$. Recall that the switching signal δ is in fact $\hat{\omega}_p$. Accordingly, only after the control law (3.4) is updated with the most recent information about ω_p, i.e., with $\hat{\omega}_p$ (and provided that this value is different from the current value of $\hat{\omega}_p$), that time instant becomes a switching instant and the corresponding $x(t_i^{\delta+})$ contributes to the left-hand side of (3.37). In case (3.37) might get violated, simply decrease τ_i's in Step 6 or cease switching according to Step 3 (i.e., the left-hand side of (3.37) remains unaltered) by using

Input-Output Triggering

the current $\hat{\omega}_p$ until (3.37) is no longer compromised (if ever). Furthermore, even from the viewpoint that $\hat{\omega}_p$ is a noisy version of ω_p, it might not be advantageous (in terms of decreasing the tracking error) to always update the control law (3.4) with the incoming information about ω_p, i.e., with $\hat{\omega}_p$. This observation is found in the following paragraph as well.

In the simulations, we choose $\omega_p(t) = (1,1)t_{[0,2.26)} + (0.6, 0.15)t_{[2.26, 9.25)} + (2,2)t_{[9.25, 12]}$, where $t_\mathcal{I}$ is the indicator function on an interval $\mathcal{I}$, i.e., $t_\mathcal{I} = t$ when $t \in \mathcal{I}$ and zero otherwise. The corresponding $\mathcal{L}_2$-gains Σ_n^δ are as follows: $\gamma_n^{(0.6, 0.15)} = 22$, $\gamma_n^{(1,1)} = 53$ and $\gamma_n^{(2,2)} = 56$. In order to illustrate Theorem 3.4 and the mechanism behind (3.35), we superpose a continuous signal $e_\omega(t) \in [-0.3, 0.3] \times [-0.3, 0.3]$, where $t \in [0, 12]$, onto the above $\omega_p(t)$, and update the control law with $\hat{\omega}_p$ being $(0.6, 0.15)$, $(1,1)$ or $(2,2)$. This way, we are able to use a fixed γ_n^δ between two switches so that the left-hand side of (3.37) does not increase unnecessarily (and because the received values are corrupted by noise anyway). In addition, the impact of changes in $\hat{y}$ on τ_i's is easier to observe. The obtained numerical results are provided in Figure 3.5. As can be seen from Figure 3.5, intersampling intervals τ_i's tend to increase as $\|x\|$ approaches the origin because M_1 and M_2 decrease. In addition, the abrupt changes of τ_i at 2.26 s and 9.25 s, visible in Figure 3.5(c), are the consequence of the abrupt changes in $\hat{\omega}_p$. In other words, τ_i's adapt to the changes in $\hat{\omega}_p$. This adaptation of τ_i's follows from (3.35) where individual gains are considered instead of the unified gain [207]. The simulation results obtained using the unified gain $\sup_{r \in \mathcal{P}} \gamma_n^r = \overline{\gamma_n} = 96$, achieved when $r = (3,3)$, and corresponding A in (3.35) are shown in Figure 3.6. Apparently, the use of the unified gains decreases τ_i's, does not allow for adaptation of τ_i, and yet does not necessarily yield stability of the closed-loop system since (3.37) does not have to hold (a similar observation is found in [207]). Consequently, the number of transmissions in the scenario depicted in Figure 3.5 is 580, while in the scenario depicted in Figure 3.6 it is 1377 (refer to Figure 3.3 as well). Finally, it should be mentioned that the oscillations of x in Figures 3.5(a) and 3.6(a) are an inherited property of the controller, and not a consequence of intermittent feedback.

3.6 Conclusions and Perspectives

In this chapter we present a methodology for input-output triggered control of nonlinear systems. Based on the currently available measurements of the output and external input of the plant, a sampling policy yielding the closed-loop system stable in some sense is devised. Using the formalism of $\mathcal{L}_p$-gains and $\mathcal{L}_p$-gains over a finite horizon, the small-gain theorem is employed to prove stability, asymptotic, and $\mathcal{L}_p$-stability (with bias) of the closed-loop system. Different types of stability are a consequence of different assumptions on the

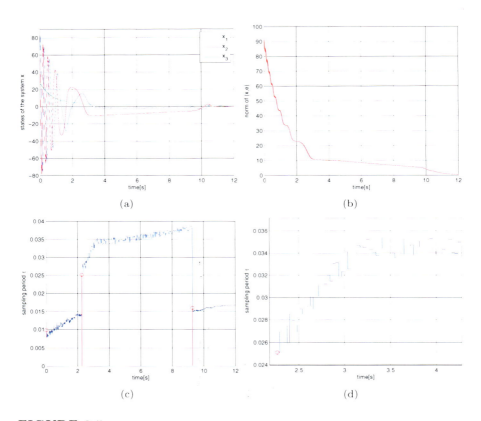

FIGURE 3.5
A realistic scenario illustrating input-output triggering: (a) states x of the tracking system, (b) norm of (x, e), (c) values of intersampling intervals τ_i's between two consecutive transmissions. Red stems indicate time instants when changes in δ happen, and (d) a detail from Figure 3.5(c).

Input-Output Triggering

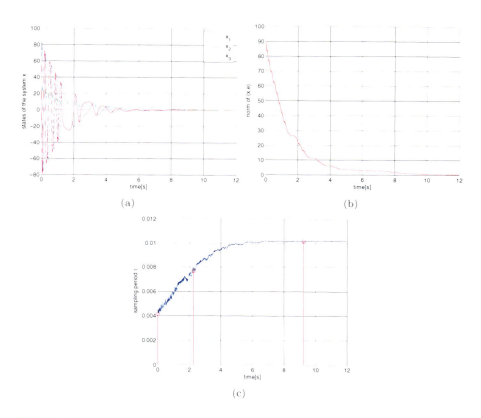

FIGURE 3.6
A realistic scenario illustrating input-output triggering using the unified gains: (a) states x of the tracking system, (b) norm of (x, e), and (c) values of intersampling intervals τ_i's between two consecutive transmissions. Red stems indicate time instants when changes in δ happen.

noise environment causing the mismatch between the actual external input and output of the plant, and the measurements available to the controller via feedback. The closed-loop systems are modeled as impulsive switched systems, and several results regarding $\mathcal{L}_p$-stability of such systems are devised. Finally, our input-output triggered sampling policy is exemplified on a trajectory tracking controller for velocity-controlled unicycles.

It is of interest to apply scattering transformation between the controller and plant in order to eliminate detrimental effects of delays (if present). Furthermore, actuators with saturation could be analyzed. In order to obtain larger intertransmission intervals, zero-order hold estimation strategies can be replaced with model-based estimation of control signals and plant outputs (as in Chapter 2). Finally, we expect our results (with slight modifications) to hold for input-to-state stability of impulsive switched systems.

3.7 Proofs of Main Results

3.7.1 Properties of Matrix Functions

Let $f : \mathbb{R} \to \mathbb{R}$ be a continuous function. The work in [168] constructs a matrix A such that $\|f(A)\| = f(\|A\|)$ for any f by requiring that matrix A is symmetric, with nonnegative entries and positive semidefinite. While symmetry of the matrix and nonnegative entries are required throughout [168], positive semidefiniteness is required only in [168, Lemma 7.1]. Exploiting the fact that f in [168, Lemma 7.1] (i.e., in Theorem 3.2 herein) is the exponential function $f(\cdot) = \exp(\cdot)$ and A is a real symmetric matrix with nonnegative entries, the following lemmas show that the positive semidefiniteness requirement of [168, Theorem 5.1] can be relaxed.

Lemma 3.1. *Suppose that $f(\cdot) = \exp(\cdot)$ and A is a real $n \times n$ symmetric matrix. The eigenvalue of A with the largest absolute value is real and nonnegative if and only if $\|f(A)\| = f(\|A\|)$.*

Proof. It is well known that real symmetric matrices can be diagonalized, i.e., $A = UDU^T$ where D is a diagonal matrix and U is an orthogonal matrix. Since the spectral norm is unitarily invariant [60], it is straightforward to show that the following equalities hold:

$$\|f(A)\| = \|UDU^T\| = \|D\| = \|\mathrm{diag}(f(\lambda_1(A)), f(\lambda_2(A)), \ldots, f(\lambda_n(A)))\| =$$
$$= \max_k |f(\lambda_k(A))|. \qquad (3.52)$$

On the other hand, using the definition of the induced matrix 2-norm $\|\cdot\|$, we have

$$f(\|A\|) = f(\max_k |\sigma_k(A)|). \qquad (3.53)$$

Input-Output Triggering 77

In addition, real symmetric matrices have the following property

$$\sigma_k(A) = |\lambda_k(A)|. \tag{3.54}$$

Using (3.54), and monotonicity and positivity of the exponential function, we conclude that expressions (3.52) and (3.53) are equal if and only if the eigenvalue of A with the largest absolute value is real and nonnegative. □

For completeness, we now write the following well-known result for symmetric matrices (see, e.g., [122, Chapter 8]).

Lemma 3.2. *If A is a symmetric matrix with nonnegative entries, then the eigenvalue of A with the largest absolute value is real and nonnegative.*

3.7.2 Proof of Theorem 3.1

Proof. We prove this theorem for the case $p \in [1, \infty)$. The proof for the case $p = \infty$ is similar.

Let us start from some initial condition $\chi(t_0)$ and apply input ω to an impulsive switched system Σ^δ given by (3.5) to obtain the state trajectory $t \mapsto \chi(t)$ and associated output $t \mapsto y(t)$. Now, we can write for every $t \geq t_0$

$$\|y[t_0, t]\|_p^p = \int_{t_0}^t \|y(s)\|^p \mathrm{d}s = \sum_{i=0}^{J-1} \int_{t_i^\delta}^{t_{i+1}^\delta} \|y(s)\|^p \mathrm{d}s + \int_{t_J^\delta}^t \|y(s)\|^p \mathrm{d}s$$

$$= \sum_{i=0}^{J-1} \|y[t_i^\delta, t_{i+1}^\delta]\|_p^p + \|y[t_J^\delta, t]\|_p^p, \tag{3.55}$$

where $J = \arg\max\{j : t_j^\delta \leq t\}$. From (3.7) we obtain

$$\|y[t_0, t]\|_p^p \leq \sum_{i=0}^{J-1} \left(\widetilde{K}(\tau_i^\delta))\|\chi(t_i^{\delta+})\| + \widetilde{\gamma}(\tau_i^\delta)\|\omega[t_i^\delta, t_{i+1}^\delta]\|_p\right)^p +$$

$$+ \left(\widetilde{K}(\tau_J^\delta))\|\chi(t_J^{\delta+})\| + \widetilde{\gamma}(\tau_J^\delta)\|\omega[t_J^\delta, t]\|_p\right)^p. \tag{3.56}$$

Using (3.8) and (3.9) yields

$$\|y[t_0, t]\|_p^p \leq \sum_{i=0}^{J-1} \left(K_M\|\chi(t_i^{\delta+})\| + \gamma_M\|\omega[t_i^\delta, t_{i+1}^\delta]\|_p\right)^p +$$

$$+ \left(K_M\|\chi(t_J^{\delta+})\| + \gamma_M\|\omega[t_J^\delta, t]\|_p\right)^p. \tag{3.57}$$

In what follows we use the following version of the *Minkowski inequality*

$$\left(\sum_{i=1}^M (a_i + b_i)^p\right)^{1/p} \leq \left(\sum_{i=1}^M a_i^p\right)^{1/p} + \left(\sum_{i=1}^M b_i^p\right)^{1/p}, \tag{3.58}$$

where $a_i, b_i \geq 0$ and $M \in \mathbb{N} \cup \{\infty\}$. Taking the pth root of (3.57) yields

$$\|y[t_0,t]\|_p \leq \left(\sum_{i=0}^{J-1} \left(K_M \|\chi(t_i^{\delta+})\| + \gamma_M \|\omega[t_i^\delta, t_{i+1}^\delta]\|_p \right)^p + \right.$$

$$\left. + \left(K_M \|\chi(t_J^{\delta+})\| + \gamma_M \|\omega[t_J^\delta, t]\|_p \right)^p \right)^{\frac{1}{p}}. \quad (3.59)$$

Applying (3.58) to the right-hand side of (3.59) with $M = J$, $a_i = K_M \|\chi(t_i^{\delta+})\|$ and $b_i = \gamma_M \|\omega[t_i^\delta, t_{i+1}^\delta]\|_p$ for $i \leq J-1$, $b_J = \gamma_M \|\omega[t_J^\delta, t]\|_p$, leads to

$$\|y[t_0,t]\|_p \leq K_M \left(\sum_{i=0}^{J} \|\chi(t_i^{\delta+})\|^p \right)^{\frac{1}{p}} +$$

$$+ \gamma_M \left(\sum_{i=0}^{J-1} \|\omega[t_i^\delta, t_{i+1}^\delta]\|_p^p + \|\omega[t_J^\delta, t]\|_p^p \right)^{\frac{1}{p}}. \quad (3.60)$$

Applying the inequality $(a+b)^{1/p} \leq a^{1/p} + b^{1/p}$, where $a, b \geq 0$, to the first term in (3.60) multiple times, and noting that, since $t_0^\delta = t_0$,

$$\sum_{i=0}^{J-1} \|\omega[t_i^\delta, t_{i+1}^\delta]\|_p^p + \|\omega[t_J^\delta, t]\|_p^p = \|\omega[t_0, t]\|_p^p$$

(as in (3.55)), we obtain

$$\|y[t_0,t]\|_p \leq K_M \left(\sum_{i=0}^{J} \|\chi(t_i^{\delta+})\| \right) + \gamma_M \|\omega[t_0, t]\|_p. \quad (3.61)$$

Applying (3.10) we obtain

$$\|y[t_0,t]\|_p \leq K_M \overline{K} \|\chi(t_0)\| + \gamma_M \|\omega[t_0, t]\|_p \quad (3.62)$$

for all $t \geq t_0$. $\square$

3.7.3 Proof of Theorem 3.2

Let $Df(t)$ denote the *left-hand derivative* of $f : \mathbb{R} \to \mathbb{R}^n$, i.e., $Df(t) = \lim_{h \to 0, h < 0} \frac{f(t+h) - f(t)}{h}$. The following two lemmas and theorem are taken from [168] and slightly modified.

Lemma 3.3. *Let $I = [t_0, t_1]$, $v \in \mathbb{R}^n$ and consider $Dv \preceq Av + d(t)$, $v(t_0) = v_0$, $\forall t \in I$, where $A \in \mathcal{A}_{n_e}^+$, $\|A\| < \infty$, and $d(t) : I \to \mathbb{R}^n$ is continuous. Then, for all $t \in I$, $v(t)$ is bounded by*

$$v(t) \preceq \exp(A(t-t_0))v_0 + \int_{t_0}^{t} \exp(A(t-s))d(s)ds.$$

Proof. Now we are ready to prove Theorem 3.2. By the hypotheses of the theorem, we have $\dot{\overline{\chi}} = \widetilde{g}(t,\chi,v) \preceq A\overline{\chi} + \tilde{y}(t,\chi,v)$ for all $t \in [t_0, t_0 + \tau]$, and the ith component of $\dot{\overline{\chi}}$ is given by:

$$\left|\tfrac{d}{dt}\chi_i(t)\right| = \left|\lim_{h \to 0, h < 0} \tfrac{\chi_i(t+h) - \chi_i(t)}{h}\right| \geq \lim_{h \to 0, h < 0} \tfrac{|\chi_i(t+h)| - |\chi_i(t)|}{h} =$$
$$= D\overline{\chi}_i(t).$$

Therefore, $D\overline{\chi} \preceq A\overline{\chi} + \tilde{y}(t)$. Using Lemma 3.3, we can write

$$\overline{\chi}(t) \preceq \exp(A(t - t_0))\overline{\chi}(t_0) + \int_{t_0}^{t} \exp(A(t - s))\tilde{y}(s)\mathrm{d}s. \tag{3.63}$$

Setting the input term $\tilde{y} \equiv 0$, we obtain $\overline{\chi}(t) \preceq \exp\left(A(t - t_0)\right)\overline{\chi}(t_0)$. Taking the norm of both sides of this inequality and using Lemmas 3.1 and 3.2 we obtain

$$\|\overline{\chi}(t)\| \leq \exp\left(\|A\|(t - t_0)\right)\|\overline{\chi}(t_0)\|. \tag{3.64}$$

Raising to the $p^{\text{th}} \in [1, \infty)$ power and integrating over $[t_0, t]$ yields

$$\|\overline{\chi}[t_0, t]\|_p^p \leq \frac{\exp(\|A\|p(t - t_0)) - 1}{p\|A\|}\|\overline{\chi}(t_0)\|^p.$$

Taking the pth root yields

$$\|\overline{\chi}[t_0, t]\|_p \leq \left(\frac{\exp(\|A\|p(t - t_0)) - 1}{p\|A\|}\right)^{\frac{1}{p}}\|\overline{\chi}(t_0)\|, \quad p \in [1, \infty). \tag{3.65}$$

The $\mathcal{L}_\infty$ bound is easily obtained by taking $\lim_{p \to \infty} \|\overline{\chi}[t_0, t]\|_p$ obtaining

$$\|\overline{\chi}[t_0, t]\|_\infty \leq \exp(\|A\|(t - t_0)) - 1)\|\overline{\chi}(t_0)\|.$$

Let us now set $\overline{\chi}(t_0) = 0$ and estimate the contribution from the input term. From (3.63) we have $\overline{\chi}(t) \preceq \int_{t_0}^{t} \exp(A(t - s))\tilde{y}(s)\mathrm{d}s$. Using Lemmas 3.1 and 3.2 we obtain

$$\|\overline{\chi}(t)\| \leq \int_{t_0}^{t} \exp(\|A\|(t - s))\|\tilde{y}(s)\|\mathrm{d}s. \tag{3.66}$$

Let us denote $\phi(s) = \exp(\|A\|s)$. Integrating the previous inequality and using Lemma 2.2 with $p = q = r = 1$ yields the $\mathcal{L}_1$-norm estimate:

$$\|\overline{\chi}[t_0, t]\|_1 \leq \|\phi[0, t - t_0]\|_1 \|\tilde{y}[t_0, t]\|_1. \tag{3.67}$$

Taking the max over $[t_0, t]$ in (3.66) and using Lemma 2.2 with $q = r = \infty$ and $p = 1$ yields the $\mathcal{L}_\infty$-norm estimate:

$$\|\overline{\chi}[t_0, t]\|_\infty \leq \|\phi[0, t - t_0]\|_1 \|\tilde{y}[t_0, t]\|_\infty. \tag{3.68}$$

We can think of (3.63) as a linear operator G mapping $\tilde{y}$ to $\overline{\chi}$ with bound for the norms $\|G\|_1 \leq \|G\|_1^*$ and $\|G\|_\infty \leq \|G\|_\infty^*$ where $\|G\|_1^*$ and $\|G\|_\infty^*$ are given by (3.67) and (3.68), respectively. Because $\|G\|_1^* = \|G\|_\infty^*$, Theorem 2.3 gives that $\|G\|_p \leq \|G\|_1^* = \|G\|_\infty^*$ for all $p \in [1, \infty]$. This yields

$$\|\overline{\chi}[t_0, t]\|_p \leq \|\phi[0, t - t_0]\|_1 \|\tilde{y}[t_0, t]\|_p, \quad p \in [1, \infty].$$

Since $\|\phi[0, t - t_0]\|_1 = \frac{\exp(\|A\|(t-t_0))-1}{\|A\|}$, we obtain

$$\|\overline{\chi}[t_0, t]\|_p \leq \frac{\exp(\|A\|(t - t_0)) - 1}{\|A\|} \|\tilde{y}[t_0, t]\|_p, \quad p \in [1, \infty]. \tag{3.69}$$

After summing up the contributions of (3.65) and (3.69), the statement of the theorem follows. □

3.7.4 Proof of Results in Section 3.4.3

The proofs of the results in Section 3.4.3 use the property over an arbitrary finite interval with constant δ introduced in the next section. After that, this property is used sequentially, over a finite horizon of arbitrary length, to obtain an $\mathcal{L}_p$-bound on (x, e).

3.7.4.1 $\mathcal{L}_p$ property over an arbitrary finite interval with constant δ

Proof. Consider the nontrivial interval $I := [\underline{t}, \underline{t} + \tau)$, $\underline{t} \geq t_0$ on which $t \mapsto \delta(t)$ is constant and with $\tau > 0$ to be defined. Let r be such that $\delta(t) = r$ for all $t \in I$. From item (ii) of Assumption 3.2, given initial condition $x(\underline{t})$, we have, for all $t \in I$,

$$\begin{aligned}\|\tilde{y}^r[\underline{t}, t]\|_p &\leq K_n^r \|x(\underline{t})\| + \gamma_n^r \|(e, e_\omega)[\underline{t}, t]\|_p \\ &\leq K_n^r \|x(\underline{t})\| + \gamma_n^r \|e[\underline{t}, t]\|_p + \gamma_n^r \|e_\omega[\underline{t}, t]\|_p.\end{aligned} \tag{3.70}$$

Item (i) of Assumption 3.2 allows us to invoke Theorem 3.2 for Σ_e^r on I. Then, we have, for all $t \in I$,

$$\|e[\underline{t}, t]\|_p \leq \widetilde{K}_e^r(\tau)\|e(\underline{t})\| + \widetilde{\gamma}_e^r(\tau)\|\tilde{y}^r[\underline{t}, t]\|_p \tag{3.71}$$

for any τ chosen as

$$\tau \in (0, \tau^{r,*}], \tag{3.72}$$

where $\widetilde{K}_e^r(\tau), \widetilde{\gamma}_e^r(\tau)$ are given as in (3.31) and $\tau^{r,*}$ is given as in (3.35). Due to the construction of $\tau^{r,*}$ and the choice of τ in (3.72), the open-loop gain of the interconnection (see Figure 3.4) from e_ω to e is $\gamma_n^r \widetilde{\gamma}_e^r(\tau)$ is less than $\kappa \in (0, 1)$. Then, combining (3.70) and (3.71), we have, for each $t \in I$,

$$\|e[\underline{t}, t]\|_p \leq \frac{\widetilde{K}_e^r(\tau)}{1 - \gamma_n^r \widetilde{\gamma}_e^r(\tau)} \|e(\underline{t})\| + \frac{\widetilde{\gamma}_e^r(\tau) K_n^r}{1 - \gamma_n^r \widetilde{\gamma}_e^r(\tau)} \|x(\underline{t})\| + \frac{\widetilde{\gamma}_e^r(\tau) \gamma_n^r}{1 - \gamma_n^r \widetilde{\gamma}_e^r(\tau)} \|e_\omega[\underline{t}, t]\|_p. \tag{3.73}$$

Next, let us use $\tau_{\max}^*$ from Remark 3.2 instead of τ in order to obtain the following upper bound for $\widetilde{K}_e^r(\tau)$ over $(0, \tau^{r,*}]$ and for any $r \in \mathcal{P}$:

$$\widetilde{K}_e^r(\tau) \leq \sup_{r \in \mathcal{P}} \left(\frac{\exp(\|A^r\| p\tau_{\max}^*) - 1}{p\|A^r\|} \right)^{\frac{1}{p}} =: \widetilde{K}_e(\tau).$$

This supremum exists due to Assumption 3.2. Likewise, $\widetilde{\gamma}_e^r(\tau)$ can be upper bounded by a constant $\widetilde{\gamma}_e(\tau)$. Next, using the detectability property in item (iii) of x from $(\widetilde{y}^r, e_\omega, e)$ (with gains K and γ), and combining (3.73) and (3.70), gives

$$\|(x,e)[\underline{t},t]\|_p \leq \widehat{K}\|(x(\underline{t}), e(\underline{t}))\| + \widehat{\gamma}\|e_\omega[\underline{t},t]\|_p \qquad (3.74)$$

for all $t \in I$, where the constants $\widehat{K}$ and $\widehat{\gamma}$ are given by

$$\widehat{K} = \sqrt{n_x + n_e} \max \left\{ K + \frac{\widetilde{\gamma}_e(\tau)\overline{K_n} + \gamma\overline{K_n} + \gamma\widetilde{\gamma}_e(\tau)\overline{K_n}}{1 - \kappa}, \right.$$

$$\left. \frac{\widetilde{K}_e(\tau) + \gamma\overline{\gamma_n}\widetilde{K}_e(\tau) + \gamma\widetilde{K}_e(\tau)}{1 - \kappa} \right\},$$

$$\widehat{\gamma} = \gamma + \frac{\widetilde{\gamma}_e(\tau)\overline{\gamma_n} + \gamma\overline{\gamma_n} + \gamma\widetilde{\gamma}_e(\tau)\overline{\gamma_n}}{1 - \kappa},$$

where we have used the bounds on K_n^r and γ_n^r given in item (ii) of Assumption 3.2. Notice that the above expressions are independent of $r \in \mathcal{P}$. □

3.7.4.2 Extending bounds to (an arbitrarily long) finite horizon

Proof. Given initial time t_0 and initial condition x_0, we use analysis on an arbitrary interval in Section 3.7.4.1 to design the sampling instants $\mathcal{T}$ to update $(\hat{y}, \hat{u})$ and, in turn, define the instants in $\mathcal{T}^\delta$ at which the signal $t \mapsto \delta(t)$ is allowed to switch. In this way, using $\underline{t} = t_0$, $x(\underline{t}) = x_0$, $e(\underline{t}) = e_0$, pick τ_0 to satisfy (3.72) to define $I_0 := [t_0, t_0 + \tau_0)$, over which $t \mapsto \delta(t)$ is constant, i.e., $\delta(t) = r_0$ for all $t \in I_0$, for some $r_0 \in \mathcal{P}$. Then, as previously explained, we obtain, for all $t \in I_0$,

$$\|(x,e)[t_0,t]\|_p \leq K^{r_0}\|(x_0, e_0)\| + \gamma^{r_0}\|e_\omega[t_0,t]\|_p. \qquad (3.75)$$

Now, at $t = t_0 + \tau_0$ a sampling event occurs, which updates (x,e) according to (3.26a), and, potentially, δ changes value. Then, using $\underline{t} = t_1$, $x(\underline{t}) = x(t_1^+)$, $e(\underline{t}) = \lim_{t \nearrow t_1} h(t) =: e(t_1)$, pick τ_1 to satisfy (3.72) to define $I_1 := [t_1, t_1 + \tau_1)$, over which $t \mapsto \delta(t)$ is constant, i.e., $\delta(t) = r_1$ for all $t \in I_1$, for some $r_1 \in \mathcal{P}$. Then, as before, we obtain, for all $t \in I_1$,

$$\|(x,e)[t_1,t]\|_p \leq K^{r_1}\|(x(t_1), e(t_1^+))\| + \gamma^{r_1}\|e_\omega[t_1,t]\|_p. \qquad (3.76)$$

Proceeding in this way for subsequent sampling intervals, in particular, for the i-th sampling time, we use $\underline{t} = t_i$, $x(\underline{t}) = x(t_i)$, $e(\underline{t}) = \lim_{t \nearrow t_i} h(t) =: e(t_i^+)$, pick τ_i to satisfy (3.72) to define $I_i := [t_i, t_1 + \tau_i)$, over which $t \mapsto \delta(t)$ is constant, i.e., $\delta(t) = r_i$ for all $t \in I_i$, for some $r_i \in \mathcal{P}$. Now, we obtain, for all $t \in I_i$,

$$\|(x,e)[t_i,t]\|_p \leq K^{r_i}\|(x(t_i), e(t_i^+))\| + \gamma^{r_i}\|e_\omega[t_i,t]\|_p. \qquad (3.77)$$

In fact, the above construction of t_i's can be performed for any $t \geq t_0$, combining (3.77) over each I_i interval with $i \in \{0, 1, 2, \ldots, N\}$, where N is such that $[0, t] \subset \cup_{i=0}^{N} I_i$ and $[0, t] \not\subset \cup_{i=0}^{N-1} I_i$. $\square$

Now we are ready to prove the results in Section 3.4.3.

3.7.4.3 Proof of Theorem 3.3

Proof. Let us now apply Theorem 3.1 with (x, e) playing the role of y and e_ω playing the role of ω. According to Theorem 3.1, we are interested only in instants $t_i \in \mathcal{T}$ that are followed by a change of the value of δ, i.e., in $t_i^\delta \in \mathcal{T}^\delta$. Recall that, due to (3.74), we have $K_M \leq \widehat{K}$ and $\gamma_M \leq \widehat{\gamma}$; hence, the suprema in (3.8) and (3.9) exist. Next, we need to verify that there exists $K_1 \geq 0$ such that $\sum_{t_i^\delta \in \mathcal{T}_0^\delta} \|(x(t_i^\delta), e(t_i^{\delta+}))\| \leq K_1 \|(x(t_0), e(t_0))\|$. Due to the perfect resets of e at $t_i^\delta \in \mathcal{T}^\delta$, given by (3.36a), and hypothesis (3.37), we infer that $K_1 = \overline{K}$ for Cases 3.1 and 3.2. In other words, (3.38) holds for Cases 3.1 and 3.2.

The case for $p = \infty$ follows similarly. $\square$

3.7.4.4 Proof of Theorem 3.4

Proof. Notice that $h(t)$ in Theorem 3.4 is more general than $h(t)$ in Remark 3.4. Consequently, the condition (ii) of Theorem 3.1 is no longer satisfied. In order to take into account the setting of Theorem 3.4, we rewrite (3.71) as follows

$$\|e[t_0, \overline{t}]\|_p \leq \widetilde{K}_e^r(\tau^{r_0,*})\|e(t_0)\| + \widetilde{\gamma}_e^r(\tau^{r_0,*})\|\widetilde{y}^r[t_0, \overline{t}]\|_p,$$

for all $\bar{t} \in I_0 = [t_0, t_0 + \tau^{r_0,*})$, and

$$\|e[t_i, \bar{t}]\|_p \leq \widetilde{K}_e^r(\tau^{r_i,*})\|h(t_i)\| + \widetilde{\gamma}_e^r(\tau^{r_i,*})\|\widetilde{y}^r[t_i, \bar{t}]\|_p$$
$$\leq \widetilde{K}_e^r(\tau^{r_i,*})\overline{K}_h + \widetilde{\gamma}_e^r(\tau^{r_i,*})\|\widetilde{y}^r[t_i, \bar{t}]\|_p$$
$$= \left(\frac{\exp(\|A^{r_i}\|p\tau^{r_i,*}) - 1}{p\|A^{r_i}\|}\right)^{\frac{1}{p}} \overline{K}_h + \widetilde{\gamma}_e^r(\tau^{r_i,*})\|\widetilde{y}^r[t_i, \bar{t}]\|_p$$
$$= \left[\int_{t_i}^{t_i + \tau^{r_i,*}} \left\|\exp\left(\|A^{r_i}\|(s - t_i)\right)\overline{K}_h\right\|^p ds\right]^{\frac{1}{p}} + \widetilde{\gamma}_e^r(\tau^{r_i,*})\|\widetilde{y}^r[t_i, \bar{t}]\|_p$$
$$\leq \left[\int_{t_i}^{t_i + \tau^{r_i,*}} \left\|\exp\left(\|A^{r_i}\|\tau^{r_i,*}\right)\overline{K}_h\right\|^p ds\right]^{\frac{1}{p}} + \widetilde{\gamma}_e^r(\tau^{r_i,*})\|\widetilde{y}^r[t_i, \bar{t}]\|_p$$
$$= \|b^{r_i}[t_i, t_i + \tau^{r_i,*}]\|_p + \widetilde{\gamma}_e^r(\tau^{r_i,*})\|\widetilde{y}^r[t_i, \bar{t}]\|_p,$$

for all $\bar{t} \in I_i = [t_i, t_i + \tau^{r_i,*})$ with $i \in \{0, 1, 2, \ldots, N\}$, where N is such that $[0, t] \subset \cup_{i=0}^N I_i$ and $[0, t] \not\subset \cup_{i=0}^{N-1} I_i$, and $b^{r_i}(t) \equiv b^{r_i} = \exp\left(\|A^{r_i}\|\tau^{r_i,*}\right)\overline{K}_h$. Notice that $\sup_{r \in \mathcal{P}} b^r \leq \overline{K}_h \exp(\overline{\eta}\tau_{\max}^*) =: b \equiv b(t)$, where $\overline{\eta}$ is defined in (ii) of Assumption 3.2 and $\tau_{\max}^*$ in Remark 3.2, is a suitable uniform choice over all I_i intervals. Next, let us use the following bounds over each interval:

$$\|e[t_i, \bar{t}]\|_p \leq \widetilde{K}_e^r(\tau^{r_i,*})\|e(t_i)\| + \widetilde{\gamma}_e^r(\tau^{r_i,*})\|\widetilde{y}^r[t_i, \bar{t}]\|_p + \|b[t_i, \bar{t}]\|_p,$$

for all $\bar{t} \in I_i = [t_i, t_i + \tau^{r_i,*})$, where $\|e(t_i)\| = 0$ for $i \in \{1, 2, ..., N\}$. Notice that the above systems are $\mathcal{L}_p$-stable with the same bias $b(t) \equiv b$ over intervals I_i.

In the same manner as before, we now apply the small-gain theorem over each interval I_i and utilize item (iii) of Assumption 3.2. Afterward, we apply Theorem 3.1 to the resulting $\mathcal{L}_p$-stable systems with bias over each interval I_i. [Notice that Theorem 3.1 readily extends to $\mathcal{L}_p$-stability with bias. By adding $\|b[t_i^\delta, t_{i+1}^\delta]\|_p$ and $\|b[t_J^\delta, t]\|_p$ to the respective terms on the right-hand side of (3.56) and following the remainder of the proof in Section 3.7.2, one arrives at $\|y[t_0, t]\|_p \leq K_M \overline{K}\|\chi(t_0)\| + \gamma_M \|\omega[t_0, t]\|_p + \|b[t_0, t]\|_p$ for all $t \geq t_0$.] In other words, following the exposition below (3.71), one readily obtains:

$$\|(x, e)[t_0, t]\|_p \leq \overline{K}\widehat{K}\|(x, e)(t_0)\| + \widehat{\gamma}\|e_\omega[t_0, t]\|_p + \|\widehat{b}[t_0, t]\|, \qquad \forall t \geq t_0,$$

where

$$\widehat{b}(t) \equiv \widehat{b} = \frac{\gamma\overline{\gamma}_n + \gamma}{1 - \kappa}\overline{K}_h \exp(\overline{\eta}\tau_{\max}^*).$$

The case of $p = \infty$ is similar. □

4
Optimal Self-Triggering

CONTENTS

4.1	Motivation, Applications and Related Works	85
4.2	Problem Statement: Performance Index Minimization	87
4.3	Obtaining Optimal Transmission Intervals	89
	4.3.1 Input-Output-Triggering via the Small-Gain Theorem .	89
	4.3.2 Dynamic Programming	90
	4.3.3 Approximate Dynamic Programming	91
	4.3.4 Approximation Architecture	91
	4.3.4.1 Desired Properties	92
	4.3.5 Partially Observable States	93
4.4	Example: Autonomous Cruise Control (Revisited)	94
4.5	Conclusions and Perspectives	95

In this chapter, we utilize the adaptive upper bounds of Chapter 3 in order to obtain stabilizing sampling instants that, at the same time, minimize a cost function pertaining to performance vs. energy trade-offs. Afterward, we revisit our Autonomous Cruise Control (ACC) example to illustrate the resulting optimal input-output triggered transmission mechanism.

4.1 Motivation, Applications and Related Works

This chapter investigates optimal intermittent feedback for nonlinear control systems. Using the currently available measurements from a plant, we develop a methodology that outputs when to update the controller with new measurements such that a given cost function is minimized. Our cost function captures trade-offs between the performance and energy consumption of the control system. The optimization problem is formulated as a Dynamic Programming (DP) problem, and Approximate Dynamic Programming (ADP) is employed to solve it. Instead of advocating a particular approximation architecture for ADP, we formulate properties that successful approximation architectures satisfy. In addition, we consider problems with partially observable states, and propose Particle Filtering (PF) to deal with partially observable states and intermittent feedback. Finally, our approach is applied to a mobile robot trajectory tracking problem.

At the moment, the research community is interested in extending intersampling intervals as much as possible without taking into account a deterioration in the performance due to intermittent feedback. In applications where energy consumption for using sensors, transmitting the obtained information, and executing control laws is relatively inexpensive compared to the slower convergence and excessive use of control power, extending intersampling intervals is not desirable. For instance, think of an airplane driven by an autopilot system designed to follow the shortest path between two points. Any deviation from the shortest path caused by intermittent feedback increases overall fuel consumption. This increase in fuel consumption is probably more costly than the cost of energy saved due to intermittent feedback. In this chapter, we encode these energy consumption trade-offs in a cost function, and design an ADP approach that yields optimal intertransmission intervals with respect to the cost function.

Similar problems to the problem considered herein are discussed in [75] and [64]. The authors in [75] balance control performance versus network cost by choosing the appropriate time delay–controller pair. The work in [64] investigates optimal control of hybrid systems based on ideas from dynamic and convex programming. While [75] associates costs with each time delay–controller pair, the work in [64] associates costs with switches between controllers. The optimization methods from [75] and [64] boil down to optimal control of switching systems (refer to [19] and [199]). However, the results of [19] and [199] are not applicable in this chapter. The authors of [199] focus on problems in which a prespecified sequence of active subsystems is given, and then seek both the optimal switching instants and the optimal continuous inputs. The work in [19] does not make assumptions about the number of switches nor the mode sequence; instead, they are determined by the solution of the problem. The authors in [19] develop their methodology for problems that include two modes, and provide directions for how to extend the methodology to problems with several distinct modes. However, this extension becomes intractable as the number of modes grows.

Motivated by [199], we adopt ADP (see [20] and [21]) as the strategy for tackling our problem. ADP is a set of methods for solving sequential decision-making problems under uncertainty by alleviating the computational burden of the infamous *curses of dimensionality* in DP [150]. In theory, DP solves a wide spectrum of optimization problems providing an optimal solution. In practice, straightforward implementations of DP algorithms are deemed computationally intractable for most of the applications. Therefore, the need for efficient ADP methods. However, comprehensive analyses and performance guarantees of these approximate methods are still unresolved (except in very special settings), and present a critical area of research. In literature, ADP is also known as *reinforcement learning* [166] and *neuro-dynamic program-*

Optimal Self-Triggering 87

ming [22]. Furthermore, ADP methods are extensively used in *operational research* [150].

The rest of the chapter is organized as follows. Section 4.2 presents the problem of optimal intermittent feedback and assumptions under which the problem is solved. The methodology brought together to solve the problem is presented in Section 4.3. The proposed methodology is verified on a trajectory tracking controller in Section 4.4. Conclusions are drawn and future challenges are discussed in Section 4.5.

4.2 Problem Statement: Performance Index Minimization

Consider a time-invariant nonlinear feedback control system consisting of a plant

$$\dot{x}_p = f_p(x_p, u, \omega_p),$$
$$y = g_p(x_p), \qquad (4.1)$$

and a controller

$$\dot{x}_c = f_c(x_c, y, \omega_c),$$
$$u = g_c(x_c), \qquad (4.2)$$

where $x_p \in \mathbb{R}^{n_p}$ and $x_c \in \mathbb{R}^{n_c}$ are the states, $y \in \mathbb{R}^{n_y}$ and $u \in \mathbb{R}^{n_u}$ are the outputs, and $\omega_p \in \mathbb{R}^{n_{\omega_p}}$ and $\omega_c \in \mathbb{R}^{n_{\omega_c}}$ are the external/exogenous inputs or disturbances of the plant and controller, respectively. Notice that y is the input of the controller, and u is the input of the plant. Let us denote the compound state of the closed-loop systems (4.1) and (4.2) by $x = (x_p, x_c)$ where $x \in \mathbb{R}^{n_x}$.

In the above control system, one tacitly assumes that the controller is fed continuously and instantaneously by the output y and the external input ω_p of the plant. However, in real-life applications this assumption is rarely fulfilled, and sometimes excessively demanding since stability of closed-loop systems can be achieved via intermittent feedback. For example, extremely fast processing units, sensors and communication devices are needed in order to emulate continuous and instantaneous feedback using digital technology. At the same time, intermittent feedback has detrimental effects on the performance of the control loop.

In order to account for the intermittent knowledge of y and ω_p by the controller, we model the links between the plant and controller as communication networks that cause intermittent exchange of information. More precisely, we introduce the output error vector e as follows:

$$e(t) := \hat{y}(t) - y(t) \qquad (4.3)$$

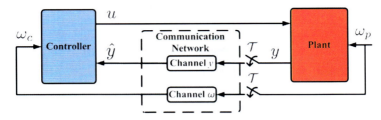

FIGURE 4.1
Diagram of a plant and controller with discrete transmission instants and communication channels giving rise to *intermittent feedback*.

where $\hat{y}$ is an estimate of y performed from the perspective of the controller, and the input error vector e_ω as follows:

$$e_\omega(t) := \hat{\omega}_p(t) - \omega_p(t) \tag{4.4}$$

where $\hat{\omega}_p$ is an estimate of ω_p from the perspective of the controller. For the sake of simplicity, we take $\hat{y}$ and $\hat{\omega}_p$ to be the most recently communicated values (or transmitted measurements) of the output and external input of the plant, i.e., we use the ZOH strategy. Now we introduce $\mathcal{T} := \{t_i : i \in \mathbb{N}_0\}$ as the set of time instants when outputs and external inputs of the plant are transmitted over communication networks. Finally, many control laws are designed such that $\omega_c = \hat{\omega}_p$. Examples are trajectory tracking controllers as in [170] and Chapter 3. An illustration of a control system indicating the communication channels that cause intermittent information is provided in Figure 4.1.

Next, we want to minimize the following cost function $V : \mathbb{R}^{n_x} \to \mathbb{R}$ that captures *performance vs. energy* trade-offs

$$V_{\tau_i}(x_0) = \underset{e_\omega}{\mathbb{E}} \left\{ \sum_{i=1}^{\infty} \gamma^i \Big[\underbrace{\int_{t_{i-1}}^{t_i} (x_p^T Q x_p + u^T R u) dt + S}_{l(x_p, u, \tau_i)} \Big] \right\} \tag{4.5}$$

over all sampling policies τ_i and for all initial conditions $x_0 \in \mathbb{R}^{n_x}$. In addition, $\gamma \in (0,1)$ is a discount factor that makes the sum (4.5) finite provided that $l(x_p, u, \tau_i)$ is bounded over all $[t_{i-1}, t_i]$ where $t_i \in \mathcal{T}$ and

$$t_i = t_{i-1} + \tau_{i-1}. \tag{4.6}$$

For clarity, we use τ_i instead of $\tau(\hat{y}(t_i), \hat{\omega}_p(t_i))$, but one has to keep in mind that, in general, intersampling intervals τ_i's depend on the most recently transmitted information from the plant, i.e., on $\hat{y}(t_i)$ and $\hat{\omega}_p(t_i)$. In addition, Q and R are positive definite matrices, and a nonnegative constant S represents the

Optimal Self-Triggering 89

cost incurred for sampling y and ω_p, transmitting $\hat{y}$ and $\hat{\omega}_p$, and updating the control signal u. In (4.5), the expectation over a stochastic signal e_ω is denoted $\underset{e_\omega}{\mathbb{E}}$.

The main problem considered herein can now be stated:

Problem 4.1. *For the system (4.1) and (4.2) with values of $\hat{\omega}_p$ and $\hat{y}$ received at t_i, $i \in \mathbb{N}_0$, find time intervals τ_i's until the next transmission instants such that (4.5) is minimized.*

We solve the above problem under the following assumptions:

Assumption 4.1. *$\hat{y}$ is corrupted by measurement noise.*

Assumption 4.2. *$\hat{\omega}_p$ is corrupted by measurement noise, and ω_p is arbitrary between two consecutive t_i's.*

Due to these assumptions, we have to deal with partially observable states (see Subsection 4.3.5 for more details).

4.3 Obtaining Optimal Transmission Intervals

This section presents the tools brought together to solve Problem 4.1 under Assumptions 4.1 and 4.2. Starting from the input-output-triggering of Chapter 3 that provides maximal stabilizing intersampling intervals $\tau_i^{\max}$'s, we find optimal τ_i^*'s for the cost function (4.5) resorting to ADP and PF.

4.3.1 Input-Output-Triggering via the Small-Gain Theorem

Building on the small-gain theorem, we develop *input-output-triggering* in Chapter 3. In other words, based on the currently available but outdated measurements of the outputs and external inputs of a plant, a simple expression for when to obtain new up-to-date measurements and execute the control law is provided. In fact, with slight modifications, our ADP approach is applicable to any self-triggered sampling policy (e.g., [6], [103], and [134]). Essentially, self-triggered sampling policies output maximal allowable intersampling intervals $\tau_i^{\max}$'s that provably yield a stable closed-loop system (4.1) and (4.2). Starting from these $\tau_i^{\max}$'s, the work presented herein finds

$$\tau_i^* \in [0, \tau_i^{\max}], \quad i \in \mathbb{N}_0 \tag{4.7}$$

that minimize (4.5). Because we know the upper bounds $\tau_i^{\max}$'s of stabilizing sampling policies, τ_i^*'s obtained in this chapter provably stabilize the plant. A comprehensive treatment of the problem of whether ADP solutions of optimal problems yield stability can be found in [206].

4.3.2 Dynamic Programming

Notice that the cost function (4.5) has the standard DP form. Let us now introduce a state transition function f that maps $x(t_{i-1})$, $u(t_{i-1})$ and $\hat{\omega}_p(t_{i-1})$ to $x(t_i)$ given some e_ω over $[t_{i-1}, t_i]$, i.e.,

$$x(t_i) = f(x(t_{i-1}), u(t_{i-1}), \tau_{i-1}, \hat{\omega}_p(t_{i-1}), e_\omega). \tag{4.8}$$

Due to intermittent feedback and the presence of nonlinearities in the plant and controller, the state transition function over τ_i's, in general, cannot be given in a closed form with, for example, a difference equation [132]. This is a typical impediment one faces when analyzing nonlinear systems under intermittent feedback. Therefore, in general, the state transition function (4.8) needs to be simulated using (4.1), (4.2), $\hat{y}(t_{i-1})$, $\hat{\omega}_p(t_{i-1})$ and e_ω over time horizon τ_{i-1}.

Next, let us assume that e_ω is a stationary stochastic process. Consequently, since we consider the infinite horizon problem (4.5) and a time-invariant control system (4.1) and (4.2), τ_i is not a function of t_i. Hence, we simply write τ instead of τ_i in the rest of the chapter. Solving the DP problem of minimizing (4.5) backward through time is combinatorially impossible since the state space x in (4.5) is uncountable. Therefore, we write the stochastic control problem of minimizing (4.5) over τ in its equivalent form known as the Bellman equation

$$V^*(z) = \inf_{\tau \in [0, \tau^{\max}]} \left(l(z, u, \tau) + \gamma \, \mathop{\mathbb{E}}_{e_\omega} \{V^*(f(z, u, \tau, \hat{\omega}_p, e_\omega))\} \right), \tag{4.9}$$

where $V^*(z)$ is called the optimal value function (or optimal cost-to-go function), and represents the cost incurred by an optimal policy τ^* when the initial condition in (4.5) is z. It is well known that V^* is the unique fixed point of (4.9). Therefore, the problem of minimizing (4.5) boils down to finding V^* in (4.9).

For notational convenience, we introduce the Bellman operator $\mathcal{M}$ as

$$\mathcal{M}g = (\mathcal{M}g)(z) = \inf_{\tau \in [0, \tau^{\max}]} \left(l(z, u, \tau) + \gamma \, \mathop{\mathbb{E}}_{e_\omega} \{g(f(z, u, \tau, \hat{\omega}_p, e_\omega))\} \right) \tag{4.10}$$

for any $g : \mathbb{R}^{n_x} \to \mathbb{R}$. Since $\gamma \in (0, 1)$, it can be shown that $\mathcal{M}$ is a contraction, i.e.,

$$\|\mathcal{M}u - \mathcal{M}v\|_s \leq \gamma \|u - v\|_s, \tag{4.11}$$

where $\|v\|_s = \sup_{z \in \mathbb{R}^{n_x}} v(z)$. The set $\mathcal{B}$ of all bounded, real-valued functions with the norm $\|\cdot\|_s$ is a Banach space. Therefore, for each initial $V^0 \in \mathcal{B}$, the sequence of value functions $V^{n+1} = \mathcal{M}V^n = \mathcal{M}^{n+1}V^0$ converges to V^*.

Two remarks are in order. First, it can be shown that the problem of finding an optimal τ^* for each state in (4.9) is non-convex. However, since τ is confined to a rather small compact set $[0, \tau^{\max}]$, we utilize gradient search

Optimal Self-Triggering 91

methods with constraints from different initial points in order to obtain τ^*. Second, the expectation $\underset{e_\omega}{\mathbb{E}}$ in (4.9) can be obtained in a closed form only for special cases. Otherwise, it can be calculated numerically by replacing the integral with a sum using a quadrature approximation. In Section 4.4, we use the Simpson formula [39].

Lastly, due to the "curses of dimensionality", solving (4.9) for $V^*(z)$ or iterating an initial V^0 is deemed intractable for most of the problems of interest; hence, we employ ADP in the next subsection where our goal is to find an approximation $\hat{V}^*$ of V^*.

4.3.3 Approximate Dynamic Programming

Among a number of methods in ADP, we choose the Value Iteration (VI) method for its simplicity and a wide spectrum of applications. Notice that $\mathcal{B}$ is an infinite dimensional vector space, meaning that it takes infinitely many parameters to describe V^*. Therefore, one introduces an approximate value function $\hat{V}^i$ of V^i where $i \in \mathbb{N}_0$. Then, the VI method iteratively applies $\mathcal{M}$ to an approximate value function $\hat{V}^0$ until $\|\hat{V}^{i+1} - \hat{V}^i\|_s < \epsilon$ where $\epsilon > 0$. Approximate value functions $\hat{V}^i$, $i \in \mathbb{N}_0$, can be represented in finite parameter approximation architectures such as Neural Networks (NNs). Note that it is not possible to obtain true value functions V^i's but only their approximations; hence, we write $\hat{V}^i$ instead of V^i. Basically, VI performs

$$\hat{V}^{i+1} = \mathcal{M}\hat{V}^i, \quad i \in \mathbb{N}_0 \tag{4.12}$$

until

$$\|\hat{V}^{i+1} - \hat{V}^i\|_s < \epsilon \tag{4.13}$$

where $\epsilon > 0$.

In order to calculate $\hat{V}^{i+1}$ in (4.12), we need to apply (4.10) over all $z \in \mathbb{R}^{n_x}$. Obviously, this is computationally impossible since $\mathbb{R}^{n_x}$ contains uncountably many points. Therefore, many ADP approaches focus on a compact subset $\mathcal{C}_x \subset \mathbb{R}^{n_x}$, choose a finite set of points $\mathcal{X} \subset \mathcal{C}_x$, and calculate $\hat{V}^{i+1}$ only for the points in $\mathcal{X}$. Afterward, the values of $\hat{V}^{i+1}$ for $\mathcal{C}_x \setminus \mathcal{X}$ are obtained via some kind of interpolation/generalization.

4.3.4 Approximation Architecture

The problem of choosing an approximation architecture that fits $\hat{V}^{i+1}$ to $\hat{V}^{i+1}(\mathcal{X})$ and, at the same time, is able to interpolate/generalize for $\hat{V}^{i+1}(\mathcal{C}_x \setminus \mathcal{X})$ appears to be crucial in order for ADP to converge. It is considered that ADP is not converging when either the stopping criterion (4.13) is never reached (refer to [55] and [100]) or $\hat{V}^*$ is not an accurate approximation of V^* [138]. The latter criterion is concerned with suboptimality of the obtained solution. In this chapter, we focus on the former deferring suboptimality analyses for future work.

The key property that has to be preserved by an approximation architecture is the contraction property (4.11) (refer to [55], [100] and [32]). In [55], the author classifies function approximators as expansion or contraction approximators. *Expansion approximators*, such as linear regressors and NNs, exaggerate changes on $\mathcal{C}_x \setminus \mathcal{X}$. *Contraction approximators* (or local averagers), such as k-nearest-neighbor, linear interpolation, grid methods and other state aggregation methods, conservatively respond to changes in $\mathcal{X}$. Therefore, on the one hand, a VI that includes a contraction approximator always converges, in the sense of (4.13), to the fixed point determined by the approximator, say $\hat{V}_{ca}^*$. However, not much can be said about the value $\|V^* - \hat{V}_{ca}^*\|_s$ (see [55] and [100]). On the other hand, a VI that includes an expansion approximator might diverge [100]. However, NNs are still a widely used approximation architecture due to their notable successes (for example, [161], [173], and [107]), adaptive architectures [187], performance guarantees under certain assumptions [32], and inventions of novel NN architectures. These novel NN architectures are also called nonparametric approximation [32] and they adapt to the training data. Examples are kernel-based NNs (refer to [23] and [63]) and recurrent NNs (refer to [167] and [63]). Almost all of the references in this subsection provide advantages and disadvantages of different approximation architectures. This fact shows the importance of the choice of approximation architecture.

A goal of this chapter is not to advocate certain architectures. Instead, based on our experience and the references above, we define properties that successful approximation architectures possess (e.g., contraction approximators and kernel-based NNs). Based on the specifics of the problem (dimensionality of the problem, availability and density of data, available processing power, memory requirements, etc.), one should choose a suitable architecture. In the example from Section 4.4, a Multilayer Perceptron (MLP) architecture is successful in fulfilling this property.

4.3.4.1 Desired Properties

Assume that $V^*(x)$ is a smooth function on $\mathcal{C}_x$, and choose a smooth function approximator. At the ith step, where $i \in \mathbb{N}_0$, randomly pick any $x' \in \mathcal{C}_x$, calculate $(\mathcal{M}\hat{V}^i)(x')$, and fit $\hat{V}^i(x')$ to $(\mathcal{M}\hat{V}^i)(x')$ obtaining $\hat{V}^{i+1}$. We are seeking an approximation architecture that satisfies the following properties

(i) $\hat{V}^{i+1}(x') = (\mathcal{M}\hat{V}^i)(x')$;

(ii) $\text{supp}(\hat{V}^{i+1} - \hat{V}^i) = \mathcal{C}_i$, where $\text{supp}(f) = \{x : f(x) \neq 0\}$ is the *support* of a function f, and $\mathcal{C}_i \subset \mathcal{C}_x$ is a convex and compact neighborhood of x; and

(iii) for any $c \in \partial \mathcal{C}_i$, where $\partial \mathcal{C}_i$ denotes the boundary of $\mathcal{C}_i$, the following

Optimal Self-Triggering

holds

$$\hat{V}^{i+1}[S] \subseteq [\hat{V}^{i+1}(c), \hat{V}^{i+1}(x')], \tag{4.14}$$

where $\hat{V}^{i+1}[S]$ is the image of the segment S connecting x' and c, in order to have $\|V^{i+1} - V^i\|_s \to 0$ as $i \to \infty$.

Remark 4.1. *Let us consider two value functions $\hat{u}^i$ and $\hat{v}^i$ in the ith step, and apply $\mathcal{M}$ at a randomly chosen x'_i. Due to (4.11), we have $\|(\mathcal{M}\hat{u}^i)(x'_i) - (\mathcal{M}\hat{v}^i)(x'_i)\| \leq \gamma \|\hat{u}^i(x'_i) - \hat{v}^i(x'_i)\|$. From property (i), we conclude that $\|\hat{u}^{i+1}(x'_i) - \hat{v}^{i+1}(x'_i)\| \leq \gamma \|\hat{u}^i(x'_i) - \hat{v}^i(x'_i)\|$. Since the approximator is smooth, we know that there exists a neighborhood $\mathcal{C}'_i \subseteq \mathcal{C}_i$ of x'_i such that $\sup_{x \in \mathcal{C}'_i} \|\hat{u}^{i+1}(x) - \hat{v}^{i+1}(x)\| \leq \sup_{x \in \mathcal{C}'_i} \|\hat{u}^i(x) - \hat{v}^i(x)\|$. This means that the nonexpansion property required in [55] is obtained locally around x'_i. The nonexpansion property from [55] is (4.11) when γ is replaced with 1. Finally, property (iii) eliminates counterexamples in which the Lebesgue measures of $\mathcal{C}'_i$, $i \in \mathbb{N}_0$, tend to zero. Consequently, generalization of the approximation architecture is ensured.*

Remark 4.2. *Property (i) is the accuracy requirement in order to preserve (4.11). Property (ii) is the "local property" found in [55], [100] and [187]. This local property is built in the activation functions of the kernel-based NNs. Property (iii) is used to ensure that $\mathcal{C}'_i$'s are not merely x'_i's. In addition, property (iii) curbs expansiveness on $\mathcal{C}_i \setminus \mathcal{C}'_i$.*

Remark 4.3. *Notice that Desired Properties imply online learning of NNs [63]. The motivation behind this choice lies in the fact that it is straightforward to check properties (i), (ii) and (iii) in online learning. Moreover, since we randomly pick points $x'_i \in \mathcal{C}_x$ in each step, we do not have to specify $\mathcal{X}$. By choosing random x'_i's, we also avoid the problem of exploration vs. exploitation [150]. On the other side, when using batch learning, properties (i), (ii) and (iii) cannot be guaranteed since NNs are expansion approximators. In fact, not until we switched to online learning in the example from Section 4.4, convergence was obtained. An extension of Desired Properties for batch learning and the problem of choosing $\mathcal{X}$ are left for future work.*

Remark 4.4. *As the stopping criterion we use the following: when $\|\hat{V}^{i+1} - \hat{V}^i\|_s < \epsilon$ for $N \in \mathbb{N}$ consecutive steps, the value iteration method has converged.*

4.3.5 Partially Observable States

Notice that the approximate value function $\hat{V}(x)$ is a function of state x. Up to this point we did not take into account that x is not available due to Assumptions 4.1 and 4.2. In other words, we are solving the DP problem (4.5) with partially observable states. More details about strategies for solving DP problems with partially observable states are found in Chapter 5 of [20].

Let us assume that the controller can access its state x_c. Consequently, the controller can calculate u at any given time. However, the controller does not have access to the state of the plant x_p but merely to $\hat{\omega}_p$ and $\hat{y}$. In other words, we are solving the DP problem (4.5) with imperfect state information, i.e., partially observable states. We circumvent this problem by introducing a Particle Filter (PF) (refer to [62] and [45]) that provides estimates $\hat{x}_p$ of the actual state x_p. More details about the problem of estimation under intermittent information can be found in [177]. It is well known that PFs are suitable for nonlinear processes, non-Gaussian and nonadditive measurement as well as process noise. We account for changes in ω_p between two consecutive sampling instants by treating e_ω as the process noise.

More precisely, we model the closed-loop system (4.1) and (4.2) as

$$x_p(t_i) = f_p^d(x_p(t_{i-1}), u(t_{i-1}), \tau_{i-1}, \hat{\omega}_p(t_{i-1}), e_\omega),$$
$$\hat{y}(t_i) = g(x_p(t_i), \nu), \qquad (4.15)$$

where f_p^d represents a discrete transition function of the plant obtained in similar fashion as (4.8), and statistics of the process noise e_ω and measurement noise ν are known and time invariant. Based on (4.15), we build a PF that extracts $\hat{x}_p$ from $\hat{\omega}_p$ and $\hat{y}$ and feeds the controller. Details of our PF implementation under intermittent feedback can be found in [177] and [176].

We deal with partially observable states by first obtaining $\hat{V}^*(x)$ for the case of perfect state information. Then, we employ PF and iterate $\hat{V}^*(x)$ using (4.12) to obtain the approximation of $\hat{V}^*(\hat{x})$ for the case of partially observable states. The motivation behind obtaining $\hat{V}^*(\hat{x})$ from $\hat{V}^*(x)$ lies in the fact that PFs need several steps to obtain a reliable estimate $\hat{x}$ starting from a poor initial estimate. Since the random choice of the state in every step in *Desired Properties* requires initialization of PFs for each step leading to a poor estimate $\hat{x}$, *Desired Properties* cannot be used for obtaining $\hat{V}^*(\hat{x})$. Basically, since $\hat{V}^*(x)$ is a close estimate of $\hat{V}^*(\hat{x})$ when $\hat{x}$ is a close estimate of x, we exploit $\hat{V}^*(x)$ to fine tune $\hat{V}^*(\hat{x})$ without need for the exploration phase.

4.4 Example: Autonomous Cruise Control (Revisited)

In this section, we apply the optimal self-triggered sampling policy to the *trajectory tracking* controller presented in Section 3.1.1. When emulating noise in (4.15), we use $e_\omega \in U([-0.3, 0.3] \times [-0.3, 0.3])$ and $\nu \in U([-0.15, 0.15] \times [-0.15, 0.15] \times [-0.15, 0.15])$, where $U(\mathcal{S})$ denotes the uniform distribution over a compact set $\mathcal{S}$.

The following coefficients were used in the cost function (4.5): $Q = 0.1I_3$, $R = 0.1I_2$, $S = 15$ and $\gamma = 0.96$. A remark is in order regarding the choice of Q, R and S. On the one hand, as we decrease S and keep Q and R fixed, the obtained sampling policy τ approaches zero. In other words, as the energy consumption for sampling, transmitting and processing decreases, the optimal intermittent feedback turns into continuous feedback. On the other hand, as S becomes greater, τ approaches τ^{max}. The above choice of Q, R and S yields $\tau \in [0.6\tau^{max}, 0.9\tau^{max}]$.

As the approximation architecture, we choose an MLP with 100 hidden neurons. In addition, we confine x to the set $\mathcal{C}_x = [-100, 100]^2 \times [-30\pi, 30\pi]$. Not until we used that many hidden neurons, properties (i), (ii) and (iii) were satisfied on $\mathcal{C}_x$. Even though activation functions in MLPs are not locally responsive, we were able to satisfy (i), (ii) and (iii). We presume the reason is low dimensionality of the considered tracking problem. For high-dimensional problems, the kernel-based NNs appear to be more suitable. In the stopping criterion from Remark 4.4 we choose $\epsilon = 1$ and $N = 10$, and obtain $\hat{V}^*(x)$ in about 300 to 400 steps depending on the initial $\hat{V}^0(x)$ and ω_p. Afterward, we obtain $\hat{V}^*(\hat{x})$ from $\hat{V}^*(x)$ using (4.12) and $\hat{x}$ fed from the particle filter. With $\epsilon = 1$ and $N = 10$, it takes about 50 simulations for $\hat{V}^*(\hat{x})$ to converge starting from $\hat{V}^*(x)$. The obtained approximation $\hat{V}^*(\hat{x})$ of $V^*(x)$ for $\omega_p = (1,1)$ is illustrated in Figure 4.2.

In the simulation included in this chapter, we choose $k_1 = 1.5$, $k_2 = 1.2$ and $k_3 = 1.1$. Figure 4.3 is obtained for the trajectory generated with $\omega_p(t) = (1,1)_{[0,1.83)} + (0.6, 0.15)_{[1.83, 8.8)} + (2,2)_{[8.8, 12]}$ where $t_\mathcal{S}$ is the indicator function on a set $\mathcal{S}$, i.e., $t_\mathcal{S} = t$ when $t \in \mathcal{S}$ and zero otherwise. In addition, Figure 4.3 shows that the intersampling interval τ tends to increase as $\|x\|$ approaches the origin.

4.5 Conclusions and Perspectives

This chapter investigates the problem of optimal input-output-triggering for nonlinear systems. We replace the traditional periodic paradigm, where up-to-date information is transmitted and control laws are executed in a periodic fashion, with optimal intermittent feedback. In other words, we develop a methodology that, based on the currently available but outdated measurements of the outputs and external inputs of a plant, provides time instants for obtaining new up-to-date measurements and execute the control law such that a given cost function is minimized. The optimization problem is formulated as a DP problem, and ADP is employed to solve it. In addition, because the investigated problems contain partially observable states, our methodology includes particle filtering under intermittent feedback. Furthermore, instead of advocating one approximation architecture over another in ADP, we for-

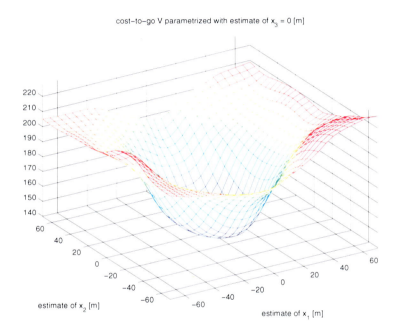

FIGURE 4.2
Approximation $\hat{V}^*(\hat{x})$ of the optimal value function $V^*(x)$ for $\omega_p = (1,1)$ depicted as a function of $\hat{x}_1 \in [-70, 70]$ and $\hat{x}_2 \in [-70, 70]$ when $\hat{x}_3 = 0$.

Optimal Self-Triggering

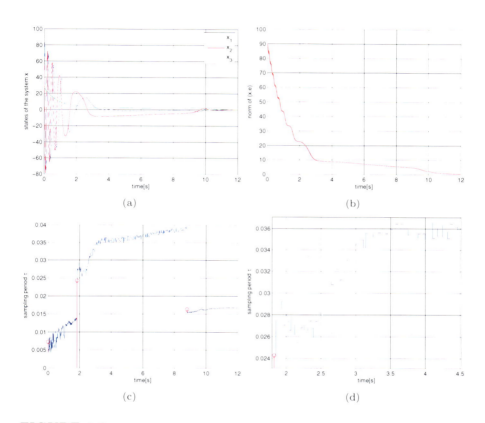

FIGURE 4.3
Illustration of the optimal input-output-triggering: (a) state x of the tracking system, (b) norm of (x, e), (c) values of sampling period τ_i between two consecutive transmissions. Red stems indicate time instants when changes in ω_p happen, and (d) a detail from Figure 4.3(c).

mulate properties that successful approximation architectures satisfy. Finally, our approach is successfully applied to a trajectory tracking controller for velocity-controlled unicycles.

In the future, the main goal is to further investigate the properties of successful approximation architectures. In addition, we plan to estimate how suboptimal the methodology developed in this chapter is.

5

Multi-Loop Networked Control Systems over a Shared Communication Channel

CONTENTS

5.1	Motivation, Applications and Related Works	100
	5.1.1 Medium Access Control	101
5.2	Markov Chains and Stochastic Stability	104
	5.2.1 Markov Chains	104
	5.2.2 Stochastic Stability	105
5.3	Problem Statement: Scheduling in Multi-Loop NCS	107
5.4	Stability and Performance	109
	5.4.1 Event-Based Scheduling Design	109
	5.4.2 Stability Analysis	112
	5.4.3 Performance and Design Guidelines	114
	5.4.4 Scheduling in the Presence of Channel Imperfections	116
5.5	Decentralized Scheduler Implementation	117
5.6	Empirical Performance Evaluation	121
	5.6.1 Optimized Thresholds λ for the Bi-Character Scheduler	121
	5.6.2 Comparison for Different Scheduling Policies	122
	5.6.3 Performance of the Decentralized Scheduler	123
	5.6.4 Performance with Packet Dropouts	125
5.7	Conclusions and Perspectives	126
5.8	Proofs and Derivations of Main Results	127
	5.8.1 Proof of Theorem 5.3	127
	5.8.2 Proof of Corollary 5.1	130
	5.8.3 Proof of Theorem 5.4	131
	5.8.4 Proof of Proposition 5.2	132
	5.8.5 Proof of Proposition 5.3	135

In this chapter, we consider multiple feedback loops being closed over a *shared communication channel*. This situation is common in practice where multiple control applications run over the same communication channel. As the transmission capacity of the shared link is limited, not all feedback loops may be closed at a time and some of the control sub-systems run in open loop. If one of those sub-systems occupies the communication channel for an excessive amount of time, the performance of the other control systems is likely to be significantly deteriorated; in the worst case, the overall system becomes unstable. In such a setting, the individual control loops are no longer operating independently but are coupled through the communication resource. Accordingly, a smart mechanism is needed which regulates the access to the communication channel (e.g., the UGES protocols from Chapter 2). In communication engineering, such a mechanism is known as *medium access control* (MAC) protocol. In this chapter we introduce different MAC policies, which in contrast to standard approaches, explicitly take the state of the individual NCS loops into account and analyze the overall behavior of the multi-loop control system under these policies. The individual control loops consist of potentially heterogeneous linear time-invariant plants with stochastic disturbances. The channel capacity is limited in terms of the transmission slots at a particular instant of time. We introduce a novel *scheduling* scheme, which dynamically prioritizes the channel access at each time step according to an error-dependent priority measure. Given local error thresholds for each control loop, the scheduling policy blocks the transmission of sub-systems with lower error values in order to free the channel for sub-systems with larger errors. The scheduler then allocates the communication resource probabilistically among the eligible sub-systems based on a prioritized measure. We prove stochastic stability of the *multi-loop NCS* under the proposed scheduler in terms of f-ergodicity of the overall network-induced error. Additionally, we derive uniform analytic performance bounds for an average cost function comprised of a quadratic error term and transmission penalty. Furthermore, results on the robustness of the scheme against channel imperfections such as packet dropouts are discussed and a decentralized implementation is presented. Simulation results show that the proposed scheme presents a significant reduction of the aggregate network-induced error variance compared to time-triggered and random access scheduling policies such as TDMA and CSMA, especially as the number of sub-systems increases.

5.1 Motivation, Applications and Related Works

Traditional digital control systems are typically associated with time-triggered control schemes and periodic sampling. The introduction of communication networks for data transmission between spatially distributed entities in large-scale systems spurs the design of more advanced sampling strategies that result in more efficient utilization of the available resources (e.g., bandwidth and energy). However, control over shared communication resources imposes several design challenges due to channel capacity limitations, network congestion, delays and dropouts [130, 73]. As many results over recent years show, e.g., [11, 169, 190, 42, 71] and references found throughout this book, it is often more beneficial to sample upon the occurrence of specific events, rather than after a fixed period of time elapses, especially when dealing with scarce resources.

Most of the available works consider an event-triggered synthesis for single-loop networked systems under limited communication resources [90, 127, 120, 112, 101, 188, 121, 37, 8, 59, 7, 169]. The main focus is on appropriate event generation to reduce the communication cost and yet achieve the required control performance including stability.

In practice, however, *multiple* networked control applications often share the resource-limited communication infrastructure for closing the loop between the sensors and the controllers; see Figure 5.1 for illustration. Relevant application domains are in industrial automation, building automation, smart transportation, smart grids, and general infrastructure systems. The communication channel typically has a limited capacity, i.e, only a limited number of transmission slots are available at a certain instant of time. If the number of control applications exceeds the number of parallel transmission slots, then inevitably some of the feedback loops may not be closed at that time instant. In such a setting, the individual feedback loops compete for the limited communication resource. In fact, their performance is coupled through limited resources: If there are feedback loops which make excessive use of the communication resources and block others, then this may lead to a performance degradation of the overall system and may even result in unstable behavior.

Compared to the single-loop scenario (see Chapters 2, 3, 4 and the references therein), multiple-loop networked control systems under communication constraints have attained little attention in the literature so far, with notable exceptions in [70, 34, 25, 26, 27, 128, 129, 155, 198, 156, 152, 9] as well as in Chapters 6, 7 and 8. The results presented in [128, 129] suggest that an event-based approach can be effectively employed as a threshold policy to govern the channel access in a networked control system with explicitly considered resource limit. Also, [33, 65, 153] conclude that event-based control and scheduling schemes outperform periodic rules in terms of resource consumption while preserving the same level of control performance in multi-loop NCSs.

However, the triggering strategy and the communication model at the level of channel access management and collision resolution have a quite intricate relationship. It is suggested in [70, 34] that event-triggering is considerably beneficial compared to the time-triggered sampling in Carrier Sense Multiple Access (CSMA)-based random channel arbitration. But with unslotted and slotted ALOHA, event-triggered sampling leads to a performance degradation [25, 26]. In summary, the interaction between the sampling mechanism and underlying communication protocol needs to be thoroughly studied.

5.1.1 Medium Access Control

Every communication channel is subject to a capacity constraint, which implies that only a limited number of networked control applications simul-

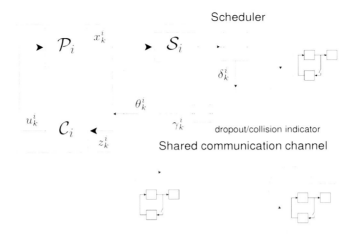

FIGURE 5.1
Multi-loop NCS operated over a shared communication channel.

taneously have access to the channel. If the number of feedback loops exceeds the number of available slots, some of the transmission requests are inevitably blocked and the corresponding sub-systems remain in open loop until they obtain access to the channel. This issue is common to any networked application; in order to resolve the access conflict, different scheduling paradigms have been developed. These mechanisms are called medium access control (MAC) protocols. The most important MAC paradigms are categorized with respect to their implementation in "centralized vs. decentralized", "time-triggered vs. event-triggered", and "deterministic vs. random (probabilistic) access" schemes.

Time division multiple access (TDMA), frequency division multiple access (FDMA) and code division multiple access (CDMA) are well-known centralized access policies. With *a priori* assigned resources to the individual data streams/applications (time slots in TDMA, code sequence in CDMA, and frequency range in FDMA), they offer collision-free transmission and high throughput even at high network load. They are called *contention-free* protocols. Moreover, quality-of-service support and bandwidth arbitration are facilitated as those methods are, in principle, capable of prioritizing channel access given either pre-determined priority order or dynamic prioritization based on the network and control systems online conditions. However, centralized strategies lack flexibility toward structural system changes and scalability, which consequently makes those approaches unfeasible to implement for structurally varying and large-scale networked systems. Centralized strategies are mostly preferred in small and medium-size NCSs without any structural changes. In addition, centralized protocols are characterized by a single point of failure, which can compromise the overall NCS performance. In a decentralized MAC protocol, the transmission time of every node is determined locally without the requirement of having global knowledge of all nodes among which the communication medium is shared. They are scalable, easy-to-install, and may flexibly react to changes in the network configuration. Decentralized policies are suitable for networked systems with a large and possibly changing number of nodes. However, collisions are inevitable within those *contention-based* protocols and need to be handled with care in the NCS design. Furthermore, collision avoidance mechanisms, e.g., the listen-before-talk scheme in *CSMA with collision avoidance* (CSMA-CA), call for all nodes to sense the channel quite often, which induces energy consumption due to idle listening, overhearing, and message overhead.

The difference between event-triggered and time-triggered scheduling approaches is that the former takes into account the current situation of the sub-systems (and the communication channel) in order to control the channel access while the latter approach schedules the communication resource according to some periodic offline rules. For example, TDMA is a time-triggered scheduling approach where the sequence of transmissions for a sub-system is determined *a priori*. The notable advantage of event-triggered approaches is the capability of real-time monitoring of the NCS and corresponding adjustment of the scheduling decisions; for instance, prioritizing the transmission requests from those nodes in urgent transmission state. Event-triggered scheduling policies consider a well-defined *event* as the trigger for a transmission rather than an elapsed fixed period of time. As many results over recent years show [11, 169, 190, 42, 71], it is often more beneficial to sample a signal within the control loop upon the occurrence of specific events, rather than after a fixed period of time elapses, especially when dealing with scarce resources. In NCSs, event-based control and scheduling schemes outperform the periodic (time-triggered) rules in terms of resource consumption while preserving the same level of control performance [33, 65, 153]. This motivates us to consider an event-triggered scheduling approach in this chapter in order to appropriately manage the resource constraints imposed on the channel capacity.

The scheduling decisions may be performed according to either deterministic or probabilistic rules. TDMA and Try-Once-Discard (TOD), already investigated in Chapter 2, are well-known deterministic mechanisms; CSMA and CSMA-CA are commonly employed probabilistic channel access managers. Deterministic rules usually render better performance than their probabilistic counterparts, but the latter are often better scalable and reconfigurable as they facilitate a decentralized implementation.

The remainder of this chapter is organized as follows. Section 5.2 presents the terminology and essential background used throughout this chapter. Section 5.3 introduces the basic problem setting and formulates the scheduling problem in multi-loop NCSs. Different event-based scheduling mechanisms are proposed in Section 5.4. Furthermore, their convergence properties are investigated and studied in the presence of transmission errors. In Section 5.5 a decentralized implementation considering state-of-the-art communication protocols is studied. Simulation results comparing the performance of the proposed scheduling approaches to others are presented in Section 5.6. Conclusions and future challenges are found in Section 5.7. Several technical results and proofs are included in Section 5.8.

5.2 Markov Chains and Stochastic Stability

We will show in the next section that the system under consideration is a homogeneous *Markov chain* evolving in $\mathbb{R}^{n_e}$. Therefore, in the following we will shortly revisit some essential notions and properties related to Markov chains evolving in uncountable state spaces as well as their convergence. The notions and properties below are mainly from the textbook [123].

5.2.1 Markov Chains

Proposition 5.1 (Markov Property). *Let $\Phi = \{\Phi_0, \Phi_1, \ldots\}$ be a Markov chain defined on $(\Omega, \mathcal{F})$ equipped with the initial measure μ. Assume $h : \Omega \to \mathbb{R}$ is a measurable and bounded function. Then for any positive integer n, the Markov property follows as*

$$\mathbb{E}_\mu\left[h\left(\Phi_{n+1}, \Phi_{n+2}, \ldots\right) | \Phi_0, \ldots, \Phi_n, \Phi_n = s\right] = \mathbb{E}_\Phi\left[h(\Phi_1, \Phi_2, \ldots)\right].$$

According to *Proposition* 5.1, the Markov property implies that the evolution of a Markov state is forgetful of all its past values except the most immediate value.

Definition 5.1 (Occupation Times and Return Times). *Let $\mathcal{S}$ be a general state space equipped with σ-algebra $\mathcal{A}(\mathcal{S})$. Then for any $A \in \mathcal{A}(\mathcal{S})$:*

1. *The occupation time, refers to the measurable function $\eta_A : \Omega \to \mathbb{Z}^+ \cup \{\infty\}$, which denotes the number of visits to set A by the Markov chain $\Phi = \{\Phi_1, \Phi_2, \ldots\}$, i.e.,*

$$\eta_A := \sum_{n=1}^{\infty} \mathbb{I}\{\Phi_n \in A\}.$$

2. *The return time refers to the measurable function $\tau_A : \Omega \to \mathbb{Z}^+ \cup \{\infty\}$, which denotes the first return time to the set A by the Markov chain $\Phi = \{\Phi_1, \Phi_2, \ldots\}$, i.e.,*

$$\tau_A := \min\{n \geq 1 : \Phi_n \in A\}.$$

Definition 5.2 (φ-Irreducible Markov Chains). *Let the Markov chain $\boldsymbol{\Phi}$ be defined on $(\mathcal{S}, \mathcal{A})$, and φ be a measure on the σ-algebra $\mathcal{A}(\mathcal{S})$. Then $\boldsymbol{\Phi}$ is said to be φ-irreducible, if for every $s \in \mathcal{S}$, $\varphi(A) > 0$ implies*

$$\mathsf{P}_s(\tau_A < \infty) > 0.$$

The measure φ is then called the irreducibility measure of the Markov chain $\boldsymbol{\Phi}$.

φ-irreducibility means that the entire state space of a φ-irreducible Markov chain can be reached via a finite number of transitions, regardless of the initial state. We will see in what follows that this property plays an essential role in stability analysis of Markov chains on uncountable state spaces.

Theorem 5.1 (ψ-Irreducible Markov Chain). *If a Markov chain $\boldsymbol{\Phi}$ defined on $(\mathcal{S}, \mathcal{A})$ is φ-irreducible, then a unique maximal irreducibility measure ψ exists on $\mathcal{A}(\mathcal{S})$ such that*

1. *$\psi \succ \varphi$;*

2. *Markov chain $\boldsymbol{\Phi}$ is φ'-irreducible for any other measure φ', if and only if $\psi \succ \varphi'$;*

3. *$\psi(A) > 0$ implies $\mathsf{P}_s(\tau_A < \infty) > 0$, for every $s \in \mathcal{S}$; and*

4. *if $\psi(A) = 0$, then $\psi(\bar{A}) = 0$, where $\bar{A} := \{z : \mathsf{P}_z(\tau_A < \infty) > 0\}$.*

Definition 5.3 (Small set). *A subset $C \in \mathcal{B}(X)$ of the measurable space $(X, \mathcal{B})$ is called ν-small if a non-trivial measure ν on $\mathcal{B}(X)$ and $k > 0$ exists such that for all $x \in C$ and $B \in \mathcal{B}(X)$,*

$$P^k(x, B) \geq \nu(B).$$

Remark 5.1. *It is comprehensively discussed in [123, Sec. 5.3.5] that for linear state-space models, all compact subsets of the state space are small sets as well.*

Definition 5.4 (Aperiodic and Strongly Aperiodic Markov Chain). *Let $\boldsymbol{\Phi}$ be a φ-irreducible Markov chain. The Markov chain $\boldsymbol{\Phi}$ is called aperiodic if the largest common period for which a d-cycle occurs equals one. The chain $\boldsymbol{\Phi}$ is called strongly aperiodic if there exists a ν-small set A such that $\nu(A) > 0$.*

5.2.2 Stochastic Stability

We mainly employ stability concepts presented in [123], where a comprehensive discussion about the stability of Markov chains in uncountable state spaces is provided. The stability of Markov chains can be investigated in terms of ergodicity. A random process is called *ergodic* if the time-average of its events over one sample sequence of transitions represents the behavior of the process over the entire state-space. Thus, ergodicity implies the existence of an invariant finite measure over the entire state-space, implying that the process returns to some sets in finite time, and does not diverge forever.

Definition 5.5 (Positive Harris Recurrence [72]). *Let the Markov chain $\Phi = (\Phi_0, \Phi_1, \ldots)$ evolve in state space X, with individual random variables measurable with respect to some known σ-algebra $\mathcal{B}(X)$. Then Φ is said to be positive Harris recurrent if:*

1. *there exists a non-trivial measure $\nu(B) > 0$ for a set $B \in \mathcal{B}$ such that for all initial states $\Phi_0 \in X$, $P(\Phi_k \in B, k < \infty) = 1$ holds, and*

2. *Φ admits a unique invariant probability measure.*

Definition 5.6 (*f*-Ergodicity [123]). *Let $f \geq 1$ be a real-valued function in $\mathbb{R}^n$. A Markov chain Φ is said to be f-ergodic, if:*

1. *Φ is positive Harris recurrent with the unique invariant probability measure π,*

2. *the expectation $\pi(f) := \int f(\Phi_k) \pi(d\Phi_k)$ is finite, and*

3. *$\lim_{k \to \infty} \|P^k(\Phi_0, .) - \pi\|_f = 0$ for every initial value $\Phi_0 \in X$, where $\|\nu\|_f = \sup_{|g| \leq f} |\nu(g)|$.*

Item 3 introduces the notion of Markov chain gradient with respect to a real-valued function of states.

Definition 5.7 (Drift for Markov Chains). *Let $V : \mathbb{R}^n \to [0, +\infty)$ be a real-valued function and Φ be a Markov chain. The drift operator Δ is defined for any non-negative measurable function V as*

$$\Delta V(\Phi_k) = \mathbb{E}[V(\Phi_{k+1})|\Phi_k] - V(\Phi_k), \quad \Phi_k \in \mathbb{R}^n. \tag{5.1}$$

The following theorem summarizes the f-ergodicity of Markov chains in general state spaces [123, Ch. 14].

Theorem 5.2 (*f*-Norm Ergodic Theorem). *Suppose that the Markov chain Φ is ψ-irreducible and aperiodic and let $f(\Phi) \geq 1$ be a real-valued function in $\mathbb{R}^n$. If a small set $\mathcal{D}$ and a non-negative real-valued function V exist such that $\Delta V(\Phi) \leq -f(\Phi)$ for every $\Phi \in \mathbb{R}^n \setminus \mathcal{D}$ and $\Delta V < \infty$ for $\Phi \in \mathcal{D}$, then the Markov chain Φ is f-ergodic.*

In summary, f-ergodicity expresses that the Markov state converges to an invariant finite-variance measure over the entire state-space. This ensures that the Markov chain is a stationary process, and guarantees that if the Markov state leaves some compact subsets of σ-algebra $\mathcal{B}(X)$, it returns to these subsets in finite time with probability one.

Lyapunov Mean Square Stability (LMSS) is another commonly employed stability concept in stochastic systems.

Definition 5.8 (Lyapunov Mean Square Stability [94]). *A linear system with state vector X_k is Lyapunov mean square stable (LMSS) if given $\varepsilon > 0$, there exists $\rho(\varepsilon)$ such that $\|X_0\|_2 < \rho$ implies*

$$\sup_{k \geq 0} \mathbb{E}\left[\|X_k\|_2^2\right] \leq \varepsilon.$$

It should be noted that Lyapunov mean square stability implies f-ergodicity but not vice versa since the latter holds not only with quadratic Lyapunov functions but also with p-power variations. A weaker notion of stability is given by the *Lyapunov Stability in Probability* (LSP).

Definition 5.9 (Lyapunov Stability in Probability [94]). *A linear system with state vector x_k is LSP if given $\varepsilon, \varepsilon' > 0$, there exists $\rho(\varepsilon, \varepsilon') > 0$ such that $|x_0| < \rho$ implies*

$$\lim_{k \to \infty} \sup \mathsf{P}\left[x_k^T x_k \geq \varepsilon'\right] \leq \varepsilon. \tag{5.2}$$

5.3 Problem Statement: Scheduling in Multi-Loop NCS

In this section, we introduce the model of a multi-loop NCS and further discuss the control and scheduling synthesis. Consider a set of N heterogeneous control loops coupled through a shared communication channel as depicted in Figure 5.1. Each individual loop consists of a discrete time linear time-invariant sub-system with stochastic disturbances (including model uncertainties) $\mathcal{P}_i$ and a controller $\mathcal{C}_i$, where the link from $\mathcal{P}_i$ to $\mathcal{C}_i$ is closed through the shared communication channel. A scheduling unit decides when a state vector $x_k^i \in \mathbb{R}^{n_{xi}}$ at time step k is to be scheduled for channel utilization, where n_i is the dimension of the i'th sub-system. The process $\mathcal{P}_i$ is modeled by the stochastic difference equation

$$x_{k+1}^i = A_i x_k^i + B_i u_k^i + w_k^i, \tag{5.3}$$

where $w_k^i \in \mathbb{R}^{n_{\omega i}}$ is i.i.d. with the standard normal distribution $\mathcal{N}(0, I)$ at each time k, while constant matrices $A_i \in \mathbb{R}^{n_{xi} \times n_{xi}}$ and $B_i \in \mathbb{R}^{n_{xi} \times n_{ui}}$ describe system and input matrices of sub-system i, respectively. The pair (A_i, B_i) is assumed to be stabilizable. The initial state x_0^i is randomly chosen from an arbitrary bounded-variance distribution. The initial state x_0 together with the noise sequence w_k generate a probability space $(\Omega, \mathcal{A}, \mathsf{P})$, where Ω is the set of all possible outcomes, $\mathcal{A}$ is a σ-algebra of events associated with probability P. The variable $\delta_k^i \in \{0, 1\}$ represents the scheduler's decision whether a sub-system i transmits at time step k or not

$$\delta_k^i = \begin{cases} 1, & x_k^i \text{ is sent through the channel} \\ 0, & x_k^i \text{ is blocked.} \end{cases}$$

Based on the scheduler's decision at a time step k, the received data at the controller side is determined as

$$z_k^i = \begin{cases} x_k^i, & \delta_k^i = 1, \\ \varnothing, & \delta_k^i = 0. \end{cases}$$

Remark 5.2. *Unless stated otherwise, we assume a loss-less channel, i.e., if a packet is transmitted, it will be surely received by the corresponding controller. Accordingly, the block "dropout/collision indicator" in Figure 5.1 is to be ignored. The case of packet dropouts due to imperfect channels is addressed in Section 5.4.4 and the case of collisions due to a decentralized implementation (contention-based protocol) in Section 5.5. Furthermore, we assume that the state is transmitted and not a measurement.*

It is assumed that the ith controller has knowledge of its own model, i.e., A_i, B_i, and the distributions of process noise, w_k^i and x_0^i, but not of the others. The control law is assumed to be a measurable and causal mapping of past observations, in particular, here we assume a linear time-invariant control law

$$u_k^i = -L_i \mathbb{E}\left[x_k^i | Z_k^i\right], \tag{5.4}$$

where $Z_k^i = \{z_0^i, \dots, z_k^i\}$ is the ith controller observation history, and L_i is a stabilizing feedback gain. If there is no transmission at time step k, a model-based estimator computes the state estimate

$$\mathbb{E}\left[x_k^i | Z_k^i\right] = (A_i - B_i L_i)\mathbb{E}\left[x_{k-1}^i | Z_{k-1}^i\right] \quad \text{if} \quad \delta_k^i = 0, \tag{5.5}$$

with $\mathbb{E}\left[x_0^i | Z_0^i\right] = \mathbf{0}_{n_{xi}}$. The estimate (5.5) is well-behaved as the stabilizing gain L_i ensures the stability of the closed-loop matrix $(A_i - B_i L_i)$. In case of a successful transmission, the transmitted state is employed to compute the control. The network-induced error $e_k^i \in \mathbb{R}^{n_{ei}}$, defined as

$$e_k^i := x_k^i - \mathbb{E}\left[x_k^i | Z_k^i\right], \tag{5.6}$$

plays a crucial role within the stability analysis as we will see in the following. Employing (5.3)–(5.5) and the definition of the network-induced error e_k^i, the dynamics of the ith-loop is written in terms of its so-called network state (x_k^i, e_k^i) as

$$x_{k+1}^i = (A_i - B_i L_i) x_k^i + \left(1 - \delta_k^i\right) B_i L_i e_k^i + w_k^i, \tag{5.7}$$

$$e_{k+1}^i = \left(1 - \delta_{k+1}^i\right) A_i e_k^i + w_k^i. \tag{5.8}$$

From (5.7) it follows that if the ith-loop is closed at time k, i.e., $\delta_k^i = 1$, the stabilizing gain L_i ensures the closed-loop matrix $(A_i - B_i L_i)$ to be Hurwitz. Moreover, (5.8) indicates that the evolution of the network induced error e_k^i is independent of the system state x_k^i and the control input u_k^i. Furthermore, we observe that stability of closed-loop system i does not imply stability of the error state e_k^i. Thus, given a stable closed-loop matrix $(A_i - B_i L_i)$, it is sufficient to show stability of e_k^i in order to show stability of the aggregate state (x_k^i, e_k^i). We show later that if A_i is unstable then the ith loop needs to be closed "often enough" over an interval to ensure stable error dynamics. This separation enables us to design the scheduler, which affects the error state e_k^i, independent of the control law u_k^i. Hence, we employ an emulation-based control strategy with the minimal requirement to stabilize the closed-loop systems in the absence of the capacity constraint.

Problem 5.1. *Design a scheduling protocol such that the network-induced error dynamics (5.8) is stabilized in a stochastic sense.*

Problem 5.2. *Investigate the performance of the multi-loop NCS under the proposed scheduling protocol and compare it to conventional approaches.*

5.4 Stability and Performance

5.4.1 Event-Based Scheduling Design

In scheduling design, the goal is to develop a medium access control scheme which utilizes limited communication resources efficiently and leads to an improved overall NCS performance compared to conventional scheduling approaches, e.g., TDMA and CSMA. Naturally, stability of the overall networked system is one of the most fundamental requirements. We assume that the communication channel is subject to a capacity constraint such that not all sub-systems can simultaneously transmit. Consequently, some of data packets are blocked and the corresponding systems operate in open loop.

Here we introduce a novel event-based error-dependent scheduling rule that dynamically prioritizes the channel access for the loops competing for channel access. Therefore we first define the overall network-induced error state $e_k \in \mathbb{R}^{n_e}$ by stacking the error vectors from all control loops as follows

$$e_k = (e_k^1, \ldots, e_k^N), \tag{5.9}$$

with $n_e = \sum_{i=1}^{N} n_{ei}$.

The following scheduling rule defines the probability of channel access for a sub-system, say i, at time step $k+1$ based on the error values e_k at one time step before as

$$P[\delta_{k+1}^i = 1 | e_k, \lambda_i] = \begin{cases} 0 & \|e_k^i\|_{Q_i}^2 \leq \lambda_i \\ 1 & \|e_k^i\|_{Q_i}^2 > \lambda_i \wedge j_\lambda \leq c \\ \frac{\|e_k^i\|_{Q_i}^2}{\sum_{j_\lambda} \|e_k^j\|_{Q_j}^2} & \|e_k^i\|_{Q_i}^2 > \lambda_i \wedge j_\lambda > c, \end{cases} \quad (5.10)$$

where $\|e_k^i\|_{Q_i}^2 := e_k^{i^T} Q_i e_k^i$ with Q_i a positive definite matrix, $c < N$ denotes the channel capacity, and j_λ is the number of loops satisfying $\|e_k^i\|_{Q_i}^2 > \lambda_i$. The scheduler is error dependent as it is informed about each sub-system's error state at each time step in order to decide the priority for channel access. Furthermore, the proposed rule possesses a probabilistic-deterministic nature and is therefore called the bi-character scheduler from now on. First, if $\|e_k^i\|_{Q_i}^2 \leq \lambda_i$ at time step k, then no transmission request associated with sub-system i is submitted (first argument in (5.10)). This feature helps to allocate the channel more efficiently by excluding the sub-systems for which a transmission is not crucial. If $j_\lambda \leq c$, then all eligible sub-systems transmit (second argument of (5.10)). Otherwise, the channel is allocated probabilistically until the capacity is reached and other transmission requests are blocked. Indeed, if $j_\lambda > c$, each eligible sub-system associated with its assigned probability takes part in a biased randomization.

Example 5.1. *Consider an NCS with $N = 4$, where two are eligible for transmission at a certain time, while $c = 1$. Assume the priorities are assigned as 0.8 and 0.2. The biased randomization is tossing an unfair coin where the probabilities of having head and tail are 80% and 20%, respectively. One sub-system will transmit based on the outcome of the single toss of the coin. Hence, it is not guaranteed that the sub-system with higher priority transmits, though it is more likely.*

Furthermore, it is clear that (5.10) is a collision-free, i.e., contention-free, policy as the scheduler selects the data packets for transmission. For the sake of brevity of presentation we assume $c = 1$ from now on, thus

$$\sum_{i=1}^{N} \delta_k^i = 1, \quad \forall k \geq 0. \quad (5.11)$$

The results provided below can easily be extended toward the general case $\sum_{i=1}^{N} \delta_k^i = c < N$, where $c > 1$.

Remark 5.3. *The bi-character scheduler (5.10) is event based. The eligibility for channel competition (deterministic part) is performed in a decentralized fashion with the tuning parameter λ_i, i.e., locally at every plant. The scheduling of the eligible systems (probabilistic policy) is performed by a central unit, i.e., the scheduling variable δ_{k+1}^i at time step $k+1$ is a function of all error states $e_k^1, \ldots, e_k^N$. To eligible sub-systems, the channel is awarded based on a biased randomization. The weight matrices Q_i directly affect the access* probability as it specifies how frequently a sub-system needs to transmit. The randomization in (5.10) does not exclude sub-systems with lower priority from transmission. This randomization provides a flexible design enabling a decentralized implementation as discussed in Section 5.5 and [115]. In addition, by slightly deviating from the current scheduler and employing the p-norm instead of the 2-norm in (5.10) with $p \gg 2$ the highest priorities have a significantly higher chance to transmit; in fact, for $p \to \infty$ they transmit almost surely as shown in [114].

In order to characterize the properties of the overall system dynamics under the scheduling protocol (5.10), we study the behavior of the overall network-induced error state $e_k \in \mathbb{R}^{n_e}$ (5.9). The scheduling law (5.10), which generates the input signal for the error state e_k according to (5.8), is a randomized policy depending only on the most recent error values, i.e., the decision on which sub-system eventually transmits at an arbitrary time step $k+1$ is correlated with the latest error state e_k. Moreover, the Gaussian noise w^i in (5.8) has a continuous everywhere-positive density function at any element e_k^i of the overall state e_k meaning that there is a non-zero probability to reach any subset of $\mathbb{R}^{n_e}$. This implies that there exists a transition probability for any event $\mathcal{E} \in \mathcal{A}$ such that

$$\mathsf{P}\left(e_{k+t} \in \mathcal{E} | e_m, m < k, e_k\right) = \mathsf{P}^t(e_{k+t} \in \mathcal{E} | e_k),$$

where $\mathsf{P}^t(e_{k+t} \in \mathcal{E})$ denotes the probability that e_k enters a set $\mathcal{E}$ after t transitions, and m is an arbitrary time index before time step k. Since the scheduling policy (5.10) is forgetful about the error states e_m, $m < k$, to decide the next transmission via δ_{k+1}^i, (5.9) is a Markov chain. It is homogeneous as the difference equation (5.8) is time invariant and noise process w_k^i is i.i.d. for every $i = \{1, \ldots, N\}$, and for any time step k. Since the noise distribution is absolutely continuous with an everywhere-positive density function, every subset of the state-space is accessible within one transition, i.e., the d-cycle is one, thus the Markov chain is aperiodic and ψ-irreducible, where ψ is a non-trivial measure on the state space $\mathbb{R}^{n_e}$.

Remark 5.4. *With the classical TOD scheduling rule*

$$\delta_{k+1}^i = \begin{cases} 0 & \|e_k^i\|_2 < \|e_k^j\|_2 \quad \text{for sub-system } j \neq i \\ 1 & \text{otherwise} \end{cases}, \quad (5.12)$$

the highest priorities transmit with probability one, it is a centralized, event-triggered, deterministic policy. The properties of the underlying Markov chain of the network-induced error e_k remain the same as for the bi-character scheduling policy (5.10). Also in this case, stability has to be shown only for e_k in order to show stability of the overall system.

Remark 5.5. *As an alternative to (5.10) we have also studied the pure prob-*

abilistic policy

$$\mathsf{P}\left[\delta_{k+1}^i = 1 \big| e_k^j, j \in \{1, \ldots, N\}\right] = \frac{\|e_k^i\|_2^p}{\sum_{j=1}^N \|e_k^j\|_2^p}, \quad (5.13)$$

where, $p \geq 1$ is an integer. Since the scheduling law (5.13) depends only on the latest error state values, the proposed rule ensures that the Markov property of the network-induced error state (5.9) remains valid as shown in [114].

5.4.2 Stability Analysis

In this section the stability of the multi-loop NCS with the proposed bi-character medium access control policy (5.10) is investigated. In order to show stability of the overall system under the scheduling policy, we only have to show that the network-induced error e_k is stabilized as derived in Section 5.3. We show that the Markov chain e_k and therefore the multi-loop NCS, arbitrated by the proposed scheduling protocol, is stochastically stable, according to the notion of f-ergodicity. In order to show stability employing the f-Norm Ergodic Theorem 5.2 we select a non-negative real-valued function $V: \mathbb{R}^{n_e} \to \mathbb{R}^+$ as

$$V(e_k) = \sum_{i=1}^{N} e_k^{i^\mathsf{T}} Q_i e_k^i. \quad (5.14)$$

Due to the characteristics of the selected function (5.14), f-ergodicity of the Markov chain (5.9) cannot always be guaranteed employing the drift ΔV over one transition step, i.e., for $k \to k+1$. We illustrate this observation for $N = 2$ by constructing the following example.

Example 5.2. *Let an NCS with two identical scalar systems compete for one channel slot. For illustration purposes, assume $Q_1 = Q_2 = 1$ and $e_k^1 = e_k^2 = \bar{e}_k > \lambda_1 = \lambda_2$ so the transmission chance for each system is $\frac{1}{2}$ according to (5.10). Employing (5.14), it is straightforward to show that the drift in (5.1) with $e_k = (e_k^1, e_k^2)$ becomes*

$$\Delta V(e_k) = \mathbb{E}[V(e_{k+1})|e_k] - V(e_k) = 2 + \|A\bar{e}_k\|_2^2 - 2\|\bar{e}_k\|_2^2.$$

For $A > \sqrt{2}$, the drift is positive, which violates the drift condition in Theorem 5.2. We show later that the ergodicity of Markov chain (5.9) under the capacity constraint (5.11) is recovered considering the drift over an interval with length equal or greater than the number of sub-systems sharing the channel.

Intuitively, only after all sub-systems have a chance to transmit, a negative drift ΔV over some interval of interest can be guaranteed. To fulfill this, we investigate the ergodicity of the Markov chain over the interval with length N. Ergodicity over an interval implies ergodicity over longer intervals (see [123,

Ch.19] for details of ergodicity over multi-step intervals). To infer f-ergodicity over the interval $[k, k+N]$, we modify the drift definition in (5.1) as

$$\Delta V(e_k, N) = \mathbb{E}[V(e_{k+N})|e_k] - V(e_k), \quad e_k \in \mathbb{R}^{n_e}. \tag{5.15}$$

Theorem 5.3. *Consider a multi-loop NCS consisting of N heterogeneous linear time-invariant stochastic sub-systems modeled as (5.3), and a transmission channel subject to the constraint (5.11), and the control, estimation and scheduling laws given by (5.4), (5.5) and (5.10), respectively. Then, for any positive λ_i's and positive definite Q_i's the Markov chain (5.9) following (5.8) is f-ergodic. Furthermore, if the gain matrix L_i (5.4) ensures the closed-loop matrix $(A_i - B_i L_i)$ to be Hurwitz $\forall\, i \in \{1, ..., N\}$, then the multi-loop NCS described in (5.3)–(5.8) with the aggregate state (x_k, e_k) under the scheduling law (5.10) is f-ergodic.*

The proof of this result employs the multi-step drift (5.15) of the Lyapunov function candidate (5.14), see Section 5.8.1.

Remark 5.6. *Theorem 5.3 assures that the Markov chain (5.9) visits compact sets $\mathcal{D}_f \subset \mathbb{R}^{n_e}$ at most every N time steps for any $\lambda_i > 0$ and $Q_i \succeq 0$. The set $\mathcal{D}_f$, however, varies with the parameters λ_i's, Q_i's, N and system matrices A_i.*

As in this particular case f-ergodicity is shown employing a quadratic Lyapunov function, it is straightforward to show Lyapunov mean square stability.

Corollary 5.1. *The multi-loop NCS described in (5.3)–(5.8) with the aggregate state (x_k, e_k) under the scheduling law (5.10) is Lyapunov mean square stable.*

The proof is presented in Section 5.8.2.

Remark 5.7. *When the deterministic TOD policy (5.12) is employed, then all the stability statements remain valid. In particular, the Markov chain underlying the network-induced error e_k is f-ergodic; the multi-loop NCS is f-ergodic and LMSS. The proof largely follows the proof arguments of Theorem 5.3, see Section 5.8.1.*

Remark 5.8. *When the pure probabilistic policy (5.13) is employed, still all the stability statements remain valid, i.e., the Markov chain underlying the network-induced error e_k is f-ergodic; the multi-loop NCS is f-ergodic and LMSS. The proof is provided in [114]; it follows the line of arguments as for Theorem 5.3, see Section 5.8.1.*

5.4.3 Performance and Design Guidelines

In this section, we investigate the performance of the bi-character scheduling policy (5.10) by obtaining analytical uniform upper bounds on the average of the per-time step cost function J_{e_k} given by

$$J_{e_k} = \sum_{j=1}^{N} {e_k^j}^\top Q_j e_k^j + \eta \delta_k^j := \sum_{j=1}^{N} \|e_k^j\|_{Q_j}^2 + \eta \delta_k^j, \quad (5.16)$$

where $\eta \geq 0$ denotes the cost of channel utilization.

Note that we solely evaluate the performance of the scheduler, i.e., there is no penalty on the signals u_k^i and x_k^i. Due to the independence of control inputs from the scheduling policy, terms involving u_k^i and x_k^i can be readily added to the cost function. However, a minimization of those terms would directly affect the design of gain L_i, which is out of the scope of this chapter.

The following lemma is essential in order to obtain the analytical upper bound for the average cost.

Lemma 5.1. *[37] Let e_k be a Markov chain with state space X. Introduce $J_{e_k}: X \to \mathbb{R}$ and a measurable function $h: X \to \mathbb{R}$, where $h(e_k) \geq 0$ for all $e_k \in X$. Define the average cost $J_{ave} = \lim_{T \to \infty} \sup \frac{1}{T} \sum_{k=0}^{T-1} \mathbb{E}[J_{e_k}]$. Then*

$$J_{\text{ave}} \leq \sup_{e_k \in X} \{J_{e_k} + \mathbb{E}[h(e_{k+1})|e_k] - h(e_k)\}.$$

Lemma 5.1 refers to one-step time transitions. Since e_k is a ψ-irreducible Markov chain evolving in the uncountable space $\mathbb{R}^{n_e}$, one can generate a Markov chain by sampling the states of the original chain at steps $\{0, N, 2N, \ldots\}$. It is straightforward to show that the generated Markov chain is ψ-irreducible, aperiodic, and time homogeneous [123, Ch. 1]. Consequently, the average cost J_{ave} can be defined over an N-step interval as follows

$$J_{\text{ave}} \leq \sup_{e_k \in X} \{J_{e_k} + \mathbb{E}[h(e_{k+N})|e_k] - h(e_k)\}. \quad (5.17)$$

Let $h(e_k) = \sum_{i=1}^{N} \|e_k^i\|_{Q_i}^2$, then from (5.16), we conclude

$$J_{\text{ave}} \leq \sup_{e_k \in X} \sum_{i=1}^{N} \left[\mathbb{E}\left[\|e_{k+N}^i\|_{Q_i}^2 | e_k\right] + \eta \delta_k^i\right]. \quad (5.18)$$

Theorem 5.4. *Consider a multi-loop NCS consisting of N heterogeneous linear time-invariant stochastic sub-systems modeled as (5.3), and a transmission channel subject to the constraint (5.11), and the control, estimation and scheduling laws given by (5.4), (5.5) and (5.10), respectively. Then the average cost $J_{ave} = \lim_{T \to \infty} \sup \frac{1}{T} \sum_{k=0}^{T-1} \mathbb{E}[J_{e_k}]$ is upper bounded by a strictly increasing function of $\|A_i\|, \text{tr}(Q_i), N,$ and η, over all initial conditions and for all $i \in \{1, \ldots, N\}$. Furthermore, the upper bound is convex with respect to λ_i for all $i \in \{1, \ldots, N\}$.*

The proof is provided in Section 5.8.3.

Remark 5.9. *Even though the convexity with respect to the local error thresholds λ_i is shown only for the upper bound, simulations suggest that the average cost (5.18) inherits this property, see also Section 5.6. Intuitively, lowering λ_i's implies that more sub-systems compete for transmission, increases the probability that sub-systems with lower priorities transmit, and hence increases the average cost. On the other hand, with higher λ_i's, more sub-systems are deterministically kept out of channel access competition and subsequently they remain in open loop.*

The obtained upper bound for the average cost function determines the size of the attractive set for the error state e_k. Theorem 5.4 indicates the influence of different system parameters, which is discussed in the following. If the degree of instability of the sub-systems increases, i.e., the matrix norm $\|A_i\|$ for at least one i increases, then also the size of the attractive set increases, which is intuitively comprehensible. If the number N of sub-systems increases while the capacity remains constant, again the size of the attractive set increases as individual sub-systems have to wait longer for channel access. In fact, the same holds true also if the ratio $\frac{c}{N}$ decreases.

As already discussed when introducing the bi-character scheduling policy (5.10), the local parameters λ_i's and Q_i's tune the deterministic threshold and assigned priorities, respectively. The weight matrices Q_i indeed determine the importance of transmissions for a certain sub-system, and therefore they are designed to assign the priorities in favor of those sub-systems with higher operational importance. Increasing them would increase the chance the corresponding sub-system has to be granted access to the channel. Setting the weight matrix very high for one certain sub-system results in that sub-system having, as expected, much more frequent transmissions than the others. Setting all the error thresholds λ_i to zero results in the pure probabilistic scheduling policy (5.13). Although it is shown that the f-ergodicity remains valid under the policy (5.13), we will see in the numerical that the performance of the bi-character policy (5.10) outperforms that of (5.13). Intuitively, setting the threshold to zero allows all sub-systems $i \in \{1, \ldots, N\}$ to take part in the channel competition at every time step. Since the channel access is supposed to be eventually granted through a randomized mechanism, even those sub-systems with negligible error values have chances to use the channel and therefore occupy the limited transmission possibilities. Hence, the channel becomes busy and other sub-systems with higher errors remain in open loop and should wait for next time steps to transmit, which leads to increase in the aggregate error variance. On the other hand, setting those thresholds very high means that more sub-systems with relatively high errors (but still below their corresponding highly set thresholds) are excluded from the channel competition and they remain in open loop until their errors exceed the corresponding thresholds. This scenario also leads to increased aggregate error variance. Therefore, there should be optimal error thresholds which maintain the balance between the aggregate error variance and giving the chance to transmit to the sub-systems which are in more stringent real-time conditions.

5.4.4 Scheduling in the Presence of Channel Imperfections

So far in this chapter, we assumed that every data packet which is awarded channel access, will be successfully received by the controller. We will now investigate the robustness of the scheduling policy with respect to data packet dropouts. Note that even in case of a centralized scheduling mechanism such as TOD (5.12) or the bi-character policy (5.10), dropouts may still happen caused by external factors such as erroneous network links, malfunctioning network hardware, weak-power wireless signals from stations located far away, environmental disturbances, or even intentionally through, e.g., the dynamic source routing protocol. In order to model packet loss, we introduce the binary variable γ_k, which indicates successful or unsuccessful transmission in each time instant

$$\gamma_k^i = \begin{cases} 1, & x_k^i \text{ successfully received} \\ 0, & x_k^i \text{ dropped.} \end{cases}$$

Thus the dynamics of the error state e_k^i depends on the sequence of variables δ_k^i and γ_k^i as

$$e_{k+1}^i = \left(1 - \theta_{k+1}^i\right) A_i e_k^i + w_k^i, \tag{5.19}$$

where $\theta_k^i = \delta_k^i \gamma_k^i$; see also Figure 5.1 for an illustration. We assume a Bernoulli packet loss, i.e., a scheduled data packet is successfully transmitted with probability p and dropped with probability $1 - p$, i.e.,

$$\mathsf{P}\left[\gamma_k^i = 1 | \delta_k^i = 1\right] = p \quad , \quad \mathsf{P}\left[\gamma_k^i = 0 | \delta_k^i = 1\right] = 1 - p. \tag{5.20}$$

Through the following proposition, we show that, under mild conditions, the TOD policy (5.12) is robust against the dropouts, i.e., the stability of the overall multi-loop NCS is preserved.

Proposition 5.2. *Consider a multi-loop NCS consisting of N heterogeneous linear time-invariant stochastic sub-systems modeled as (5.3), and a transmission channel subject to the constraint (5.11), and the control, estimation and scheduling laws given by (5.4), (5.5) and (5.12), respectively. Assume that the scheduled data packets are successfully transmitted by probability p as described in (5.20) with*

$$p > 1 - \frac{1}{\sum_{i=1}^{N} \|A_i\|_2^{2N}}.$$

Then the network-induced error state (5.19) is f-ergodic.

The proof is provided in Section 5.8.4. f-ergodicity of the overall multi-loop NCS with the aggregate state (x_k, e_k) readily follows from the assumption of the closed-loop matrix $(A_i - B_i L_i)$ to be Hurwitz $\forall\, i \in \{1, ..., N\}$.

The result indicates that the transmission needs to be successful often enough depending on the degree of instability of the sub-systems and the number of sub-systems: The higher the degree of instability of the sub-systems is, the lower the packet loss probability is allowed to be. Also, the larger the number N of sub-systems sharing the single transmission channel is, the higher the transmission success rate is required to be, which is intuitive.

We have derived analogous results for the proposed bi-character scheduling policy (5.10). In particular, if we can deterministically bound the packet loss occurrences during an interval $[k, k+N]$ by m, i.e., $m < N$, then the network-induced error state (5.19) is f-ergodic, see [116] for a formal proof.

5.5 Decentralized Scheduler Implementation

The proposed bi-character scheduling approach (5.10) is intrinsically centralized: The event triggers, in our case the weighted error norms from all sub-systems involved in channel access competition, are provided to the scheduler at every time step. Therefore, the scheduling protocol has the form of a centralized control unit where all sub-systems are connected and update their event information in a timely manner. This approach, as discussed earlier, might not always be desirable due to a long distance between the sending stations, high amount of data which needs to be transmitted and analyzed by the centralized scheduling unit, extra transmission costs, and even privacy issues. Decentralized control and scheduling mechanisms with local decision-making units are appropriate when such issues are relevant.

In this section, we discuss the approximate decentralized implementation of the centralized bi-character scheduling policy (5.10), i.e., the ith sub-system is provided with only local information A_i, B_i, W_i, λ_i, Z_k^i and the distribution of x_0^i, where λ_i is the local error threshold and $Z_k^i = \{z_0^i, \ldots, z_k^i\}$. First, note that the deterministic scheduling process is performed locally within each loop, i.e., in a decentralized fashion, because the condition $\|e_{k'}^i\| \leq \lambda_i$ is checked locally in order to decide whether the sub-system is eligible for the channel competition. Therefore, the decentralized design, described in what follows, approximates the probabilistic part of the scheduling policy (5.10), as the deterministic part is inherently decentralized.

In order to introduce the decentralized scheduling mechanism, it is essential to have a closer look at the communication channel. First of all, it is assumed that the operational time scale of the communication channel, denoted as *micro time slots*, is much finer than that of the local control systems, denoted as *macro time slots*. This implies that between each two macro time slots $k\tau_{\text{macro}} \to (k+1)\tau_{\text{macro}}$, the micro time slots are distributed as $\{k\tau_{\text{macro}}, k\tau_{\text{macro}} + \tau, \ldots, k\tau_{\text{macro}} + (h-1)\tau, (k+1)\tau_{\text{macro}}\}$, with τ and τ_{macro} denoting the duration of the micro and the macro time slots, respectively, and the integer h denoting the number of micro time slots within one control period, i.e., one macro time slot. Considering the current communication and telecommunication standards in terms of bandwidth, time resolution, and speed in comparison to many control applications, it is reasonable to assume $h \gg 1$. One data packet can be transmitted through the communication channel starting from every micro time slot $\{1, \ldots, h\}$. We assume that a data packet will be received by the corresponding controller before the next macro time slot starts.

Every sub-system i knows its latest local error e_k^i as well as λ_i at time step k when the sub-system is about to decide whether to compete for channel access in time step $k+1$. A system which is eligible for channel transmission at some time $k'+1$, locally inspects whether the channel is free. We introduce integer random variables $\nu_k^i \in \{0, 1, \ldots, (h-1)\}$, so-called *waiting times*, for each eligible sub-system. The integer ν_k^i denotes the number of micro slots that sub-system i waits before listening to the communication channel at time $kh + \nu_k^i$, where the time from now on is expressed in terms of the number of micro time slots. If the channel is free, sub-system i transmits. Otherwise, sub-system i backs off and does not attempt to transmit over $\{\tau_{\text{macro}}, k\tau_{\text{macro}} + \tau, \ldots, k\tau_{\text{macro}} + (h-1)\tau\}$ again. We propose for every eligible sub-system i, that the waiting time ν_k^i is chosen randomly from a finite-variance concave probability mass function (*pmf*) with the error-dependent mean

$$\mathbb{E}[\nu_k^i] = \frac{1}{\|e_k^i\|_{Q_i}^2}. \qquad (5.21)$$

The waiting times $\nu_k^i \in \{0, 1, \ldots, (h-1)\}$ are chosen for every eligible sub-system according to its corresponding local mass function. Note that they are independent from each other, i.e., the design is indeed decentralized. The concavity of the local probability mass functions emphasizes the prioritized character of the decentralized scheduling rule, as it ensures the random waiting times are chosen with higher probabilities around the mean (5.21).

The decentralized implementation of the scheduling policy is accordingly represented by

$$\mathrm{P}[\delta_{k+1}^i = 1 | e_k^i] = \begin{cases} 0 & \|e_k^i\|_{Q_i}^2 \leq \lambda_i \\ \mathrm{P}[\nu_k^i < \nu_k^j] & \forall j \in \mathcal{G}_k, j \neq i, \end{cases} \qquad (5.22)$$

where $\mathcal{G}_k$ denotes the set of all eligible sub-systems at time instant k, i.e., $\mathcal{G}_k = \{i | \|e_k^i\|_{Q_i}^2 > \lambda_i\}$.

Similar to the centralized scheme (5.10), there is no guarantee that the sub-system with the highest error is awarded the channel even though that sub-system has more chance. In Figure 5.2, the probabilistic channel access procedure is schematically shown for a specific realization with two sub-systems. For illustrative purposes, $h=11$ even though in reality the time duration τ of the communication channel is often much finer.

As the proposed policy is decentralized, there is a possibility of collision,

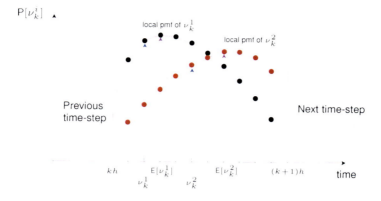

FIGURE 5.2
Two sub-systems compete for channel access by randomly choosing their waiting times ν_k^1 and ν_k^2, according to their corresponding local probability mass functions with error-dependent means $\mathbb{E}[\nu_k^1]$ and $\mathbb{E}[\nu_k^2]$) (purple arrows). The blue arrows illustrate a possible realization of the waiting times, here sub-system 1 won the competition as its waiting time is lower than for sub-system 2.

namely if at least two subsystems choose exactly identical waiting times. As all sub-systems are blocked from transmission within the corresponding macro time slot then, a collision has the same effect as a packet dropout. Accordingly, the error dynamics can be represented using the packet drop/collision indicator variable γ_k^i by (5.19). The probability of collision obviously depends on the choice of the local probability mass functions, the number of micro time slots h, and other system parameters such as the number of sub-systems N, the scheduler parameters λ_i, Q_i, and the sub-system dynamics. The implementation of the proposed decentralized scheduling scheme is illustrated in Figure 5.3.

Of course, scheduler parameters should be chosen such that collisions happen rarely. However, there is still a non-zero probability that collisions happen at all time instants over the interval $[k, k+N]$. This leads to the conclusion that there is a non-zero probability that all sub-systems operate in open loop as all transmission attempts have failed due to collisions. Therefore, the stability of NCSs under such a decentralized policy cannot be shown in terms of f-ergodicity. Instead, stability is shown in terms of Lyapunov Stability in Probability [94].

Proposition 5.3. *Consider a multi-loop NCS consisting of N heterogeneous linear time-invariant stochastic sub-systems modeled as (5.3), and a transmission channel subject to the constraint (5.11), and the control and estimation laws given by (5.4) and (5.5), respectively. Under the scheduling law (5.22)*

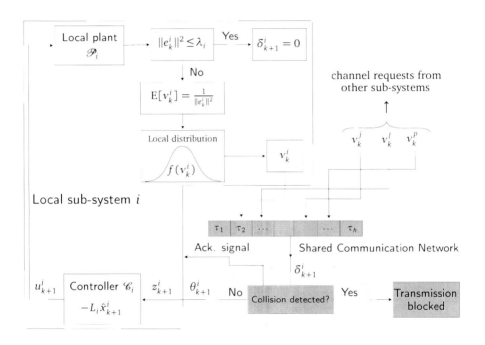

FIGURE 5.3
Architecture for the decentralized implementation of the scheduler.

with the waiting times chosen according to a finite-variance concave probability mass function with mean (5.21), the overall system is Lyapunov stable in probability.

The proof is provided in Section 5.8.5. A similar decentralized scheduling policy with the adaptation of the error thresholds depending on communication channel feedback is investigated in [113].

5.6 Empirical Performance Evaluation

In this section, the performance of the proposed event-triggered scheduler (5.10) is investigated and compared with TDMA and idealized CSMA policies. We also demonstrate that the deterministic feature of the scheduler yields performance improvements compared to the pure probabilistic policy (5.13).

To make the simulations more general, we assume a heterogeneous multi-loop NCS comprised of two classes of sub-systems. The first class consists of control loops with unstable plants, the second class of control loops with stable processes. The system parameters (5.3) are $A_1 = 1.25$, $B_1 = 1$ for the first class and $A_2 = 0.75$, and $B_2 = 1$ for the second class. The initial condition is $x_0^1 = x_0^2 = 0$, and the random disturbance is given by $w_k^i \sim \mathcal{N}(0, 1)$. We consider N loops with an equal number of loops belonging to each of the two classes. In order to stabilize the sub-systems, we choose a deadbeat control law in (5.4), i.e., $L_i = A_i$ for each class $i \in \{1, 2\}$ and a model-based observer given by (5.5). For the purpose of simplicity, we select $Q_i = 1$ for each class $i \in \{1, 2\}$. The results are obtained from Monte Carlo simulations over a horizon of 5×10^5 time steps.

5.6.1 Optimized Thresholds λ for the Bi-Character Scheduler

For illustrative purposes, we assume that the local error thresholds are equal for all sub-systems. Therefore, we drop the subscript i and simply denote the thresholds λ in the remainder of this section. The simulation results presented in Figure 5.4 demonstrate how the variance of the network-induced error (5.9) change with respect to the local error thresholds λ for multi-loop NCSs with different numbers of connected sub-systems under the scheduling policy (5.10). We observe that indeed the aggregate error variance is a convex function of the error threshold λ; see also Section 5.4.3. In consequence, there is an optimal error threshold that minimizes the aggregate error variance. We also observe, that the optimal value of λ monotonically increases with an increasing number N of control loops. This is expected since the competition to access the sole

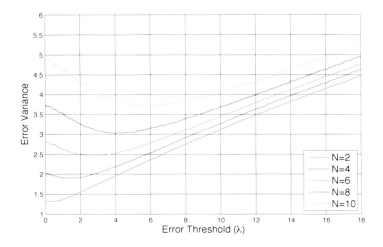

FIGURE 5.4
Error variance over local error thresholds for multi-loop NCSs with different number of control loops.

communication channel intensifies by increasing the number of sub-systems while keeping the number of available simultaneous transmission possibilities equal (here $c = 1$).

5.6.2 Comparison for Different Scheduling Policies

In the following we will study the performance of the proposed error-dependent scheduling laws (5.10), (5.12), and (5.13) in comparison to classical schemes as TDMA and CSMA, and a uniform event-based policy as proposed in [128]. Therefore we use the same multi-loop NCS setup as before. As error thresholds λ, we consider the optimal ones as derived in the previous paragraph as a function of the number N of sub-systems; see also Figure 5.4.

Figure 5.5 provides comparisons in terms of average network-induced error variance between the different scheduling policies for different numbers of sub-systems $N \in \{2, 4, 6, 8, 10, 20\}$ subject to the constraint (5.11). Note that for $N > 2$, we have more unstable systems than available transmission slots, i.e., at least one unstable sub-system is in open loop at every time step. The chosen optimal thresholds λ for different N are also presented in Figure 5.5.

To have a fair comparison, we derived the optimal TDMA pattern by brute force search over a finite time window (finding the optimal TDMA pattern is NP-hard). We search for the patterns resulting in minimum average error variance over the considered window for NCS setups with $N \in \{2, 4, 6, 8, 10, 20\}$. The optimal search is, however, extremely time-consuming, while the pattern changes sensitively w.r.t. the system parameters. The search for $N = 4$ over 9 time steps lasts nearly 11 hours on a 3.90 GHz 4690 Core i5 CPU, indicating the lack of scalability of this approach. The CSMA is a non-prioritized randomized channel arbitration; the transmission probability for a sub-system i is computed *a priori* also depending on system parameters for a fair comparison as $\frac{A_i^2}{\sum_{j=1}^N A_j^2}$ at each time step. The uniform event-based policy proposed in [128] disregards the error-dependent prioritization in policy (5.10) and instead employs uniform channel access probabilities. Hence, the scheduler randomly selects one sub-system among the eligible ones with uniform probability. Within the CSMA scheme, the sub-systems are selected randomly for channel access with a uniform probability, independent of the sub-system state.

As observable in Figure 5.5, the TOD scheduling policy (5.12) outperforms all other policies, which is to be expected as always the sub-system with the largest error is granted channel access. However, this is a fully centralized approach and only suitable for small-scale multi-loop NCSs. The fully decentralized, i.e., scalable, probabilistic policy (5.13) performs astonishingly well, for small N close to the TDMA scheme, and for larger N even better than TDMA. As expected, the proposed policy (5.10) performs better than the purely probabilistic one, as only sub-systems with larger errors compete for channel access due to the eligibility criterion. The CSMA scheme, which is also fully decentralized, performs worst: it results in an acceptable performance only up to $N = 4$. For $N = 6$ the variance takes values of magnitude 10^4. This is also expected as CSMA is a static protocol which may result in a long non-transmission period for unstable systems. The performance of the uniform event-based policy from [128] is between the optimal TDMA and the CSMA scheme. The proposed error-dependent scheduling laws (5.10), (5.12), and (5.13) all show a rather graceful performance degradation with an increasing number of sub-systems (while the capacity remains constant at $c = 1$).

5.6.3 Performance of the Decentralized Scheduler

Now, the performance of the proposed decentralized scheduling (5.22) is compared to the proposed error-dependent scheduling laws (5.10), (5.13) and the classical schemes TDMA and CSMA. Therefore we use the same multi-loop NCS setup as before. As error thresholds λ we consider the optimal ones as derived in the previous paragraph as a function of the number N of sub-systems, see also Figure 5.4 and Table 5.1. The number of micro time slots in the communication system is $h = 150$. The waiting times are chosen randomly from a Poisson distribution with mean (5.21) for each sub-system at every time instant.

In a decentralized scheduling policy, there is an inevitable chance of collision as there is the possibility that two or more sub-systems choose the same waiting time. The collision probability increases with a higher number N of

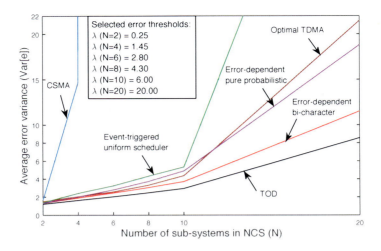

FIGURE 5.5
Comparison of the average error variance vs. the number of control loops for different scheduling policies.

control loops. This is validated by the empirically derived number of collisions presented in Table 5.1.

Number of plants (N)	2	4	6	8	10
Error threshold (λ)	0.25	1.45	2.80	4.30	6.00
Collisions in 2×10^5 samples	252	972	2094	3535	5009
Collisions (%)	0.126	0.486	1.047	1.767	2.504

TABLE 5.1
Optimal error thresholds λ and the number of collisions depending on the number of control loops.

Figure 5.6 shows the performance of the different scheduling policies in terms of the aggregate network-induced error variance for multi-loop NCSs with a different number of subsystems $N \in \{2, 4, 6, 8, 10\}$ subject to the constraint (5.11). Here in the TDMA scheme, every sub-system transmits according to an *a priori* defined schedule every N time steps. With the CSMA scheme (ICSMA) every sub-system has a chance of transmission of $\frac{1}{N}$ at each time step. For comparison we also consider the case where all sub-systems transmit at every time step by relaxing the capacity constraint to $c = N$, i.e., every subsystem transmits at every time step. This is the (non-achievable) lower bound for error variance.

The ICSMA protocol results in an acceptable performance only for $N = 2$, while the bi-character approach (5.10) outperforms TDMA. As expected, for the pure probabilistic scheduler (5.13) the error variance increases compared

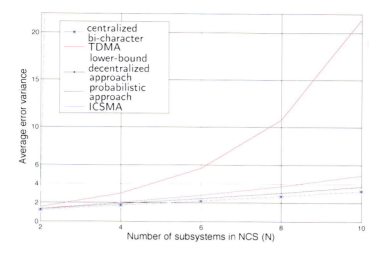

FIGURE 5.6
Comparison of the average error variance vs. the number of control loops for different scheduling policies.

to the bi-character scheduler, since those subsystems with small errors now have non-zero chance to utilize the channel.

5.6.4 Performance with Packet Dropouts

In this section we investigate the performance of the multi-loop NCS under the TOD scheduling policy (5.12) and with packet dropouts. To simulate the effect of packet dropouts, a fixed success probability p is assigned to each sub-system i and the binary random variable γ_k^i is then determined as the outcome of the Bernoulli distribution such that $\gamma_k^i = 1$ with probability p and $\gamma_k^i = 0$ with probability $1 - p$. Otherwise the simulation conditions, such as system parameters and the two classes of sub-systems (stable and unstable), are chosen as in Section 5.6.2. For comparison the multi-loop NCS is also simulated under the optimal TDMA scheme, the classical CSMA scheme and the uniform event-based strategy from [128]; see also Section 5.6.2 for details on those scheduling schemes.

Figure 5.7 provides the comparison in terms of the average aggregate network-induced error variance for the multi-loop NCS with a different number of sub-systems $N \in \{2, 4, 6, 8, 10, 20\}$, subject to the capacity constraint (5.11), i.e., $c = 1$. The TOD scheduler without packet dropout ($p = 100\%$) is clearly superior to the three other scheduling approaches (optimal TDMA, uniform event-based, CSMA). The performance deteriorates gracefully as the number of sub-systems increases. For dropout probabilities of 10% (($p = 90\%$) and 20% ($p = 80\%$) the aggregate error variance is still well bounded even

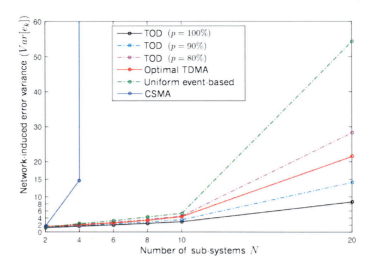

FIGURE 5.7
Comparison of the average error variance vs. the number of control loops for different scheduling policies with packet dropout.

for $N = 20$. This indicates that the derived bound (5.43) is rather conservative: They compute to $p > \{0.6374, 0.9175, 0.9771, 0.9930, 0.9977, 0.9999\}$ for $N = \{2, 4, 6, 8, 10, 20\}$. Naturally with increasing packet dropout probability, the performance deteriorates. Still, the TOD scheduler outperforms all others even with a dropout probability of 10%. As previously discussed, under these conditions the CSMA protocol results in an acceptable variance only up to $N = 4$.

5.7 Conclusions and Perspectives

In this chapter, we address the problem of scheduling in multi-loop NCS over shared capacity-limited communication channels. We propose different event-based scheduling policies which all have in common that they determine the priority for channel access depending on the state of the sub-systems. We show stability of the overall multi-loop NCS in a probabilistic sense and provide conditions for stability under packet dropouts. In practice, the choice for a centralized scheme or a decentralized scheme largely depends on the network size. The performance for centralized scheduling policies is clearly superior over decentralized strategies. However, this comes at the cost of additional overhead also in terms of communication and is suitable only for small to medium-size multi-loop NCSs. Decentralized strategies inevitably result in collisions, i.e., packet dropouts caused internally. The performance could come arbitrarily close to the centralized case if the communication system operates on a significantly faster time scale than the control sub-systems. The proposed state-dependent scheduling policies clearly outperform the classical MAC paradigms such as TDMA and CSMA. There are many open challenges which need to be addressed in the future, such as the consideration of non-linear dynamics for the sub-systems and an optimization-based approach to state-dependent scheduling.

5.8 Proofs and Derivations of Main Results

Before proceeding to the proof of the main theorem of this chapter, we express e^i_{k+N} as function of a previous error value $e^i_{k+r_i}$ at a certain time step $k+r_i$, with $r_i \in [0, N-1]$:

$$e^i_{k+N} = \prod_{j=r_i+1}^{N} \left(1 - \delta^i_{k+j}\right) A_i^{N-r_i} e^i_{k+r_i} \qquad (5.23)$$
$$+ \sum_{r=r_i}^{N-1} \left[\prod_{j=r+2}^{N} \left(1 - \delta^i_{k+j}\right) A_i^{N-r-1} w^i_{k+r}\right].$$

5.8.1 Proof of Theorem 5.3

Proof. To study stability, we let the NCS operate from time step k to $k+N-1$ under the policy (5.10). Then, time $k+N$ is scheduled considering all possible scenarios that might happen over $[k, k+N-1]$. To that end, we define at every time $k' \in [k, k+N]$ two time-varying disjoint sets $S^1_{k'}$ and $S^2_{k'}$, containing all excluded sub-systems and all eligible sub-systems for transmission, respectively. Therefore, for every $i \in \{1, \ldots, N\}$ at a time step k' we have $i \in S^1_{k'}$ if $\|e^i_{k'}\|^2_{Q_i} \leq \lambda_i$ and $i \in S^2_{k'}$ if $\|e^i_{k'}\|^2_{Q_i} > \lambda_i$. Clearly, $S^1_{k'} \cup S^2_{k'} = N$. Inclusion in either set $S^1_{k'}$ or $S^2_{k'}$ depends on both transmission occurrence and the noise process. Having this said, we discern three complementary and disjoint cases for a sub-system i. Sub-system i:

c_1: has either transmitted or not within the past $N-1$ time steps, and $i \in S^1_{k+N-1}$, i.e., $\|e^i_{k+N-1}\|^2_{Q_i} \leq \lambda_i$;

c_2: has transmitted at least once within the past $N-1$ time steps, i.e., $\delta^i_{k'} = 1$ at a certain time step $k' \in [k, k+N-1]$, and $i \in S^2_{k+N-1}$, i.e., $\|e^i_{k+N-1}\|^2_{Q_i} > \lambda_i$;

c_3: has not transmitted within the past $N-1$ time steps, i.e., $\delta^i_{k'}=0$ for all time steps $k' \in [k, k+N-1]$, and $i \in S^2_{k+N-1}$, i.e., $\|e^i_{k+N-1}\|^2_{Q_i} > \lambda_i$.

Each sub-system is characterized by exactly one of the cases in transition from k to $k+N-1$. For each of the three cases, the N-step drift (5.15) is split into partial drifts:

$$\Delta V(e^{i \in c_j}_k, N) = \sum_{c_j} \mathbb{E}\big[\|e^i_{k+N}\|^2_{Q_i} | e_k\big] - V(e^{c_j}_k), \tag{5.24}$$

where $V(e^{c_j}_k) = \sum_{c_j} \|e^i_k\|^2_{Q_i}$ for each case $c_j \in \{c_1, c_2, c_3\}$. We investigate (5.24) for each case c_j in order to invoke Theorem 5.2. For all sub-systems $i \in c_1$ we know $\delta^i_{k+N} = 0$, which is followed from $\|e^i_{k+N-1}\|^2_{Q_i} \leq \lambda_i$. Therefore

$$\sum_{c_1} \mathbb{E}\big[\|e^i_{k+N}\|^2_{Q_i} | e_k\big] = \sum_{c_1} \mathbb{E}\big[\|A_i e^i_{k+N-1} + w^i_{k+N-1}\|^2_{Q_i} | e_k\big]$$
$$\leq \sum_{c_1} \lambda_i \|A_i\|^2_2 + \mathrm{tr}(Q_i), \tag{5.25}$$

where the inequality (5.51) is ensured by the Cauchy–Schwarz inequality. Define $f_{c_1} = \epsilon_1 V(e^{c_1}_k) - \xi^{b^+}_1$, with $\xi^{b^+}_1 = \sum_{c_1} \lambda_i \|A_i\|^2_2 + \mathrm{tr}(Q_i)$ and $\epsilon_1 \in (0, 1]$. Then, we can find a small set $\mathcal{D}_1$ and constant ϵ_1 such that $f_{c_1} \geq 1$ and $\Delta V(e^{c_1}_k, N) \leq \sum_{c_1} \lambda_i \|A_i\|^2_2 + \mathrm{tr}(Q_i) - V(e^{c_1}_k) \leq -f_{c_1}$.

For a sub-system $i \in c_2$, assume that a transmission occurred at time step $k+r_i$, where $r_i \in [0, N-1]$, i.e., $\delta^i_{k+r_i} = 1$. By statistical independence of the noise sequence w^i_{k+r} and error $e^i_{k+r_i-1}$, from (5.52) we have

$$\sum_{c_2} \mathbb{E}\big[\|e^i_{k+N}\|^2_{Q_i} | e_k\big] \leq \sum_{c_2} \sum_{r=r_i}^{N} \mathrm{tr}(Q_i) \|A^{N-r}_i\|^2_2. \tag{5.26}$$

Define $f_{c_2} = \epsilon_2 V(e^{c_2}_k) - \xi^{b^+}_2$, where $\xi^{b^+}_2$ is the right side of the inequality in (5.53) and $\epsilon_2 \in (0, 1]$. Then we can find a small set $\mathcal{D}_2 \subset \mathbb{R}^{n_e}$ and ϵ_2 such that $f_{c_2} \geq 1$ and $\Delta V(e^{c_2}_k, N) \leq \sum_{c_2} \sum_{r=r_i}^{N} \mathrm{tr}(Q_i) \|A^{N-r}_i\|^2_2 - V(e^{c_2}_k) \leq -f_{c_2}$.

The sub-systems $i \in c_3$ are eligible to compete for channel access at time step $k+N$ because they belong to S^2_{k+N-1}. To infer f-ergodicity, we split the third case c_3 into two complementary and disjoint sub-cases as follows:

$l^{c_3}_1$ Sub-system i has not transmitted over the past $N-1$ time steps, but $i \in S^1$ at least once over $k, k+N-2$,

$l^{c_3}_2$ Sub-system i has not transmitted over the past $N-1$ time steps, and $i \in S^2$ for all the $N-1$ time steps.

Recall that the sub-systems $i \in c_3$ belong to S^2_{k+N-1}. For sub-case $l^{c_3}_1$, suppose that $k+r_i$ was the last time for which $i \in S^1_{k+r_i}$, which implies

$\|e_{k+r_i}^i\|_{Q_i}^2 \leq \lambda_i$. Knowing that $\delta_{k'}^i = 0$ for $i \in c_3$ until time step $k+N$, we reach

$$\sum_{l_1^{c_3}} \mathbb{E}[\|e_{k+N}^i\|_{Q_i}^2 | e_k] \leq$$
$$\sum_{l_1^{c_3}} \left[\lambda_i \|A_i^{N-r_i}\|_2^2 + \sum_{r=r_i}^{N-1} \text{tr}(Q_i) \|A_i^{N-r-1}\|_2^2 \right]. \quad (5.27)$$

Define $f_{l_1^{c_3}} = \epsilon_{l_1^{c_3}} V(e_k^{l_1^{c_3}}) - \xi_{l_1^{c_3}}^{b^+}$, where $\xi_{l_1^{c_3}}^{b^+}$ is given in (5.54), with $\epsilon_{l_1^{c_3}} \in (0,1]$. Thus, we can find small set $\mathcal{D}_{l_1^{c_3}} \subset \mathbb{R}^{n_e}$ and $\epsilon_{l_1^{c_3}}$ such that $f_{l_1^{c_3}} \geq 1$, and $\Delta V(e_k^{l_1^{c_3}}, N) \leq -f_{l_1^{c_3}}$.

In sub-case $l_2^{c_3}$, sub-systems are all in $S_{k'}^2$ since $\|e_{k'}^i\|_{Q_i}^2 > \lambda_i$ for all $k' \in [k, k+N-1]$. From (5.52) with $r'=0$:

$$\sum_{l_2^{c_3}} \mathbb{E}\left[\|e_{k+N}^i\|_{Q_i}^2 | e_k \right]$$
$$\leq \sum_{l_2^{c_3}} \left[\sum_{r=1}^{N} \text{tr}(Q_i) \|A_i^{N-r}\|_2^2 \right] + \|A_i^N\|_2^2 \|e_k^i\|_{Q_i}^2$$
$$\leq \sum_{l_2^{c_3}} \left[\sum_{r=1}^{N} \text{tr}(Q_i) \|A_i^{N-r}\|_2^2 \right] + \|A_i^N\|_2^2 V(e_k^{l_2^{c_3}}). \quad (5.28)$$

Unlike the other cases, the upper bound in (5.55) depends on the initial value via the last term. As the considered cases cannot happen all together, let us calculate the probability for sub-case $l_2^{c_3}$ to happen according to the policy (5.10). Recall that the interval of interest has length N. Therefore, if one system, say j, is not awarded channel access during the entire interval $[k, k+N]$, then there exists another sub-system, say i, which is awarded channel access more than once. Let $k+\bar{r}$ denote the most recent step at which sub-system i has transmitted, i.e., $\delta_{k+\bar{r}}^i = 1$. The probability that i transmits again at $k+N$, i.e., $\delta_{k+N}^i = 1$, in the presence of system j with no prior transmission at all time steps $\tilde{k}$, is computed as

$$\mathsf{P}[\delta_{k+N}^i = 1 | \delta_{k+\bar{r}}^i = 1, \delta_{\tilde{k}}^j = 0, \|e_{\tilde{k}}^j\|_{Q_j}^2 > \lambda_j]$$
$$= \mathbb{E}\left[\mathsf{P}[\delta_{k+N}^i = 1 | e_k] | \delta_{k+\bar{r}}^i = 1, \delta_{\tilde{k}}^j = 0, \|e_{\tilde{k}}^j\|_{Q_j}^2 > \lambda_j \right]$$
$$= \mathbb{E}\left[\frac{\|e_{k+N-1}^i\|_{Q_i}^2}{\sum_{j \in c_2} \|e_{k+N-1}^j\|_{Q_j}^2 + \sum_{j \in c_3} \|e_{k+N-1}^j\|_{Q_j}^2} \bigg| z_{i,j} \right]$$
$$\leq \mathbb{E}\left[\frac{\|\sum_{r=\bar{r}}^{N-1} A_i^{N-r-1} w_{k+r-1}^i\|_{Q_i}^2}{\sum_{j \in c_2} \lambda_j + \sum_{j \in l_1^{c_3}} \lambda_j + \sum_{j \in l_2^{c_3}} \|e_k^j\|_{Q_j}^2} \bigg| z_{i,j} \right]$$
$$\leq \frac{\sum_{r=\bar{r}}^{N-1} \text{tr}(Q_i) \|A_i^{N-r-1}\|_2^2}{\sum_{j \in c_2} \lambda_j + \sum_{j \in l_1^{c_3}} \lambda_j + \sum_{j \in l_2^{c_3}} \|e_k^j\|_{Q_j}^2} = \mathsf{P}_{l_2^{c_3}}, \quad (5.29)$$

where $z_{i,j}$ abbreviates the conditions of the expectation, and the worst-case scenario is considered which entails $\|e_{k'}^{l_2^{c_3}}\|_{Q_j}^2 \leq \|e_{k'+1}^{l_2^{c_3}}\|_{Q_j}^2$. From (5.29) one infers that the probability of subsequent transmission for a certain sub-system, in the presence of sub-systems with large errors and no prior transmission, can be made arbitrarily close to zero by tuning λ_j's and Q_j's. Intuitively, the chance of transmission for a sub-system $j \in l_2^{c_3}$ can be increased by increasing λ_j of sub-systems $j \notin l_2^{c_3}$, i.e., $j \in \{c_2, l_1^{c_3}\}$.

Incorporating the occurrence probabilities for each case $\{c_1, c_2, c_3\}$, the N-step drift (5.1) can be expressed as

$$\Delta V(e_k, N) = \sum_{j=1}^{3} \sum_{i \in c_j} \mathsf{P}_{c_j} \mathbb{E}\left[\|e_{k+N}^i\|_{Q_i}^2 | e_k\right] - V(e_k^{c_j}), \quad (5.30)$$

where P_{c_j} represents the probability of each case $\{c_1, \ldots, c_3\}$ happening, and $\sum_{j=1}^{3} \mathsf{P}_{c_j} = 1$. Recalling the right-hand side of (5.55), and applying the probability (5.29) for the case $l_2^{c_3}$ according to (5.30), we can find the upper bound for the partial drift as follows:

$$\Delta V(e_k^{j \in l_2^{c_3}}, N) \leq \left[\mathsf{P}_{l_2^{c_3}} \sum_{l_2^{c_3}} \|A_j^N\|_2^2 - 1\right] V(e_k^{l_2^{c_3}}) + \xi_{l_2^{c_3}}^{b+},$$

with $\xi_{l_2^{c_3}}^{b+} = \mathsf{P}_{l_2^{c_3}} \sum_{l_2^{c_3}} \left[\sum_{r=1}^{N} \text{tr}(Q_j) \|A_j^{N-r}\|_2^2\right]$. Define $f_{l_2^{c_3}} = \epsilon_{l_2^{c_3}} V(e_k^{l_2^{c_3}}) - \xi_{l_2^{c_3}}^{b+}$, with $\epsilon_{l_2^{c_3}} \in (0, 1]$. We can find small set $\mathcal{D}_{l_2^{c_3}}$ and $\epsilon_{l_2^{c_3}}$ s. t. $f_{l_2^{c_3}} \geq 1$, and $\Delta V(e_k^{l_2^{c_3}}, N) \leq -f_{l_2^{c_3}}$.

So far, the partial drifts for all cases $c_1 - c_3$ are shown to be convergent. Therefore, the small set $\mathcal{D}_f \subset \mathbb{R}^{n_e}$ can be selected as $\mathcal{D}_f = \mathcal{D}_1 \cup \mathcal{D}_2 \cup \mathcal{D}_{l_1^{c_3}} \cup \mathcal{D}_{l_2^{c_3}}$ ensuring that real-valued functions $f(e_k) \geq 1$ exist such that $\Delta V(e_k, N) \leq -\min f(e_k)$ for $e_k \in \mathbb{R}^{n_e} \setminus \mathcal{D}_f$, and $\Delta V(e_k, N) < \infty$ for $e_k \in \mathcal{D}_f$. This confirms that the condition in Theorem 5.2 holds for the overall drift defined in (5.30), which readily proves that the error Markov chain (5.9) is f-ergodic. The f-ergodicity of the overall system (5.7) with the aggregate state (x_k, e_k) follows directly from the triangular structure of (5.7) and the stabilizing gains L_i. □

5.8.2 Proof of Corollary 5.1

Proof. An LTI system with state vector X_k is said to possess Lyapunov mean square stability (LMSS) if, given $\varepsilon > 0$, there exists $\rho(\varepsilon)$ such that $\|X_0\|_2 < \rho$ implies [94]:

$$\sup_{k \geq 0} \mathbb{E}\left[\|X_k\|_2^2\right] \leq \varepsilon.$$

Assuming $X_k = (x_k, e_k)$, the LMSS of the overall NCS state is achieved if $\sum_{c_j} \mathsf{P}_{i \in c_j} \mathbb{E}\left[\|e_k^i\|_2^2\right] \leq \varepsilon$ for all $c_j \in \{c_1, c_2, c_3\}$, given the stabilizing control gains L_i exist for all $i \in \{1, \ldots, N\}$. Uniform upper bounds for $\mathbb{E}[\|e_k^{i \in c_j}\|_2^2]$ for cases $\{c_1, c_2, l_1^{c_3}\}$ are derived in (5.51)–(5.54) over N-step intervals, considering P_{c_j} are upper bounded with one. For the case $l_2^{c_3}$, the uniform upper bound can be obtained by employing (5.55) and (5.29). The bound can obtained from the expression (5.38) assuming $Q_j = I$. □

5.8.3 Proof of Theorem 5.4

Proof. We derive uniform upper bounds for the average cost (5.18) for each case c_1, c_2 and c_3 considering the same multi-loop NCS model as assumed in Theorem 5.3. For the cases c_1, c_2, and $l_1^{c_3}$, the upper bounds derived in (5.51)–(5.54) are valid and only the communication cost should be added. Therefore, we derive the following for the average cost (5.18):

$$J_{\text{ave}}^{i \in c_1} \leq \sum_{c_1} \lambda_i \|A_i\|_2^2 + N\eta + \text{tr}(Q_i), \tag{5.31}$$

$$J_{\text{ave}}^{i \in c_2} \leq N\eta + \sum_{c_2} \sum_{r=r_i}^{N} \text{tr}(Q_i) \|A_i^{N-r}\|_2^2, \tag{5.32}$$

$$J_{\text{ave}}^{i \in l_1^{c_3}} \leq \sum_{l_1^{c_3}} \left[\lambda_i \|A_i^{N-r_i}\|_2^2 + \sum_{r=r_i}^{N-1} \|A_i^{N-r-1}\|_2^2 \, \text{tr}(Q_i) \right], \tag{5.33}$$

where $N\eta$ upper-bounds the cost of communication. For sub-systems in $l_2^{c_3}$, the non-uniform upper bound (5.55) is valid. Therefore, according to (5.18), we have

$$J_{\text{ave}}^{i \in l_2^{c_3}} \leq \sup_{e_k} \sum_{l_2^{c_3}} \|A_i^N\|_2^2 \|e_k^i\|_{Q_i}^2 + \sum_{r=1}^{N} \text{tr}(Q_i) \|A_i^{N-r}\|_2^2.$$

We rewrite the average cost (5.18) for different cases as $J_{\text{ave}} = \mathsf{P}_{c_1} J_{\text{ave}}^{i \in c_1} + \mathsf{P}_{c_2} J_{\text{ave}}^{i \in c_2} + \mathsf{P}_{l_1^{c_3}} J_{\text{ave}}^{i \in l_1^{c_3}} + \mathsf{P}_{l_2^{c_3}} J_{\text{ave}}^{i \in l_2^{c_3}}$. Since $\delta_k^i = 0$ for all sub-systems $i \in l_2^{c_3}$ and for all $\bar{k} \in [k+N]$, $J_{\text{ave}}^{i \in l_2^{c_3}} \leq \sup_{e_k \in \mathbb{R}^{n_e}} \sum_{i \in l_2^{c_3}} \mathbb{E}\left[\|e_{k+N}^i\|_{Q_i}^2 | e_k\right]$. Given the probability of occurrence for the sub-case $l_2^{c_3}$ in (5.29), and considering the worst-case scenario, we conclude

$$\mathsf{P}_{l_2^{c_3}} J_{\text{ave}}^{i \in l_2^{c_3}} \leq \frac{\sum_{r=r_q}^{N} \text{tr}(Q_q) \|A_q^{N-r}\|_2^2}{\sum_{i \in c_2} \lambda_i + \sum_{i \in l_1^{c_3}} \lambda_i} \sum_{i \in l_2^{c_3}} \sum_{r=1}^{N} \text{tr}(Q_i) \|A_i^{N-r}\|_2^2$$

$$+ \sum_{r=r_q}^{N} \text{tr}(Q_q) \|A_q^{N-r}\|_2^2 \sum_{i \in l_2^{c_3}} \|A_i^N\|_2^2. \tag{5.34}$$

Summing up the uniform upper bounds (5.35)–(5.38) for all cases $\{c_1, c_2, c_3\}$, we obtain the overall upper bound for the average cost function (5.18) independent of the initial state. The overall upper bound is strictly increasing w.r.t. $\|A_i\|$, N, $\text{tr}(Q_i)$, and η, for all $i \in \{1, \ldots, N\}$. Since the error thresholds λ_i's are determined locally for each sub-system i, and moreover the obtained overall upper bound for the average cost is the summation of local functions which are convex w.r.t. their corresponding local threshold λ_i, the overall upper bound is convex w.r.t. the λ_i's, and the proof then readily follows.

We derive the upper bounds for the average cost (5.18) for each case c_1, c_2 and c_3 considering the same multi-loop NCS model as assumed in Theorem 5.3. For the cases c_1 and c_2, we notice that the upper bounds derived in (5.51) and (5.53) are valid. Therefore, from (5.18), we derive

$$J_{\text{ave}}^{i \in c_1} \leq \sum_{c_1} \lambda_i \|A_i\|_2^2 + n_\delta \eta + \text{tr}(Q_i), \tag{5.35}$$

$$J_{\text{ave}}^{i \in c_2} \leq n_\delta \eta + \sum_{c_2} \sum_{r=r_i}^{N} \text{tr}(Q_i) \|A_i^{N-r}\|_2^2, \tag{5.36}$$

where n_δ represents the number of transmissions. For sub-systems in c_3 both upper bounds in (5.54) and (5.55) are still valid. Therefore, for both $l_1^{c_3}$ and $l_2^{c_3}$, we have

$$J_{\text{ave}}^{i \in l_1^{c_3}} \leq \sum_{l_1^{c_3}} \left[\lambda_i \|A_i^{N-r_i}\|_2^2 + \sum_{r=r_i}^{N-1} \|A_i^{N-r-1}\|_2^2 \, \text{tr}(Q_i) \right] \tag{5.37}$$

$$J_{\text{ave}}^{i \in l_2^{c_3}} \leq \sup_{e_k} \sum_{l_2^{c_3}} \|A_i^N\|_2^2 \|e_k^i\|_{Q_i}^2 + \sum_{r=1}^{N} \text{tr}(Q_i) \|A_i^{N-r}\|_2^2.$$

The average cost in (5.18) can be rewritten for different cases as $J_{\text{ave}} = \mathsf{P}_{c_1} J_{\text{ave}}^{i \in c_1} + \mathsf{P}_{c_2} J_{\text{ave}}^{i \in c_2} + \mathsf{P}_{l_1^{c_3}} J_{\text{ave}}^{i \in l_1^{c_3}} + \mathsf{P}_{l_2^{c_3}} J_{\text{ave}}^{i \in l_2^{c_3}}$. With probability of occurrence for $l_2^{c_3}$ in (5.29) still valid, and considering the worst-case scenario, we have for $l_2^{c_3}$

$$\mathsf{P}_{l_2^{c_3}} J_{\text{ave}}^{j \in l_2^{c_3}} \leq \sum_{l_2^{c_3}} \|A_j\|_2^2 \sum_{r=\bar{r}}^{N-1} \text{tr}(Q_i) \|A_i^{N-r-1}\|_2^2 + \tag{5.38}$$

$$\frac{\left[\sum_{l_2^{c_3}} \sum_{r=1}^{N} \text{tr}(Q_j) \|A_j^{N-r}\|_2^2 \right] \left[\sum_{r=\bar{r}}^{N-1} \text{tr}(Q_i) \|A_i^{N-r-1}\|_2^2 \right]}{\sum_{c_2} \lambda_j + \sum_{l_1^{c_3}} \lambda_j}.$$

As (5.38) is a uniform bound along with the upper bounds for cases $\{c_1, c_2, l_1^{c_3}\}$, the result readily follows. □

5.8.4 Proof of Proposition 5.2

Proof. To address stability, we consider the operation of the NCS over an interval with length N, i.e., from the initial time step k to $k+N$ under the TOD policy (5.12). Then, according to (5.15), we calculate the expectation of error at the final time step $k+N$ considering all possible transitions in the entire state-space $\mathbb{R}^{n_e}$ over the interval $[k, k+N]$. To that end, we divide the set of all sub-systems into two complementary and disjoint sets G and $\bar{G}$, where $|G| + |\bar{G}| = N$, where $|\cdot|$ denotes the cardinality operator. Set G contains the sub-systems which have been awarded channel access at least once over the entire interval $[k, k+N]$, either the transmission has been successful or not. The set $\bar{G}$ includes the sub-systems which have never been awarded channel access over the same interval. Thus, if a sub-system is awarded channel access but the scheduled packet is dropped, it belongs to the set G. Assume m sub-systems belong to the set $\bar{G}$. According to the *law of iterated expectations* and from (5.52), we conclude for a sub-system $i \in G$ that

$$\begin{aligned}
&\mathbb{E}\left[\|e_{k+N}^i\|_2^2 | e_k, \gamma_{k+r_i}^i\right] \\
&= p\mathbb{E}\left[\|e_{k+N}^i\|_2^2 | e_k, \gamma_{k+r_i}^i = 1\right] + (1-p)\mathbb{E}\left[\|e_{k+N}^i\|_2^2 | e_k, \gamma_{k+r_i}^i = 0\right] \\
&\leq p\sum_{r=r_i-1}^{N-1} \mathsf{tr}(W_i)\|A_i^{N-r-1}\|_2^2 + (1-p)\mathbb{E}\left[\|A_i^N e_k^i + \sum_{r=0}^{N-1} A_i^{N-r-1} w_{k+r}^i\|_2^2 | e_k\right] \\
&\leq \mathsf{tr}(W_i)\left(p\sum_{r=r_i-1}^{N-1}\|A_i^{N-r-1}\|_2^2 + (1-p)\left[\|A_i^N\|_2^2\|e_k^i\|_2^2 + \sum_{r=0}^{N-1}\|A_i^{N-r-1}\|_2^2\right]\right),
\end{aligned}$$
(5.39)

in which $k + r_i$ represents the latest time sub-system i is awarded channel access.

Sub-systems $j \in \bar{G}$ have never been awarded channel access over the past N time steps, i.e., $\delta_{\bar{k}}^j = 0$ for all $\bar{k} \in [k, k+N]$. If $m > 0$, then there exists at least one sub-system $q \in G$ which has transmitted more than once over the past N time steps, either successfully or unsuccessfully. Consider the two latest successful or unsuccessful transmissions of sub-system q happened at time steps $k+r_q$ and $k+r_{qq}$, where $r_q < r_{qq} \leq N$. This implies $\|e_{k+r_{qq}-1}^q\|_2 > \|e_{k+r_{qq}-1}^j\|_2$ for all $j \in \bar{G}$, according to (5.12). For a sub-system $j \in \bar{G}$ the following inequality is straightforward to show

$$\mathbb{E}\left[\|e_{k+N}^j\|_2^2 | e_k\right] \leq \|A_j^{N-r_{qq}+1}\|_2^2 \mathbb{E}\left[\|e_{k+r_{qq}-1}^q\|_2^2 | e_k\right] + \sum_{r=r_{qq}-1}^{N-1} \mathsf{tr}(W_j)\|A_j^{N-r-1}\|_2^2.$$
(5.40)

Considering the dropout possibility $1-p$ associated with the scheduled transmission of sub-system q at time step $k + r_q$, the expectation of error norm at time $k + r_{qq} - 1$ for sub-system q becomes

$$\begin{aligned}
&\mathbb{E}\left[\|e_{k+r_{qq}-1}^q\|_2^2 | e_k, \gamma_{k+r_q}^q\right] \\
&= p\mathbb{E}\left[\|e_{k+r_{qq}-1}^q\|_2^2 | e_k, \gamma_{k+r_q}^q = 1\right] + (1-p)\mathbb{E}\left[\|e_{k+r_{qq}-1}^q\|_2^2 | e_k, \gamma_{k+r_q}^q = 0\right] \\
&\leq p\sum_{r=r_q-1}^{r_{qq}-2} \mathsf{tr}(W_q)\|A_q^{r_{qq}-r-2}\|_2^2 \\
&\quad + (1-p)\mathbb{E}\left[\|A_q^{r_{qq}-1} e_k^q + \sum_{r=0}^{r_{qq}-2} A_q^{r_{qq}-r-2} w_{k+r}^q\|_2^2 | e_k\right].
\end{aligned}$$

Substituting the above expression in (5.40), the following upper bound for $\mathbb{E}\left[\|e_{k+N}^j\|_2^2|e_k\right]$ follows

$$\mathbb{E}\left[\|e_{k+N}^j\|_2^2|e_k\right] \leq p\|A_j^{N-r_{qq}+1}\|_2^2 \sum_{r=r_q-1}^{r_{qq}-2} \operatorname{tr}(W_q)\|A_q^{r_{qq}-r-2}\|_2^2$$
$$+ (1-p)\|A_j^{N-r_{qq}+1}\|_2^2 \mathbb{E}\left[\|A_q^{r_{qq}-1}e_k^q + \sum_{r=0}^{r_{qq}-2} A_q^{r_{qq}-r-2} w_{k+r}^q\|_2^2|e_k\right]$$
$$+ \sum_{r=r_{qq}-1}^{N-1} \operatorname{tr}(W_j)\|A_j^{N-r-1}\|_2^2$$
$$\leq \zeta_{>0} + (1-p)\|A_j^{N-r_{qq}+1}\|_2^2 \|A_q^{r_{qq}-1}\|_2^2 \|e_k^q\|_2^2, \tag{5.41}$$

where

$$\zeta_{>0} = p\|A_j^{N-r_{qq}+1}\|_2^2 \sum_{r=r_q-1}^{r_{qq}-2} \operatorname{tr}(W_q)\|A_q^{r_{qq}-r-2}\|_2^2 + \sum_{r=r_{qq}-1}^{N-1} \operatorname{tr}(W_j)\|A_j^{N-r-1}\|_2^2$$
$$+ (1-p)\|A_j^{N-r_{qq}+1}\|_2^2 \sum_{r=0}^{r_{qq}-2} \operatorname{tr}(W_q)\|A_q^{r_{qq}-r-2}\|_2^2.$$

Having (5.39) and (5.41), we compute the N-step drift definition (5.15) as follows

$$\Delta V(e_k, N) = \sum_{i \in G} \mathbb{E}\left[\|e_{k+N}^i\|_2^2|e_k\right] + \sum_{j \in \bar{G}} \mathbb{E}\left[\|e_{k+N}^j\|_2^2|e_k\right] - V(e_k)$$
$$\leq \sum_{i \in G} \left[p \sum_{r=r_i-1}^{N-1} \operatorname{tr}(W_i)\|A_i^{N-r-1}\|_2^2\right]$$
$$+ \sum_{i \in G} \left[(1-p)\left[\|A_i^N\|_2^2\|e_k^i\|_2^2 + \sum_{r=0}^{N-1} \operatorname{tr}(W_i)\|A_i^{N-r-1}\|_2^2\right]\right]$$
$$+ \sum_{j \in \bar{G}} \zeta_{>0} + (1-p)\|A_j^{N-r_{qq}+1}\|_2^2\|A_q^{r_{qq}-1}\|_2^2\|e_k^q\|_2^2 - V(e_k).$$

Then, we can rewrite the N-step drift operator as

$$\Delta V(e_k, N) \leq \tau_{>0} + \sum_{i \in G}\left[(1-p)\left[\|A_i^N\|_2^2\|e_k^i\|_2^2\right]\right] \tag{5.42}$$
$$+ \sum_{j \in \bar{G}}(1-p)\|A_j^{N-r_{qq}+1}\|_2^2\|A_q^{r_{qq}-1}\|_2^2\|e_k^q\|_2^2 - V(e_k),$$

where

$$\tau_{>0} = \zeta_{>0} + \sum_{i \in G}\left[p \sum_{r=r_i-1}^{N-1} \operatorname{tr}(W_i)\|A_i^{N-r-1}\|_2^2\right]$$
$$+ \sum_{i \in G}\left[(1-p) \sum_{r=0}^{N-1} \operatorname{tr}(W_i)\|A_i^{N-r-1}\|_2^2\right].$$

Now, we re-express (5.42), using the fact that $V(e_k) > \|e_k^i\|_2^2$, for every $i \in \{1, \ldots, N\}$

$$\Delta V(e_k, N) \leq \tau_{>0} + (1-p)V(e_k) \left[\sum_{j \in \bar{G}} \|A_j^{N-r_{qq}+1}\|_2^2 \|A_q^{r_{qq}-1}\|_2^2 + \sum_{i \in G} \|A_i^N\|_2^2 \right] - V(e_k).$$

In order to achieve a decreasing drift function, it is essential to satisfy the following inequality

$$(1-p) \left[\sum_{j \in \bar{G}} \|A_j^{N-r_{qq}+1}\|_2^2 \|A_q^{r_{qq}-1}\|_2^2 + \sum_{i \in G} \|A_i^N\|_2^2 \right] - 1 < 0.$$

Considering the worst-case scenario, i.e., $A_j > A_i$ for all the m sub-systems $j \in \bar{G}$ and $N - m$ sub-systems $i \in G$, the lower bound for the success probability p can be derived as follows

$$p > 1 - \frac{1}{\sum_{j \in \bar{G}} \|A_j\|_2^{2N} + \sum_{i \in G} \|A_i^N\|_2^2}.$$

The above lower bound depends on the time-variant sets G and $\bar{G}$, thus we employ the inequality $\sum_{i=1}^N \|A_i\|_2^{2N} > \sum_{j \in \bar{G}} \|A_j\|_2^{2N} + \sum_{i \in G} \|A_i^N\|_2^2$, leading to the time-invariant lower bound on p as

$$p > 1 - \frac{1}{\sum_{i=1}^N \|A_i\|_2^{2N}}. \quad (5.43)$$

Having (5.43) satisfied, we define $f_2 : \mathbb{R}^{n_e} \to \mathbb{R}$ as $f_2 = \epsilon_2 V(e_k) - \tau_{>0}$, with $\epsilon_2 \in (0, 1]$. Therefore, small set $\mathcal{D}_2$ and ϵ_2 can be found such that $f_2 \geq 1$, and $\Delta V(e_k, N) \leq -f_2$, for $e_k \notin \mathcal{D}_2$. This readily proves the f-ergodicity of (5.9). $\square$

5.8.5 Proof of Proposition 5.3

Here we show Lyapunov Stability in Probability (LSP) [94] of the overall system; see also Definition 5.9. Before presenting the main proof, we show in the following lemma that LSP of the overall system can be shown by solely considering the error state e_k.

Lemma 5.2. *Consider a system dynamics given by (5.7). The condition in Definition 5.9) is equivalent to*

$$\lim_{k \to \infty} \sup \mathsf{P}\left[e_k^T e_k \geq \xi' \right] \leq \xi, \quad (5.44)$$

where $\xi' > 0$ and the constant ξ fulfills $0 \leq \xi \leq \varepsilon$.

Proof. The system state x_k^i for each loop i evolves as

$$x_{k+1}^i = (A_i - B_i L_i)x_k^i + (1 - \theta_{k+1}^i)B_i L_i e_k^i + w_k^i. \tag{5.45}$$

As already discussed, the evolution of the error e_k^i is independent of the system state x_k^i within each individual control loop. Furthermore, by assumption, the emulative control law (5.4) ensures the closed-loop matrix $(A_i - B_i L_i)$ is Hurwitz. Together with the assumption that x_0^i has a bounded variance distribution, it follows that the system state x_k^i is converging. In addition, the disturbance process w_k^i is i.i.d. according to $\mathcal{N}(0, I)$, and is bounded in probability. Thus, showing $\lim_{k\to\infty} \sup \mathsf{P}\left[e_k^{i\mathsf{T}} e_k^i \geq \xi_i'\right] \leq \xi_i$ ensures the existence of constants ε_i and $\varepsilon_i' > 0$ such that $\lim_{k\to\infty} \sup \mathsf{P}\left[x_k^{i\mathsf{T}} x_k^i \geq \varepsilon_i'\right] \leq \varepsilon_i$. As individual loops operate independently, we take the aggregate NCS state (x_k, e_k). Then, the existence of ξ and $\xi' > 0$ such that $\lim_{k\to\infty} \sup \mathsf{P}\left[e_k^\mathsf{T} e_k \geq \xi'\right] \leq \xi$, implies existence of ε and $\varepsilon' > 0$ such that $\lim_{k\to\infty} \sup \mathsf{P}\left[x_k^\mathsf{T} x_k \geq \varepsilon'\right] \leq \varepsilon$, and the proof readily follows. □

As expected values are more straightforward in pursuing further analysis than probabilities, we employ *Markov's inequality* for $\xi' > 0$

$$\mathsf{P}\left[e_k^\mathsf{T} e_k \geq \xi'\right] \leq \frac{\mathbb{E}\left[e_k^\mathsf{T} e_k\right]}{\xi'}. \tag{5.46}$$

This confirms that showing the error is uniformly bounded in expectation ensures finding appropriate ξ and $\xi' > 0$ such that (5.44) is satisfied for arbitrary $\rho(\xi', \xi)$. Introducing positive definite matrices Q_i, we focus on deriving an upper bound for the expectation of weighted quadratic error norm

$$\mathbb{E}\left[e_k^\mathsf{T} Q e_k\right] = \sum_{i=1}^{N} \mathbb{E}\left[e_k^{i\mathsf{T}} Q_i e_k^i\right] = \sum_{i=1}^{N} \mathbb{E}\left[\|e_k^i\|_{Q_i}^2\right], \tag{5.47}$$

where $Q = \mathrm{diag}(Q_i)$. This modifies the condition (5.44) as

$$\lim_{k\to\infty} \sup \mathsf{P}\left[e_k^\mathsf{T} Q e_k \geq \bar{\xi}'\right] \leq \bar{\xi}. \tag{5.48}$$

Due to the capacity constraint (5.11), the boundedness of (5.47) cannot always be shown over one step transition, as already illustrated in Section 5.4.2. Following the same arguments, we look over the interval $[k, k+N]$, with arbitrary initial state e_k in order to show LSP.

Proof. We assume that the NCS operates from some time instant k until $k+N-1$ and we predict the error evolution considering all the possible scenarios under the introduced scheduling policy over the interval $[k, k+N-1]$. Then, looking at time step $k+N$, we show that the aggregate error state e_{k+N} fulfills (5.47). We divide the subsystems $i \in \{1, \ldots, N\}$ at each time step $k' \in [k, k+N-1]$ into two disjoint sets as

$$i \in \begin{cases} \mathcal{G}_{k'} & \|e_{k'}^i\|_{Q_i}^2 > \lambda_i \\ \bar{\mathcal{G}}_{k'} & \|e_{k'}^i\|_{Q_i}^2 \leq \lambda_i, \end{cases} \tag{5.49}$$

where $\mathcal{G}_{k'} \cup \bar{\mathcal{G}}_{k'} = N$. According to (5.10), subsystems belonging to $\mathcal{G}_{k'}$ are considered for transmission at time $k'+1$. Note that not only a transmission results in error decrement for a subsystem, but the disturbance process might also decrease the error. Therefore, the inclusion in either set $\mathcal{G}_{k'}$ or $\bar{\mathcal{G}}_{k'}$ depends on both transmission occurrence and disturbance $w^i_{k'}$. To take this into account, we discern three complementary and mutually exclusive cases, covering the entire state space that the Markov chain e_k evolves, at time step $k+N-1$ as:

Subsystem i:

c_1: has either successfully transmitted or not within the past $N-1$ time steps, and is in set $\bar{\mathcal{G}}_{k+N-1}$, i.e.,

$$i \in \bar{\mathcal{G}}_{k+N-1} \;\Rightarrow\; \|e^i_{k+N-1}\|^2_{Q_i} \leq \lambda_i;$$

c_2: has successfully transmitted at least once within the past $N-1$ time steps, and is in set $\mathcal{G}_{k+N-1}$, i.e.,

$$\exists k' \in [k, k+N-1] : \theta^i_{k'} = 1 \text{ and } \|e^i_{k+N-1}\|^2_{Q_i} > \lambda_i;$$

c_3: has not successfully transmitted within the past $N-1$ time steps, and is in set $\mathcal{G}_{k+N-1}$, i.e.,

$$\forall k' \in [k, k+N-1] : \theta^i_{k'} = 0 \text{ and } \|e^i_{k+N-1}\|^2_{Q_i} > \lambda_i.$$

We study the boundedness of error expectation over the interval $[k, k+N]$ for cases c_1–c_3. Since, the cases are complementary and mutually exclusive, we can calculate the probability for each case to happen, and express (5.47) as

$$\sum_{i=1}^{N} \mathbb{E}\left[\|e^i_{k+N}\|^i_{Q_i}\right] = \sum_{c_l} \mathsf{P}_{c_l} \mathbb{E}\left[\|e^i_{k+N}\|^i_{Q_i} | c_l, \mathsf{P}_{c_l}\right], \tag{5.50}$$

where $\sum_{l=1}^{3} \mathsf{P}_{c_l} = 1$.

Suppose that some subsystems i belong to c_1. Since $i \in \bar{\mathcal{G}}_{k+N-1}$, it follows from (5.49) that $\|e^i_{k+N-1}\|^2_{Q_i} \leq \lambda_i$. Thus, they are not eligible for transmission at time $k+N$, i.e., $\delta^i_{k+N} = 0$. Then, it follows from (5.8) and (5.50)

$$\sum_{i \in c_1} \mathbb{E}\left[\|e^i_{k+N}\|^2_{Q_i} | e_k\right] = \sum_{i \in c_1} \mathbb{E}\left[\|A_i e^i_{k+N-1} + w^i_{k+N-1}\|^2_{Q_i} | e_k\right]$$

$$\leq \sum_{c_1} \|A_i\|^2_2 \mathbb{E}\left[\|e^i_{k+N-1}\|^2_{Q_i} | e_k\right] + \mathbb{E}\left[\|w^i_{k+N-1}\|^2_{Q_i}\right]$$

$$\leq \sum_{c_1} \|A_i\|^2_2 \lambda_i + \mathbb{E}\left[\|w^i_{k+N-1}\|^2_{Q_i}\right]. \tag{5.51}$$

This fulfills the condition (5.48) with $\bar{\xi}' > \sum_{c_1} \|A_i\|^2_2 \lambda_i + \mathbb{E}\left[\|w^i_{k+N-1}\|^2_{Q_i}\right]$, and $\bar{\xi} = \frac{\sum_{c_1} \mathbb{E}\left[\|e^i_{k+N}\|^2_{Q_i} | e_k\right]}{\bar{\xi}'} < 1$.

For some $i \in c_2$, let a successful transmission occur at time step $k+r'_i$, where $r'_i \in [1, N-1]$, i.e., $\theta^i_{k+r'_i} = 1$. We express e^i_{k+N} as a function of the error at time $k+r'_i-1$ as

$$e^i_{k+N} = \prod_{j=r'_i}^{N}(1-\theta^i_{k+j})A_i^{N-r'_i+1}e^i_{k+r'_i-1}$$
$$+\sum_{r=r'_i}^{N}\left[\prod_{j=r+1}^{N}(1-\theta^i_{k+j})A_i^{N-r}w^i_{k+r-1}\right], \quad (5.52)$$

where we define $\prod_{N+1}^{N}(1-\theta^i_{k+j}) := 1$. The first term of the above equality vanishes as $\theta^i_{k+r'_i} = 1$. By statistical independence of w^i_{k+r-1} and θ^i_{k+j}, it follows from (5.52)

$$\sum_{c_2}\mathbb{E}[\|e^i_{k+N}\|^2_{Q_i}|e_k]$$
$$=\sum_{c_2}\mathbb{E}\left[\|\sum_{r=r'_i}^{N}\prod_{j=r+1}^{N}[1-\theta^i_{k+j}]A_i^{N-r}w^i_{k+r-1}\|^2_{Q_i}\right]$$
$$\leq \sum_{c_2}\sum_{r=r'_i}^{N}\mathbb{E}[\|A_i^{N-r}w^i_{k+r-1}\|^2_{Q_i}]. \quad (5.53)$$

In fact, we disregard the scheduling process in the last inequality. We are allowed to do so since $\prod_{j=r+1}^{N}\left[1-\theta^i_{k+j}\right] \leq 1$. Hence, the condition (5.48) is satisfied considering $\bar{\xi}'$ is chosen to be larger than (5.53), and $\bar{\xi} = \frac{\sum_{c_2}\mathbb{E}[\|e^i_{k+N}\|^2_{Q_i}|e_k]}{\bar{\xi}'} < 1$.

The subsystems in the third case are eligible for channel access at time $k+N$. To infer (5.50), we split the third case, c_3, into two complementary and disjoint sub-cases as follows:

$l_1^{c_3}$ system i has not transmitted within the past $N-1$ time steps, but has been in $\bar{\mathcal{G}}$ at least once, the last occurred at a time $k+r'_i$, with $r'_i \in [0,\ldots,N-2]$.

$l_2^{c_3}$ system i has not transmitted within the past $N-1$ time steps, and has been in $\mathcal{G}$ for all time $[k, k+N-1]$.

Recall that the subsystems $i \in c_3$ belong to $\mathcal{G}_{k+N-1}$. For sub-case $l_1^{c_3}$, $k+r'_i$ is the last time for which $i \in \bar{\mathcal{G}}_{k+r'_i}$, which in turn implies $\|e^i_{k+r'_i}\|^2_{Q_i} \leq \lambda_i$. Knowing that $\theta^i_{k'} = 0$ for $i \in c_3$ until time step $k+N$, we reach

$$\sum_{l_1^{c_3}}\mathbb{E}[\|e^i_{k+N}\|^2_{Q_i}|e_k] \leq$$
$$\sum_{l_1^{c_3}}\left[\|A_i^{N-r'_i}\|^2_2\lambda_i + \sum_{r=r'_i}^{N-1}\mathbb{E}[\|A_i^{N-r-1}w^i_{k+r}\|^2_{Q_i}]\right]. \quad (5.54)$$

The condition (5.48) is met by choosing $\bar{\xi}'$ larger than the uniform upper bound (5.54), and $\bar{\xi} = \frac{\sum_{l_1^{c_3}}\mathbb{E}[\|e^i_{k+N}\|^2_{Q_i}|e_k]}{\bar{\xi}'} < 1$.

The subsystems $j \in l_2^{c_3}$ have always been candidates for channel access, i.e., $j \in \mathcal{G}_{[k,k+N-1]}$. Hence, $\|e^j_{k'}\|^2_{Q_j} > \lambda_j$ for all $k' \in [k, k+N-1]$. From (5.8), we conclude

$$\sum_{l_2^{c_3}} \mathbb{E}\left[\|e^j_{k+N}\|^2_{Q_j} | e_k\right]$$
$$\leq \sum_{l_2^{c_3}} \mathbb{E}\left[\|A_j e^j_{k+N-1}\|^2_{Q_j} | e_k\right] + \mathbb{E}\left[\|w^j_{k+N-1}\|^2_{Q_j}\right]. \quad (5.55)$$

Expression (5.55) is not uniformly bounded since the term e^j_{k+N-1} in (5.55) is not bounded according to (5.49). However, as the considered cases cannot happen together, we calculate the probability for sub-case $l_2^{c_3}$ to happen according to the scheduling policy (5.10). Note that in deriving (5.51), (5.53) and (5.54) it was not necessary to consider the probability of the corresponding cases to happen. To calculate the probability of happening we need to consider the possibility of collisions. As collisions may happen at all time steps, there is a non-zero probability that all the scheduled packets collide. This means that all subsystems operate in open loop at all time steps. We investigate two collision scenarios: 1) there has been at least one successful transmission over the interval $[k, k+N-1]$, and 2) there has been no successful transmission over the entire interval. We assume that if a collision is reported, then the channel is not awarded to any subsystem.

Investigating the first scenario, we assume that whenever a collision is detected and consequently all the subsystems have to operate in open loop, a virtual control loop has successfully transmitted instead of a real subsystem. This means at the time the collision occurs, N real subsystems and one virtual one share the communication channel and the channel is awarded to the virtual subsystem. The virtual loops have the same discrete LTI dynamics as in (5.3). As the worst-case situation, let the channel experience $m < N-1$ collisions in the interval $[k, k+N-1]$. Thus, at time $k+N$ we have N real and m virtual subsystems, where all virtual ones have transmitted. Since we have $N+m$ subsystems, we need to extend our interval to $[k, k_m]$, where $k_m = k+N+m$. Consideration of virtual loops is merely to justify the longer interval and plays no more role in the analysis. If a subsystem $j \in l_2^{c_3}$ transmits at k_m, then its error becomes bounded in expectation. Otherwise, if j has never transmitted, then there exists another subsystem, say i, which has transmitted more than once. Let $k+\bar{r}$ denote the most recent step in which $\theta^i_{k+\bar{r}} = 1$ for $\bar{r} \leq N+m-1$. The probability that subsystem i re-transmits at the final time step k_m, in the presence of the subsystem $j \in l_2^{c_3}$, can be expressed as

$$\mathsf{P}[\theta^i_{k_m} = 1 | \theta^i_{k+\bar{r}} = 1, \theta^j_{k'} = 0 \quad \forall k' \in [k, k_m]]$$
$$= \mathsf{P}[\nu^j_{k_m-1} > \nu^i_{k_m-1} | \theta^i_{k+\bar{r}} = 1, \theta^j_{k'} = 0 \quad \forall k' \in [k, k_m]]$$
$$\leq \frac{\mathbb{E}[\nu^j_{k_m-1} | \theta^j_{k'} = 0 \quad \forall k' \in [k, k_m]]}{s^i_{k_m-1}\tau},$$

where the last expression follows from *Markov's inequality* considering the positive constant waiting time $\nu^i_{k_m-1} = s^i_{k_m-1}\tau$ corresponds to subsystem i is given, and $s^i_{k_m-1} \in \{1,\ldots,h-1\}$. The latest error value of subsystem j at time k_m-1 is required to be given in order to have an expectation of the waiting times $\nu^j_{k_m-1}$. Therefore, having the last expression conditioned on $e^j_{k_m-1}$, we have from the *law of iterated expectation* that

$$\frac{\mathbb{E}[\nu^j_{k_m-1}|\theta^j_{k'}=0 \quad \forall k' \in [k,k_m]]}{s^i_{k_m-1}\tau}$$

$$= \frac{\mathbb{E}[\mathbb{E}[\nu^j_{k_m-1}|e^j_{k_m-1}]|\theta^j_{k'}=0 \quad \forall k' \in [k,k_m]]}{s^i_{k_m-1}\tau}$$

$$= \frac{1}{s^i_{k_m-1}\tau \|e^j_{k_m-1}\|^j_{Q_j}}, \tag{5.56}$$

where the last equality follows from (5.21). The expression (5.56) confirms that having large error values correspond to subsystems $j \in l_2^{c_3}$ with no prior transmission reduces the probability of re-transmission of a subsystem $i \notin l_2^{c_3}$.

Having (5.50) extended for the interval $[k, k_m]$ and considering the expression (5.55) for the expectation of error for a subsystem $j \in l_2^{c_3}$, we employ (5.56) as follows

$$\sum_{l_2^{c_3}} \mathsf{P}_{l_2^{c_3}} \mathbb{E}\left[\|e^j_{k_m}\|^2_{Q_j}|e_k\right]$$

$$= \sum_{l_2^{c_3}} \mathsf{P}[\nu^j_{k_m-1} > \nu^i_{k_m-1}] \mathbb{E}\left[\|e^j_{k_m}\|^2_{Q_j}|e_k\right]$$

$$\leq \sum_{l_2^{c_3}} \frac{\|A_j\|^2_2 \|e^j_{k_m-1}\|^2_{Q_j}}{s^i_{k_m-1}\tau \|e^j_{k_m-1}\|^2_{Q_j}} + \frac{\mathbb{E}\left[\|w^j_{k_m-1}\|^2_{Q_j}\right]}{s^i_{k_m-1}\tau \|e^j_{k_m-1}\|^2_{Q_j}}$$

$$\leq \sum_{l_2^{c_3}} \frac{\|A_j\|^2_2}{s^i_{k_m-1}\tau} + \frac{\mathbb{E}\left[\|w^j_{k_m-1}\|^2_{Q_j}\right]}{\lambda_j s^i_{k_m-1}\tau}, \tag{5.57}$$

where the last inequality follows from knowing that $\|e^j_{k_m-1}\|^2_{Q_j} > \lambda_j$ for every subsystem $j \in l_2^{c_3}$. Since (5.57) is uniformly bounded, (5.48) holds by selecting $\bar{\xi} = \frac{\sum_{l_2^{c_3}} \mathbb{E}[\|e^j_{k_m}\|^2_{Q_j}|e_k]}{\bar{\xi}'} < 1$ and $\bar{\xi}'$ larger than (5.57), over the interval $[k, k_m]$. Expression (5.57) can be made small by tuning λ_j's and Q_j's but not arbitrarily, due to its first term. It confirms, despite having unstable plants and sparse capacity which might cause a subsystem with large error wait for transmission, the expected error remains bounded. Moreover, (5.57) is derived considering the couplings between the subsystems which occurs in the communication channel.

It was shown earlier that condition (5.47) holds at time $k+N$ within each case c_1-c_2 and $l_1^{c_3}$, which implies that they stay bounded in expectation over longer finite horizons. Thus, rewriting (5.50) over the extended interval $[k, k_m]$

yields

$$\sum_{i=1}^{N+m} \mathbb{E}\left[\|e^i_{k_m}\|^2_{Q_i}\right] = \sum_{c_l} \mathsf{P}_{c_l} \mathbb{E}\left[\|e^i_{k_m}\|^2_{Q_i}|c_l, \mathsf{P}_{c_l}\right]$$
$$\leq \sum_{c_1} \mathbb{E}\left[\|e^i_{k_m}\|^2_{Q_i}|c_1\right] + \sum_{c_2} \mathbb{E}\left[\|e^i_{k_m}\|^2_{Q_i}|c_2\right]$$
$$+ \sum_{l_1^{c_3}} \mathbb{E}\left[\|e^i_{k_m}\|^2_{Q_i}|l_1^{c_3}\right] + \sum_{l_2^{c_3}} \mathsf{P}_{l_2^{c_3}} \mathbb{E}\left[\|e^i_{k_m}\|^2_{Q_i}|l_2^{c_3}, \mathsf{P}_{l_2^{c_3}}\right] < \bar{\varsigma},$$

where $\bar{\varsigma}$ sums up the finite uniform upper bounds for cases c_1-c_3, assuming at least one successful transmission over $[k, k_m]$. This ensures the error Markov chain e_k satisfies (5.47), which in turn affirms the overall NCS possesses LSP.

The second collision scenario prevents employing the probability $\mathsf{P}_{l_2^{c_3}}$ to show that (5.47) holds for sub-case $l_2^{c_3}$, since no transmission happens over $[k, k+N]$. To infer (5.48), we calculate the probability that at least two subsystems select identical waiting times at every time step, leading to an all-time step collision scenario. As all subsystems operate in open loop, we calculate (5.48) for all $i \in \{1, \ldots, N\}$. Assume a subsystem i has selected $\nu^i_{k'} = s^i_{k'}\tau$ at some time k', for $s^i_{k'} \in \{1, \ldots, h-1\}$. The probability that a subsystem j has identical waiting time $\nu^j_{k'} = s^i_{k'}\tau$ can be calculated. We know

$$\mathbb{E}[\nu^j_{k'}] = \sum_{m=1}^{h-1} m\tau . \mathsf{P}(\nu^j_{k'} = m\tau).$$

Therefore, we conclude

$$\mathsf{P}(\nu^j_{k'} = s^i_{k'}\tau) = \frac{1}{s^i_{k'}\tau}\left[\mathbb{E}[\nu^j_{k'}] - \sum_{m=1, m \neq s^i_{k'}}^{h-1} m\tau . \mathsf{P}(\nu^j_{k'} = m\tau)\right]$$
$$< \frac{1}{s^i_{k'}\tau}\left[\frac{1}{\lambda_j} - \sum_{m=1, m \neq s^i_{k'}}^{h-1} m\tau . \mathsf{P}(\nu^j_{k'} = m\tau)\right] \leq \frac{1}{s^i_{k'}\tau\lambda_j}.$$

Extending this for every pair of subsystems i and j which collides at every time step $k' \in [k, k+N]$, we find the upper bound for the probability of having successive collisions as

$$\mathsf{P}\left[\sum_{i=1}^{N} \theta^i_{k'} = 0, \forall k' \in [k, k+N]\right] \leq \prod_{k'=k}^{k+N} \sum_{i=1}^{N} \sum_{j=1, j \neq i}^{N} \frac{1}{s^i_{k'}\tau\lambda_j}.$$

From (5.52), if no subsystem transmits, we can choose $\bar{\xi}' = \sum_{i=1}^{N} \|A_i^N e^i_k + \sum_{r=1}^{N} A_i^{N-r} w^i_{k+r-1}\|^2_{Q_i} > 0$. Then, we have

$$\sup_{e_k} \mathsf{P}\left[\sum_{i=1}^{N} \|e^i_{k+N}\|^2_{Q_i} \geq \bar{\xi}'\right] < \prod_{k'=k}^{k+N} \sum_{i=1}^{N} \sum_{j=1, j \neq i}^{N} \frac{1}{s^i_{k'}\tau\lambda_j}, \qquad (5.58)$$

for an arbitrary $\rho(\bar{\xi}', \bar{\xi})$ such that $\sum_{i=1}^{N} \|e^i_k\|^2_{Q_i} < \rho$ and LSP of the overall NCS is readily obtained according to (5.48). □

Part II
MULTI-AGENT APPLICATIONS

6

Topology-Triggering of Multi-Agent Systems

CONTENTS

6.1	Motivation, Applications and Related Works	146
6.2	Initial-Condition-(In)dependent Multi-Agent Systems and Switched Systems	149
	6.2.1 Switched Systems and Average Dwell Time	150
	6.2.2 Graph Theory	151
6.3	Problem Statement: Transmission Intervals Adapting to Underlying Communication Topologies	151
6.4	Topology-Triggering and Related Performance vs. Lifetime Trade-Offs	153
	6.4.1 Designing Broadcasting Instants	154
	6.4.2 Switching Communication Topologies	158
	6.4.2.1 Switching without Disturbances	158
	6.4.2.2 Switching with Disturbances	160
6.5	Example: Output Synchronization and Consensus Control with Experimental Validation	161
	6.5.1 Performance vs. Lifetime Trade-Offs	163
	6.5.2 Experimental Setup	164
	6.5.3 Energy Consumption	165
	6.5.4 Experimental Results	166
6.6	Conclusions and Perspectives	168
6.7	Proofs and Derivations of Main Results	170
	6.7.1 From Agent Dynamics to Closed-Loop Dynamics	170
	6.7.2 Introducing Intermittent Data Exchange	171
	6.7.3 Proof of Proposition 6.1	172
	6.7.4 Proof of Theorem 6.2	172
	6.7.5 Proof of Theorem 6.3	174
	6.7.6 Proof of Theorem 6.4	175

In this chapter, our interest is to reduce requirements posed on communication, sensing, processing and energy resources in Multi-Agent Networks (MANs) without compromising the objectives of the MANs. Consequently, the hardware expenses and energy consumption are driven down whilst MAN multitasking, such as inter-network collaboration, is facilitated. Scenarios, in which agents of one network need to achieve a common goal, call for the study of decentralized cooperative control schemes. In order to accomplish this goal in an uncertain and noisy setting and to detect changes in the communication topology, agents have to exchange information. Because each transmission and reception of information necessitates energy, communication should be induced only when the goal completion can no longer be guaranteed in order to prolong the MAN mission. To that end, we devise an information exchange mechanism which draws upon ideas of self-triggered communication. The proposed mechanism is inspected both theoretically and experimentally (employing off-the-shelf wireless sensor platforms) for performance vs. lifetime trade-offs using a single-integrator consensus case study. Our mechanism is applicable to heterogeneous agents with exogenous disturbances (or modeling uncertainties), to switching communication topologies and to both initial-condition-dependent and initial-condition-independent long-term cooperative behaviors. The investigated stability notions include $\mathcal{L}_p$-stability and Input-to-State Stability (with respect to a set). This chapter does not consider delays, model-based estimators or noisy measurements. Noisy measurements are readily included through $\mathcal{L}_p$-stability with bias as illustrated in Chapters 7 and 8. Delays and estimators are also considered in Chapters 7 and 8.

6.1 Motivation, Applications and Related Works

The emergence of new technologies provides networks of increasingly accomplished agents. Such agents may be mobile and may possess significant processing, sensing, communication as well as memory and energy storage capabilities (refer to Figure 6.1). At the same time, aspirations to satisfy ever-growing demands of the industrial and civil sectors bring about novel engineering paradigms such as Cyber-Physical Systems [148] and the Internet of Things [125]. The essence of these paradigms is fairly similar—to extend even further the concepts of heterogeneity, safety, decentralization, scalability, reconfigurability, and robustness of MANs by laying more burden on the agents. Everyday examples of MANs are cooperative multi-robot systems [159] and Wireless Sensor Networks (WSNs) [1].

According to [148, 93, 125], we are heading toward densely deployed MANs that coexist side by side, sharing the same physical environment as illustrated in Figure 6.2. In order to realize the full potential of MANs, one typically allows each agent of a network to interact with any agent of neighboring networks. In other words, the existence of specialized agents (e.g., gateways) is no longer required to achieve inter-network collaboration. Neighboring MANs may need to share information (e.g., measurements, intentions) as well as resources and services (e.g., storage space, energy supplies, processing power,

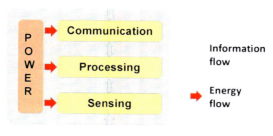

FIGURE 6.1
Block scheme of an agent indicating information and energy flows.

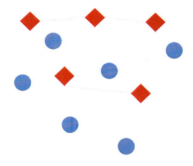

FIGURE 6.2
Two decentralized co-located MANs.

Internet or routing services). However, this interaction must not compromise the objectives of each individual network. Commonly investigated objectives of stand-alone networks are consensus attainment [159, 43, 143] and output synchronization [202, 57, 175].

Agents usually have limited and costly resources at their disposal which renders agent resource management of critical importance. Recently, several authors have proposed an avenue to agent resource management by means of decentralized *event-triggered* and *self-triggered* control/sensing/communication schemes (refer to [202, 57, 144] and [193, 43, 143, 178, 175], respectively). In event-triggered schemes, one defines a desired performance and a transmission of up-to-date information is induced when an event representing the unwanted performance occurs. In this chapter, the desired performance is closed-loop stability. In self-triggered approaches, currently available information is used to compute the next transmission instant, i.e., to *predict* the occurrence of the triggering event. Therefore, self-triggering enables agents to predict time intervals of idleness over which energy conservation modes can be activated [154, 194, 99]. Alternatively, agents can freely engage in other activities (e.g., inter-network collaboration) prior to the (pre)computed transmission instant. As a caveat, the performance of event- and self-triggered control schemes is more susceptible to disturbances, modeling uncertainties and noise due to longer open-loop-control intervals (refer to previous chapters, especially Chapter 4). While Chapter 4 and the references therein bring theoretical studies of performance vs. resource consumption trade-offs for plant-controller control problems, the work presented herein investigates theoretically and experimentally these trade-offs for MANs. To the best of our knowledge, these trade-offs for MANs are yet to be addressed both theoretically and experimentally.

Since many works regarding event- and self-triggering are still being published, it is clear that this area has not matured yet and further research endeavors are vital. In fact, most of the proposed methodologies are applicable to a rather specific application-dependent setting (e.g., specific agent and controller dynamics as well as specific topology and inter-agent coupling), which often hinders equitable comparisons between methodologies and impedes their transferability to a different setting. Thus, generalizing and unifying frameworks are needful. To that end, we devise a unifying framework for general heterogeneous agents with (possibly dynamic and output-feedback) local controllers in the presence of exogenous disturbances (or modeling uncertainties) and switching topologies. We point out that, except for [202], all of the aforementioned works consider state-feedback controllers. The unifying feature of our framework is manifested in the fact that both initial-condition-dependent and initial-condition-independent long-term behaviors of MANs can be analyzed. For example, initial-condition-dependent long-term behaviors are investigated in [202, 43, 143, 175] while initial-condition-independent long-term behaviors are addressed in [193, 57, 144, 178].

The principal requirement in our framework is $\mathcal{L}_p$-stability (with respect to a set) of the closed-loop system as discussed in Chapter 2. In other words, we do not impose specific requirements on the agent and controller dynamics per se, or on the underlying communication topology. Basically, when given local controllers do not yield the closed-loop system $\mathcal{L}_p$-stable (with respect to a set), one can look for another topology or design alternative controller. In this regard, our framework is akin to [193] and [144]. Nevertheless, the requirements in [193] and [144] are imposed locally on each agent (i.e., $\mathcal{L}_p$-stability and Input-to-State Stability properties, respectively) and on agent interactions (i.e., weak coupling). In addition, self-triggered counterparts of [144, 193] are still to be devised. Furthermore, [202] requires proportional controllers and passive agents, which in turn implies that the number of inputs equals the number of outputs for all agents, while [43] investigates single-integrator agents with proportional control laws. The authors in [57] impose the Hurwitz condition on local dynamics (rather than on the closed-loop dynamics as we do herein) and the diagonal dominant matrix structure on the nominal closed-loop dynamics. In [143], the authors tailor several ternary controllers for self-triggered practical consensus of single-integrator agents and show that those controllers possess some desirable robustness features. As opposed to [143], our methodology, as well as the methodologies in [193, 202, 144], aims at devising triggering conditions for a variety of existing control schemes that are designed on the premise of continuous information flows (i.e., an emulation-based approach). As far as communication topologies are concerned, only [202] considers time-varying topologies (though balanced ones). Furthermore, [43] and [143] consider undirected fixed topologies. On the other hand, our framework encompasses directed switching topologies. It is worth mentioning that the approaches in [43] and [193] are not Zeno-free. Lastly, unlike [202, 57, 43, 144, 143], our work is able to consider external disturbances (or modeling uncertainties).

The remainder of the chapter is organized as follows. Section 6.2 presents the notation and terminology used throughout this chapter. Section 6.3 formulates the self-triggered information exchange problem for MANs comprised of heterogeneous linear systems. Consequently, we state problems related to feasibility of self-triggered mechanisms as well as to the related theoretical and experimental MAN performance vs. lifetime analyses. Our self-triggered mechanism is devised in Section 6.4. Theoretical performance vs. lifetime trade-offs and experimental results for a single-integrator consensus case study are in Section 6.5. Conclusions and future challenges are found in Section 6.6. Several technical results and proofs are included in Section 6.7.

6.2 Initial-Condition-(In)dependent Multi-Agent Systems and Switched Systems

Consider a nonlinear impulsive system

$$\Sigma \begin{cases} \dot{x} = f(x,\omega) \\ y = g(x,\omega) \end{cases} t \in \bigcup_{i \in \mathbb{N}_0} [t_i, t_{i+1}), \\ x(t^+) = h(x(t)) \quad t \in \mathcal{T}, \end{cases} \tag{6.1}$$

where $x \in \mathbb{R}^{n_x}$ denotes the state, $\omega \in \mathbb{R}^{n_\omega}$ reflects the external disturbance (or modeling uncertainties), and $y \in \mathbb{R}^{n_y}$ denotes the output of the system. We assume enough regularity on f and h to guarantee existence of the solutions given by right-continuous functions $t \mapsto x(t)$ on $[t_0, \infty)$ starting from x_0 at $t = t_0$. Jumps of the state x (or impulses) occur at each $t \in \mathcal{T} := \{t_i : i \in \mathbb{N}\}$. The value of the state after the jump is given by $x(t^+) = \lim_{t' \searrow t} x(t')$ for each $t \in \mathcal{T}$. For a comprehensive discussion of impulsive systems, refer to [12].

In the following definitions, we use the set

$$\mathcal{B}_y := \{y \in \mathbb{R}^{n_y} | \exists b \in \mathcal{B} \text{ such that } y = g(b, \mathbf{0}_{n_\omega})\}, \tag{6.2}$$

where $\mathcal{B} \subseteq \mathbb{R}^{n_x}$.

Definition 6.1. *(Uniform Global Exponential Stability w.r.t. a set)* For $\omega \equiv 0_{n_\omega}$, the system Σ is Uniformly Globally Exponentially Stable (UGES) w.r.t. a set $\mathcal{B}$ if there exist $k, l > 0$ such that $\|x(t)\|_\mathcal{B} \leq k \exp(-l(t-t_0))\|x(t_0)\|_\mathcal{B}$ for all $t \geq t_0$ and for any $x(t_0)$.

Definition 6.2. *(Input-to-State Stability w.r.t. a set)* The system Σ is Input-to-State Stable (ISS) w.r.t. a set $\mathcal{B}$ if there exist a class-$\mathcal{KL}$ function β and a class-$\mathcal{K}_\infty$ function γ such that, for any $x(t_0)$ and every input ω, the corresponding solution $x(t)$ satisfies $\|x(t)\|_\mathcal{B} \leq \beta(\|x(t_0)\|_\mathcal{B}, t - t_0) + \gamma(\|\omega[t_0, t]\|_\infty)$.

Definition 6.3. *($\mathcal{L}_p$-stability w.r.t. a set)* Let $p \in [1, \infty]$. The system Σ is $\mathcal{L}_p$-stable from ω to y w.r.t. a set $\mathcal{B}$ with gain $\gamma \geq 0$ if there exists $K \geq 0$ such that $\|y[t_0, t]\|_{p, \mathcal{B}_y} \leq K\|x(t_0)\|_\mathcal{B} + \gamma\|\omega[t_0, t]\|_p$ for any $t \geq t_0$, $x(t_0)$ and ω.

Definition 6.4. *($\mathcal{L}_p$-detectability w.r.t. a set)* Let $p \in [1, \infty]$. The state x of Σ is $\mathcal{L}_p$-detectable from (ω, y) to x w.r.t. a set $\mathcal{B}$ with gain $\gamma_d \geq 0$ if there exists $K_d \geq 0$ such that $\|x[t_0, t]\|_{p, \mathcal{B}} \leq K_d \|x(t_0)\|_\mathcal{B} + \gamma_d \|y[t_0, t]\|_{p, \mathcal{B}_y} + \gamma_d \|\omega[t_0, t]\|_p$ for any $t \geq t_0$, $x(t_0)$ and ω.

Definitions 6.1 and 6.2 are motivated by [88] and [108], while Definitions 6.3 and 6.4 are motivated by [168]. Notice that K, γ, K_d and γ_d in Definitions 6.3 and 6.4 are not unique.

6.2.1 Switched Systems and Average Dwell Time

Consider a family of systems (6.1) indexed by the parameter ρ taking values in a set $\mathcal{P} = \{1, 2, \ldots, m\}$. Let us define a right-continuous and piecewise constant function $\sigma : [t_0, \infty) \to \mathcal{P}$ called a *switching signal* [105]. The role of σ is to specify which system is active at any time $t \geq t_0$. The resulting *switched system* investigated herein is given by

$$\Sigma_\sigma \begin{cases} \dot{x} = f_\sigma(x, \omega) \\ y = g_\sigma(x, \omega) \end{cases} t \in \bigcup_{i \in \mathbb{N}_0} [t_i, t_{i+1}), \\ x(t^+) = h_\sigma(x(t)) \quad t \in \mathcal{T}. \end{cases} \quad (6.3)$$

For each switching signal σ and each $t \geq t_0$, let $N_\sigma(t, t_0)$ denote the number of discontinuities, called *switching times*, of σ on the open interval (t_0, t). We say that σ has *average dwell time* τ_a if there exist $N_0, \tau_a > 0$ such that

$$N_\sigma(t, t_0) \leq N_0 + \frac{t - t_0}{\tau_a} \quad (6.4)$$

for every $t \geq t_0$. For a comprehensive discussion, refer to [105] and [74]. In this chapter, different values of σ correspond to different topologies L, while state jump instants t_i's indicate time instants at which an exchange of information takes place.

Notice that our definitions of impulsive and switched systems do not explicitly rule out accumulations of jumping and switching instants in finite time as is typically done in the literature (e.g., [105], [74] [53], [159, Chapter 2]). A priori exclusions of these phenomena are not in the essence of self-triggered communication. In fact, valid self-triggered communication policies must guarantee that communication instants do not accumulate in finite time, which is known as Zeno behavior (refer to Remark 6.3). Consequently, self-triggering eliminates the problem of arbitrary fast switching because changes in the communication topology are irrelevant while information is not being exchanged.

6.2.2 Graph Theory

A *directed graph*, or digraph, is a pair $\mathcal{G} = (\mathcal{V}, \mathcal{E})$, where $\mathcal{V} = \{1, \ldots, N\}$ is a nonempty set of *nodes* (or vertices) and $\mathcal{E} \subset \mathcal{V} \times \mathcal{V}$ is the set of the corresponding *edges*. When the edge (i, j) belongs to $\mathcal{E}$, it means that there is an information flow from the node i to the node j. We do not allow *self-loops*, i.e., edges that connect a vertex to itself. When both (i, j) and (j, i) belong to $\mathcal{E}$, we say that the *link* between i and j is *bidirectional*. Otherwise, the link between i and j is *unidirectional*. The set of *neighbors* of the node i is $\mathcal{N}_i = \{j \in \mathcal{V} : (j, i) \in \mathcal{E}\}$, which is all nodes that the node i can obtain information from. A *path* in a graph is a sequence of vertices such that from each of its vertices there is an edge to the next vertex in the sequence. A *cycle* in $\mathcal{G}$ is a directed path with distinct nodes except for the starting and ending node. An *inclusive cycle* for an edge is a cycle that contains the edge on its path. A *directed tree* is a directed graph in which every node has exactly one parent except for one node. A *subgraph* $\mathcal{G}^s = (\mathcal{V}^s, \mathcal{E}^s)$ of $\mathcal{G}$ is a graph such that $\mathcal{V}^s \subseteq \mathcal{V}$ and $\mathcal{E}^s \subseteq \mathcal{E} \cap (\mathcal{V}^s \times \mathcal{V}^s)$. A *directed spanning tree* $\mathcal{G}^s$ of $\mathcal{G}$ is a subgraph of $\mathcal{G}$ such that $\mathcal{G}^s$ is a directed tree and $\mathcal{V}^s = \mathcal{V}$. A graph $\mathcal{G}$ *contains a directed spanning tree* if a directed spanning tree is a subgraph of $\mathcal{G}$.

Given a graph $\mathcal{G}$, the graph Laplacian matrix $L \in \mathbb{R}^{|\mathcal{V}| \times |\mathcal{V}|}$ is defined as

$$L = [l_{ij}], \quad l_{ij} = \begin{cases} -1, & j \in \mathcal{N}_i \\ |\mathcal{N}_i|, & j = i \\ 0, & \text{otherwise} \end{cases}.$$

6.3 Problem Statement: Transmission Intervals Adapting to Underlying Communication Topologies

Under the premise of continuous information exchange between neighboring agents, the closed-loop dynamics of a large class of MAN control problems can be written as the following switched linear system:

$$\dot{x} = A_\sigma^{\text{cl}} x + \omega,$$
$$y = C^{\text{cl}} x, \qquad (6.5)$$

where x, y and ω are stack vectors comprised of (possibly translated) agents' states, outputs and exogenous disturbances (or modeling uncertainties), respectively. In other words, when N agents are considered, it follows that $x := (\xi_1, \xi_2, \ldots, \xi_N) - \xi_\sigma^{\text{p}}$, $y := (\zeta_1, \zeta_2, \ldots, \zeta_N) - \zeta_\sigma^{\text{p}}$ and $\omega := (\omega_1, \omega_2, \ldots, \omega_N)$, where ξ_i, ζ_i and ω_i are the ith agent state, output and exogenous disturbance, respectively. Because of the translation of $(\xi_1, \xi_2, \ldots, \xi_N)$ by a particular solution ξ_σ^{p}, the equilibrium manifold of the closed-loop dynamics includes the origin when $\omega \equiv \mathbf{0}_{n_\omega}$. The associated output is ζ_σ^{p}. Appendix 6.7.1 delineates steps relating (6.5) with linear heterogeneous agents and control laws commonly found in the literature. Notice that, for a finite number of agents N, there can be at most 2^{N^2-N} different topologies (i.e., individual systems constituting the switched system) as self-loops are not allowed. Accordingly, $|\mathcal{P}| = m$, where $m \leq 2^{N^2-N}$. Relevant graph theory concepts are provided in Section 6.2.2.

Assumption 6.1. *All individual systems in (6.5) are characterized by the same equilibrium manifold*

$$\mathcal{B} := \text{Ker}(A_\rho^{\text{cl}}) = \text{Ker}(A_\varrho^{\text{cl}}) \qquad \forall \rho, \varrho \in \mathcal{P}.$$

The above assumption can be achieved by adapting d in (6.33) as the topology changes. Apparently, $\mathcal{B}$ represents the set of equilibrium points when $\omega \equiv \mathbf{0}_{n_\omega}$. Consequently,

$$\mathcal{B}_y := \{y \in \mathbb{R}^{n_y} | x \in \mathcal{B} \text{ such that } y = C^{\text{cl}} x\}.$$

Assumption 6.2. *For each $\rho \in \mathcal{P}$, all eigenvalues of A_ρ^{cl} have nonpositive real parts. In addition, the eigenvalues with zero real parts are located in the origin and $\mathcal{A}(A_\rho^{\text{cl}}) = \mathcal{G}(A_\rho^{\text{cl}})$ for each $\rho \in \mathcal{P}$.*

Remark 6.1. *In case $\mathcal{B} \neq \mathbf{0}_{n_x}$, that is, $\mathcal{A}(A_\rho^{\text{cl}}) = \mathcal{G}(A_\rho^{\text{cl}}) \neq 0$ for each $\rho \in \mathcal{P}$, the MAN long-term behavior depends on the agents' initial conditions. Basically, $\text{Ker}(A_\rho^{\text{cl}})$ is nontrivial, spanned by the eigenvectors corresponding to the zero eigenvalue and represents the equilibrium manifold. Such matrices are typically found in coordinated tasks that look primarily for an agreement while the actual value of the agreement/consensus point depends on the agents' initial conditions. When $\mathcal{A}(A_\rho^{\text{cl}}) = \mathcal{G}(A_\rho^{\text{cl}}) = 0$, $\rho \in \mathcal{P}$, then A_ρ^{cl} is Hurwitz and the MAN long-term behavior is independent of the agents' initial conditions as there is merely one attractor, the sole equilibrium point.*

Definition 6.5. *Suppose we have a closed-loop system given by (6.5). We say that the MAN achieves its objective if $y \to \mathcal{B}_y$ as $t \to \infty$.*

Remark 6.2. *The above definition includes both output and state synchronization/consensus problems (consult [202, 178] and [57, 144, 193, 43, 143, 175], respectively). The latter is characterized by $C^{\text{cl}} = \mathbf{I}_{n_x}$.*

Recall that (6.5) is obtained on the premise of continuous information exchange among neighbors. However, continuous information flows in real-life applications are often not achievable as discussed in Chapter 1 and throughout this book. Nevertheless, MAN cooperation is still being achieved in realistic settings. This observation indicates that there is some built-in robustness of the closed-loop dynamics (6.5) with respect to intermittent information exchange. By thriving on this built-in robustness, self-triggering extends agents' lifetime and allows the agents to engage in inter-network activities. However, self-triggering consensus degrades network performance due to greater susceptibility to modeling uncertainties and exogenous disturbances as demonstrated in this book.

Problem 6.1. *Modify the closed-loop dynamics (6.5) such that the hardware limitations are taken into account and design a self-triggered data exchange mechanism.*

Problem 6.2. *Theoretically investigate MAN performance vs. lifetime trade-offs when the designed self-triggered mechanism is employed.*

Problem 6.3. *Experimentally validate the obtained theoretical results.*

6.4 Topology-Triggering and Related Performance vs. Lifetime Trade-Offs

In order to account for intermittent data exchange, the closed-loop system (6.5) can be modified as follows:

$$\dot{x} = A_\sigma^{\text{cl}} x + B_\sigma^{\text{cl}} e + \omega, \qquad (6.6)$$

where B_σ^{cl} is a matrix of appropriate dimensions and e is an auxiliary error signal with the following dynamics

$$\dot{e} = -C^{\text{cl}} \dot{x}. \qquad (6.7)$$

Further details regarding B_σ^{cl} and e are not needed to follow the subsequent exposition. Nevertheless, the interested reader is referred to Section 6.7.2, where B_σ^{cl} and e are constructed for common cooperative control problems.

From (6.6), we infer that the underlying communication topology captured in $\sigma(t)$, i.e., in the graph Laplacian matrix L_σ, plays an instrumental role in cooperative control. Consequently, agents need to discover the underlying communication topology in a decentralized manner. To that end, we advocate the approach from [14] due to its finite-time-convergence property and applicability to directed graphs. According to [14], the following assumption needs to be placed on the communication topology.

Assumption 6.3. *All unidirectional links have an inclusive cycle.*

After discovering a certain L_ρ, where $\rho \in \mathcal{P}$, agents utilize that knowledge to synchronize broadcasting/receiving instants in order to avoid lost packages. Namely, only those agents that have no common receivers and are not receivers themselves at a particular time instant are allowed to broadcast their states. By grouping such agents via Algorithm 1 into subsets $\mathcal{P}_\rho^i$, an agent partition is obtained. The input to the algorithm is L_ρ and the outputs are subsets $\mathcal{P}_\rho^i$. The number of nonempty $\mathcal{P}_\rho^i$'s is $T_\rho \leq N$ and empty $\mathcal{P}_\rho^i$'s are pruned. Figure 6.3 illustrates the partition obtained for

$$L_\rho = \begin{bmatrix} 2 & -1 & 0 & -1 & 0 \\ 0 & 1 & -1 & 0 & 0 \\ 0 & 0 & 1 & 0 & -1 \\ -1 & -1 & 0 & 2 & 0 \\ 0 & 0 & -1 & -1 & 2 \end{bmatrix}. \tag{6.8}$$

Algorithm 1 The algorithm developed for graph partitioning. Taking L_ρ, where $\rho \in \mathcal{P}$, as the input, the algorithm outputs a partition $\{\mathcal{P}_\rho^1, \ldots, \mathcal{P}_\rho^N\}$.

$\mathcal{P}_\rho^i \leftarrow \{\emptyset\}$ for all $i \in \{1, \ldots, N\}$; $k \leftarrow 0$
for $i = 1$ to N **do**
    if $i \notin \mathcal{P}_\rho^m$ for every $m \in \{1, \ldots, N\}$ **then**
        $k \leftarrow k + 1$
        $\mathcal{P}_\rho^k \leftarrow \mathcal{P}_\rho^k \cup \{i\}$
        for $j = i + 1$ to N **do**
            if $L_\rho(i).L_\rho(j) = \mathbf{0}_N$ for all $i \in \mathcal{P}_\rho^k$ **then**
                $\mathcal{P}_\rho^k \leftarrow \mathcal{P}_\rho^k \cup \{j\}$
            end if
        end for
    end if
end for

6.4.1 Designing Broadcasting Instants

Let us now design stabilizing broadcasting instants for each $\mathcal{P}_\rho^i$. For simplicity, let us consider the following TDMA scheduling protocol, which is similar in spirit to the Round Robin (RR) protocol from Chapter 2, illustrated in Figure 6.4.

Protocol 6.1. *The agents from $\mathcal{P}_\rho^{[(i+1) \mod N]+1}$ broadcast their outputs τ_ρ seconds after the agents from $\mathcal{P}_\rho^{[i \mod N]+1}$ have broadcast their outputs, where $a \mod b$ denotes the remainder of the division of a by b for $a, b \in \mathbb{N}$.*

The impact of broadcasting agents' outputs is as follows:

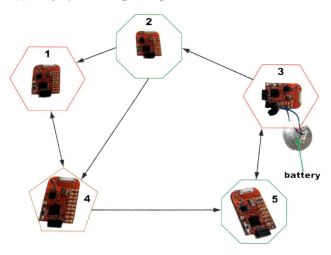

FIGURE 6.3
The graph partition $\mathcal{P}_\rho^1 = \{1,3\}$, $\mathcal{P}_\rho^2 = \{2,5\}$ and $\mathcal{P}_\rho^3 = \{4\}$ obtained via Algorithm 1. Accordingly, $T_\rho = 3$. In order not to clutter this figure with a battery for each node, only Node 3 is connected to a battery.

Property 6.1. *If the ith agent broadcasts at time t, the corresponding components of e reset to zero while other components remain unchanged, i.e.,*

$$\left. \begin{aligned} e_{(i-1)n_\zeta+1}^+(t) = \ldots = e_{in_\zeta}^+(t) &= 0, \\ e_j^+(t) &= e_j(t), \end{aligned} \right\} \quad (6.9)$$

for all $j \in \{1,\ldots,Nn_\zeta\} \setminus \{(i-1)n_\zeta+1,\ldots,in_\zeta\}$, where the set difference is denoted $\setminus$.

Due to the extensions of [168] presented in Chapter 3, our framework is applicable to the larger class of *uniformly persistently exciting scheduling protocols* [168] and not merely to Protocol 6.1. However, we do not pursue that direction herein in order not to obfuscate the main points of the present chapter.

As can be inferred from Protocol 6.1, agents know when they should "hear" from their neighbors, which is a prominent feature of self-triggering. When an agent does not "hear" from a neighbor in the precomputed time interval, that agent induces the *topology discovery* algorithm [14] in order to keep track of changes in the communication topology. Upon discovering a new topology, an up-to-date partition is obtained via Algorithm 1, the corresponding TDMA scheduling commences, and so on. Consequently, τ_ρ from Protocol 6.1 adapts to the new topology in order to preserve MAN stability. As a result, one could label our self-triggered control scheme as *topology-triggering*. However, we do not insist on the term topology-triggering in order not to inundate the area with excessive terminology. For instance, [43] and [143]

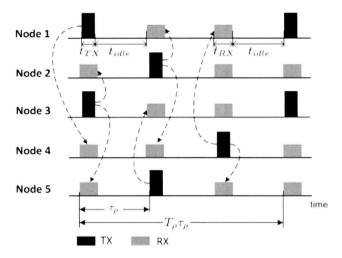

FIGURE 6.4
An illustration of the considered TDMA scheduling for the partition depicted in Figure 6.3. The abbreviation TX stands for *transmission* while RX stands for *reception*. Apparently, our TDMA scheduling prevents limitations (i)–(iv) from Section 6.3 for a sufficiently large τ_ρ, i.e., $\tau_\rho \geq \max\{t_{TX}, t_{RX}\}$, for each $\rho \in \mathcal{P}$. On the other hand, τ_ρ has to be sufficiently small in order to preserve closed-loop stability.

devise state-triggering, Chapter 3 designs input-output-triggering while [135] develops team-triggering.

We are still left to compute values of τ_σ that stabilize the closed-loop system (6.6)–(6.7) with intermittent data exchange. To that end, let us interconnect dynamics (6.6) and (6.7) and employ the small-gain theorem [88]. First, we upper bound the output error dynamics (6.7) for a fixed topology, i.e., $\sigma(t) \equiv \rho$ for some $\rho \in \mathcal{P}$, as follows:

$$\bar{e} = \overline{-C^{\mathrm{cl}}(A_\rho^{\mathrm{cl}} x + B_\rho^{\mathrm{cl}} e + \omega)} \preceq A_\rho^* \bar{e} + \tilde{y}_\rho(x, \omega), \qquad (6.10)$$

where

$$A_\rho^* = [a_{ij}^*] := \max\{|c_{ij}^*|, |c_{ji}^*|\}, \qquad (6.11)$$

$$\tilde{y}_\rho(x, \omega) := \overline{-C^{\mathrm{cl}}(A_\rho^{\mathrm{cl}} x + \omega)}. \qquad (6.12)$$

In (6.11), we use $-C^{\mathrm{cl}} B_\rho^{\mathrm{cl}} = [c_{ij}^*]$. Observe that (6.10) is preparing the terrain for application of Theorem 3.2. Notice that $A_\rho^* \in \mathcal{A}_{n_c}^+$ and $\tilde{y}_\rho : \mathbb{R}^{n_x} \times \mathbb{R}^{n_\omega} \to \mathbb{R}_+^{n_c}$ is a continuous function. With this choice of A_ρ^* and $\tilde{y}_\rho$, the upper bound (6.10) holds for all $(x, e, \omega) \in \mathbb{R}^{n_x} \times \mathbb{R}^{n_c} \times \mathbb{R}^{n_\omega}$ and all $t \in \mathbb{R}$.

Theorem 6.1. *Suppose that Protocol 6.1 is implemented and $\sigma(t) \equiv \rho$, where*

$\rho \in \mathcal{P}$. In addition, suppose that $\tau_\rho \in (0, \tau_\rho^*)$, where $\tau_\rho^* := \frac{\ln(2)}{\|A_\rho^*\|T_\rho}$. Then, the error system (6.7) is $\mathcal{L}_p$-stable from $\tilde{y}_\rho$, given by (6.12), to e for any $p \in [1, \infty]$ with gain

$$\gamma_\rho^e = \frac{T_\rho \exp(\|A_\rho^*\|(T_\rho - 1)\tau_\rho)(\exp(\|A_\rho^*\|\tau_\rho) - 1)}{\|A_\rho^*\|(2 - \exp(\|A_\rho^*\|T_\rho\tau_\rho))}, \tag{6.13}$$

and constant

$$K_\rho^e = \frac{1}{2 - \exp(\|A_\rho^*\|T_\rho\tau_\rho)} \left(\frac{\exp(p\|A_\rho^*\|T_\rho\tau_\rho) - 1}{p\|A_\rho^*\|} \right)^{\frac{1}{p}}. \tag{6.14}$$

Proof. See the proof of Theorem 3.2. □

Next, take (e, ω) to be the input and $\tilde{y}_\rho$, given by (6.12), to be the output of the dynamics (6.6). For future reference, this system is termed a *nominal system*. Due to Assumption 6.2, [175, Theorem 1] yields $\mathcal{L}_p$-stability w.r.t $\mathcal{B}$ of the system (6.6) with input (ω, e) and any output $\mathring{y}_\rho$ linear in x and ω. In addition, [175, Theorem 1] provides expressions for an associated constant and $\mathcal{L}_p$-gain. The following proposition shows that one can employ [175, Theorem 1] to infer $\mathcal{L}_p$-stability of (6.6) for the (nonlinear) output $\tilde{y}_\rho$ given by (6.12).

Proposition 6.1. *Suppose that the system (6.6) with input (ω, e) and (linear) output $\mathring{y}_\rho := -C^{\mathrm{cl}}(A_\rho^{\mathrm{cl}} x + \omega)$ is $\mathcal{L}_p$-stable w.r.t $\mathcal{B}$ with some constant K_ρ and gain γ_ρ. Then, the system (6.6) with input (ω, e) and (nonlinear) output $\tilde{y}_\rho$, given by (6.12), is $\mathcal{L}_p$-stable w.r.t $\mathcal{B}$ with the same constant K_ρ and gain γ_ρ.*

Apparently, systems (6.6) and (6.7) are interconnected according to Figure 6.5. We point out that $\tilde{y}_\rho$ is an auxiliary signal used to interconnect (6.6) and (6.7), but does not exist physically. According to the small-gain theorem, the open loop gain $\gamma_\rho \gamma_\rho^e$ must be strictly less than unity in order for this interconnection to be $\mathcal{L}_p$-stable from ω to $(\tilde{y}_\rho, e)$ w.r.t. $(\mathcal{B}_y, \mathbf{0}_{n_e})$ and, due to $\mathcal{L}_p$-detectability of the nominal system (see the proof of Theorem 6.2 for more), $\mathcal{L}_p$-stable from ω to (x, e) w.r.t. $(\mathcal{B}, \mathbf{0}_{n_e})$.

Theorem 6.2. *If the interbroadcasting interval τ_ρ in (6.13) is such that $\gamma_\rho \gamma_\rho^e < 1$, then the MAN objective in the sense of Definition 6.5 is $\mathcal{L}_p$-stable from ω to (x, e) w.r.t. $(\mathcal{B}, \mathbf{0}_{n_e})$ for given $p \in [1, \infty]$.*

Remark 6.3. *Notice that $\gamma_\rho^e(\tau_\rho)$ in (6.13) is a monotonically increasing function of $\tau_\rho \in [0, \tau_\rho^*)$. In addition, notice that $\gamma_\rho^e(0) = 0$. Due to [175, Theorem 1], we know that $\gamma_\rho < \infty$. Since our goal is to design τ_ρ such that $\gamma_\rho \gamma_\rho^e(\tau_\rho) < 1$, we first find τ_ρ' such that $\gamma_\rho \gamma_\rho^e(\tau_\rho') = 1$, and then compute $\tau_\rho = \kappa \tau_\rho'$, where $\kappa \in (0, 1)$. Due to monotonicity of $\gamma_\rho^e(\tau_\rho)$, the obtained τ_ρ' is strictly positive; hence, $\tau_\rho = \kappa \tau_\rho'$ is strictly positive. Consequently, the unwanted Zeno behavior [53] is avoided. In other words, our approach does not yield continuous feedback, which is impossible to implement in digital technology. Since we are interested in obtaining the interbroadcasting interval τ_ρ as large as possible, we choose κ as large as possible (e.g., $\kappa = 0.999$).*

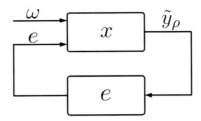

FIGURE 6.5
Interconnection of the nominal and the error dynamics.

Remarks 2.5 and 2.6 are applicable to Theorem 6.2 as well.

6.4.2 Switching Communication Topologies

Before proceeding further, we point out that scenarios with $\mathcal{B} = \mathbf{0}_{n_x}$ are less involved and do not entail all the steps delineated below (when compared to scenarios with $\mathcal{B} \neq \mathbf{0}_{n_x}$). Essentially, scenarios with $\mathcal{B} = \mathbf{0}_{n_x}$ boil down to [178, Section V].

6.4.2.1 Switching without Disturbances

The previous subsection establishes $\mathcal{L}_p$-stability of the interconnection in Figure 6.5. In what follows, we demonstrate that this interconnection is UGES w.r.t. $(\mathcal{B}, \mathbf{0}_{n_e})$ when $\omega \equiv \mathbf{0}_{n_\omega}$. To that end, we introduce the following substitution (i.e., change of coordinates)

$$z = S_\rho x,$$

where S_ρ is an invertible matrix with real entries that transforms A_ρ^{cl} into the Real Jordan Form (refer to [175, Section IV.A] for more details). According to [175, Section IV.B], $\mathcal{B}$ is spanned by the last $\mathcal{A}$ columns of S_ρ^{-1}. In addition, the *complementary space* of $\mathcal{B}$, denoted $\mathcal{B}^c$, is spanned by the first $n_x - \mathcal{A}$ columns of S_ρ^{-1}. Having that said, it is convenient to label the first $n_x - \mathcal{A}$ entries of z as z_r and the last $\mathcal{A}$ entries of z as z_q, i.e.,

$$z := (z_r, z_q).$$

As a result, we can write

$$x = S_\rho^{-1} z = S_\rho^{-1}\big((z_r, \mathbf{0}_{\mathcal{A}}) + (\mathbf{0}_{n_x - \mathcal{A}}, z_q)\big).$$

Thus,

$$\|x\|_{\mathcal{B}} \leq \big\| \underbrace{S_\rho^{-1}(z_r, \mathbf{0}_{\mathcal{A}})}_{\in \mathcal{B}^c} + \underbrace{S_\rho^{-1}(\mathbf{0}_{n_x - \mathcal{A}}, z_q)}_{\in \mathcal{B}} \big\|_{\mathcal{B}} = \big\| S_\rho^{-1}(z_r, \mathbf{0}_{\mathcal{A}}) \big\|_{\mathcal{B}}$$
$$\leq \big\| S_\rho^{-1}(z_r, \mathbf{0}_{\mathcal{A}}) \big\| \leq \big\| S_\rho^{-1} \big\| \|z_r\|. \tag{6.15}$$

Notice that z_r is the state of the reduced system associated with the nominal system (see [175, Section IV.B]). Since our choice of the nominal system output $\tilde{y}_\rho$, stated in (6.12), yields $\mathcal{B}_{\tilde{y}_\rho} = \mathbf{0}_{n_{\tilde{y}_\rho}}$, the reduced system output equals $\tilde{y}_\rho$. Thus, the inputs (e, ω) and the output $\tilde{y}_\rho$ of the reduced and nominal systems are exactly the same. The reduced system is basically

$$\begin{aligned} \dot{z}_r &= A_\sigma^{\mathrm{cl,r}} z_r + B_\sigma^{\mathrm{cl,r}} e + B_\sigma^{\mathrm{r}} \omega, \\ \tilde{y}_\rho &= \overline{C_\sigma^{\mathrm{r}} z_r + \omega}. \end{aligned} \quad (6.16)$$

For explicit expressions of $A_\sigma^{\mathrm{cl,r}}, B_\sigma^{\mathrm{cl,r}}, B_\sigma^{\mathrm{r}}$ and C_σ^{r} see [175, Section IV.B]. In other words, the reduced system is merely a different state-space realization (though a lower-dimensional realization) of the input-output mapping of the nominal system. In addition, notice that (6.7) is in fact

$$\dot{e} = -C^{\mathrm{cl}}\left(A_\sigma^{\mathrm{cl}} S_\sigma^{-1}(z_r, z_q) + B_\sigma^{\mathrm{cl}} e + \omega\right) = -C^{\mathrm{cl}}\left(A_\sigma^{\mathrm{cl}} S_\sigma^{-1}(z_r, \mathbf{0}_\mathcal{A}) + B_\sigma^{\mathrm{cl}} e + \omega\right).$$

Since the results invoked below are stated in terms of the Euclidean norm, the reduced system allows us to utilize those results at once (i.e., we use $\|z_r\|$ rather than $\|x\|_\mathcal{B}$ and relate those two norms via (6.15)).

Theorem 6.3. *Suppose that the conditions of Theorem 6.2 hold and $\omega \equiv \mathbf{0}_{n_\omega}$. In addition, assume that L_ρ, $\rho \in \mathcal{P}$, is fixed. Then, the interconnection (6.16)–(6.7) is UGES. Consequently, the closed-loop system (6.6)–(6.7) is UGES w.r.t. $(\mathcal{B}, \mathbf{0}_{n_e})$.*

To shorten the notation, we introduce $\chi := (z_r, e)$. According to Theorem 6.3, each subsystem in $\mathcal{P}$ is UGES (i.e., UGES w.r.t. $(\mathcal{B}, \mathbf{0}_{n_e})$ when considering x instead of z_r as the nominal system state). Let us now apply [12, Theorem 15.3] to each subsystem in $\mathcal{P}$. From (6.49) we infer that the flow and jump maps are Lipschitz continuous and are zero at zero. In addition, jump times t_i's are predefined (i.e., time-triggered according to Protocol 6.1 and do not depend on the actual solution of the system as long as the topology is fixed), and such that $0 < t_1 < t_2 < \ldots < t_i$ and $\lim_{i \to \infty} t_i = \infty$ hold. Consequently, all conditions of [12, Theorem 15.3] are met. From [12, Theorem 15.3], we know that there exist functions $V_\rho : \mathbb{R} \times \mathbb{R}^{n_{z_r}+n_e} \to \mathbb{R}$, $\rho \in \mathcal{P}$, that are right-continuous in t and Lipschitz continuous in χ, and satisfy the following inequalities

$$\begin{aligned} c_{1,\rho}\|\chi\|^2 &\leq V_\rho(t,\chi) \leq c_{2,\rho}\|\chi\|^2, & t \geq t_0, & \quad (6.17) \\ D_\rho^+ V_\rho(t,\chi) &\leq -c_{3,\rho}\|\chi\|^2, & t \notin \mathcal{T}, & \quad (6.18) \\ V_\rho(t^+,\chi^+) &\leq V_\rho(t,\chi), & t \in \mathcal{T}, & \quad (6.19) \end{aligned}$$

for all $\chi \in \mathbb{R}^{n_{z_r}+n_e}$, where $c_{1,\rho}$, $c_{2,\rho}$ and $c_{3,\rho}$ are positive constants. These constants are readily obtained once k and l from Definition 6.1 are known (see the proof of [12, Theorem 15.3]). In the above inequalities, $D_\rho^+ V_\rho(t,\chi)$

denotes the upper right derivative of function V_ρ with respect to the solutions of the ρth system. The upper right derivative of V_ρ is given by

$$D_\rho^+ V_\rho(t,\chi) := \limsup_{h \to 0, h > 0} \left(\frac{1}{h} [V_\rho(t+h, \chi(t+h)) - V_\rho(t, \chi(t))] \right),$$

where $\chi(t)$, $t \geq t_0$, denotes the trajectory of the ρth system. We now rewrite (6.17) and (6.18) as follows

$$c_1 \|\chi\|^2 \leq V_\rho(t,\chi) \leq c_2 \|\chi\|^2, \qquad t \geq t_0, \qquad (6.20)$$
$$D_\rho^+ V_\rho(t,\chi) \leq -2\lambda_0 V_\rho(t,\chi), \qquad t \notin \mathcal{T}, \qquad (6.21)$$
$$V_\rho(t,\chi) \leq \mu V_\varrho(t,\chi), \qquad t \geq t_0, \qquad (6.22)$$

for all $\chi \in \mathbb{R}^{n_{z_r} + n_e}$ and all $\rho, \varrho \in \mathcal{P}$, where

$$c_1 = \min_{\rho \in \mathcal{P}} c_{1,\rho} > 0, \qquad c_2 = \max_{\rho \in \mathcal{P}} c_{2,\rho} > 0,$$
$$\lambda_0 = \min_{\rho \in \mathcal{P}} \frac{c_{3,\rho}}{2 c_{1,\rho}} > 0, \qquad \mu = \max_{\rho, \varrho \in \mathcal{P}} \frac{c_{2,\rho}}{c_{1,\varrho}} > 0.$$

Notice that $\mu > 1$ in the view of interchangeability of ρ and ϱ in (6.22). Following ideas from [105] and [74], we obtain the following result:

Theorem 6.4. *Consider the family of m systems for which (6.19), (6.20), (6.21) and (6.22) hold. Then the resulting switched system is UGES for every switching signal σ with average dwell time*

$$\tau_a > \frac{\ln \mu}{2 \lambda_0} \qquad (6.23)$$

and N_0 arbitrary.

Corollary 6.1. *The MAN objective in the sense of Definition 6.5 is UGES w.r.t. $(\mathcal{B}, \mathbf{0}_{n_e})$ for every switching signal σ with the average dwell time (6.23) and N_0 arbitrary.*

Proof. This proof is akin to the last part of the proof for Theorem 6.3. □

6.4.2.2 Switching with Disturbances

Let us now examine the MAN properties when $\omega \not\equiv \mathbf{0}_{n_\omega}$. Notice that impulsive switched systems can be interpreted as time-varying impulsive systems. From Theorem 6.4 we infer that the corresponding state transition matrix $\Phi(t, t_0)$ satisfies

$$\|\Phi(t, t_0)\| \leq k \exp(-l(t - t_0)), \qquad (6.24)$$

where $k = \sqrt{\frac{c_2}{c_1} \mu^{N_0}}$ and $l = \lambda$ for some $\lambda \in (0, \lambda_0)$. For the explicit form of state transition matrices of linear time-varying impulsive systems, refer to [12,

Chapter 3]. From the corresponding variation of constants formula (see [12, Chapter 3])

$$\chi(t) = \Phi(t,t_0)\chi(0) + \int_{t_0}^{t} \Phi(t,s)[B_\sigma^{r\top} - C^{cl\top}]^\top \omega(s) ds,$$

and (6.24), we obtain

$$\|\chi(t)\| \le k \exp(-l(t-t_0))\|\chi(0)\| + bk \int_{t_0}^{t} \exp(-l(t-s))\|\omega(s)\| ds,$$

where $\max_{\rho \in \mathcal{P}} \left\|[B_\rho^r - C^{cl}]\right\| \le b$. Since $t \ge t_0$, $\int_{t_0}^{t} \exp(-l(t-s)) ds \le 1/l$ for any t_0. Using [160, Theorem 12.2], we infer that the switched system of interest is *uniformly bounded-input bounded-state* stable which in turn implies ISS (refer to [159, Theorem 2.35 & Remark 2.36] for more details). Likewise, $\mathcal{L}_p$-stability from ω to (z_r, e) is obtained following the lines of [180]. We conclude the above discussion in the following theorem.

Theorem 6.5. *The MAN objective in the sense of Definition 6.5 is ISS and $\mathcal{L}_p$-stable w.r.t. $(\mathcal{B}, \mathbf{0}_{n_e})$ from ω to (x, e) for every switching signal σ with the average dwell time (6.23) and N_0 arbitrary.*

Remark 6.4. *Recall that changes of the topology in $[t_i, t_{i+1})$, where $t_i, t_{i+1} \in \mathcal{T}$, remain unnoticed until t_{i+1} (or even later). Therefore, if $\min_{\rho \in \mathcal{P}} \tau_\rho \ge \tau_a$, then we effectively have that the switched system of interest is UGES for any switching signal. Obviously, we want to obtain τ_ρ's as large as possible. This is yet another motivation for developing self-triggered control policies.*

Remark 6.5. *The above result is similar to the well-known result of [91]. The difference is that [91] considers continuous communication among agents. In fact, the main result of [91] is a special case of ours when $\tau_\rho \to 0$ for all $\rho \in \mathcal{P}$. Therefore, Theorem 6.5 generalizes the main result of [91] to more realistic networking artifacts.*

6.5 Example: Output Synchronization and Consensus Control with Experimental Validation

In this section we employ the results of Section 6.4 to examine performance vs. lifetime trade-offs for the single-integrator consensus problem (refer to [175] and the references therein). Afterward, our theoretical predictions are experimentally verified using a set of wireless sensor platforms. We point out that other theoretical trade-offs (e.g., those involving T_ρ and $\|L_\rho\|$), performance metrics (e.g., time-efficiency and energy-efficiency as in [83]) or MAN tasks can be analyzed along the same lines. Alas, specifics of each of these analyses hinder unifying trade-off results in the spirit of Section 6.4. Accordingly, this section focuses on a subset of the cooperative tasks from Section 6.4.

As demonstrated in [175], the single-integrator consensus problem leads to $A_\sigma^{\text{cl}} = B_\sigma^{\text{cl}} = -k(L_\sigma \otimes \mathbf{I}_{n_\xi})$ and $C^{\text{cl}} = \mathbf{I}_{n_x}$, where k is a positive control gain and $\otimes$ denotes the Kronecker product. Among several other topologies that satisfy Assumption 6.3, we work out the methodology from Section 6.4 on L_ρ, given by (6.8), and

$$L_\varrho = \begin{bmatrix} 1 & -1 & 0 & 0 & 0 \\ 0 & 1 & -1 & 0 & 0 \\ 0 & 0 & 1 & 0 & -1 \\ -1 & 0 & 0 & 1 & 0 \\ 0 & 0 & -1 & -1 & 2 \end{bmatrix}.$$

Algorithm 1 yields $\mathcal{P}_\varrho^1 = \{1,3\}$, $\mathcal{P}_\varrho^2 = \{2,4\}$ and $\mathcal{P}_\varrho^3 = \{5\}$. When $n_\xi = 1$, the corresponding substitution matrices are

$$S_\rho = \begin{bmatrix} 0.0166 & -0.3979 & 0.06 & 0.1306 & 0.1906 \\ 0.1558 & 0.1322 & -0.2684 & 0.1243 & -0.1440 \\ -1.4166 & 0.1979 & -0.4600 & 1.0694 & 0.6094 \\ 1.3333 & 0 & 0 & -1.3333 & 0 \\ 0.0667 & 0.2 & 0.4 & 0.1333 & 0.2 \end{bmatrix},$$

$$S_\varrho = \begin{bmatrix} -0.1994 & -0.3032 & 0.1994 & 0.1141 & 0.1891 \\ 0.2025 & -0.1625 & -0.2025 & 0.2051 & -0.0426 \\ 0.5327 & -0.3635 & -0.5327 & -0.7807 & 1.1442 \\ -0.5 & 0.5 & 0 & 0.5 & -0.5 \\ 0.1667 & 0.1667 & 0.3333 & 0.1667 & 0.1667 \end{bmatrix},$$

yielding the reduced closed-loop matrices in (6.16)

$$A_\rho^{\text{cl,r}} = k \begin{bmatrix} -1.1226 & 0.7449 & 0 & 0 \\ -0.7449 & -1.1226 & 0 & 0 \\ 0 & 0 & -2.7549 & 0 \\ 0 & 0 & 0 & -3 \end{bmatrix},$$

$$A_\varrho^{\text{cl,r}} = k \begin{bmatrix} -0.7672 & 0.7926 & 0 & 0 \\ -0.7926 & -0.7672 & 0 & 0 \\ 0 & 0 & -2.4656 & 0 \\ 0 & 0 & 0 & -2 \end{bmatrix},$$

$$B_\rho^{\text{cl,r}} = k \begin{bmatrix} 0.0974 & 0.5451 & -0.2672 & -0.0540 & -0.3213 \\ -0.1873 & 0.1479 & 0.2566 & -0.2369 & 0.0197 \\ 3.9026 & -0.5451 & 1.2672 & -2.9460 & -1.6787 \\ -4 & 0 & 0 & 4 & 0 \end{bmatrix},$$

$$B_\varrho^{\text{cl,r}} = k \begin{bmatrix} 0.3134 & 0.1038 & -0.3134 & 0.0751 & -0.1789 \\ 0.0027 & 0.3650 & -0.0027 & -0.2478 & -0.1172 \\ -1.3134 & 0.8962 & 1.3134 & 1.9249 & -2.8211 \\ 1 & -1 & 0 & -1 & 1 \end{bmatrix},$$

$$B_\rho^{\text{r}} = \begin{bmatrix} 0.0166 & -0.3979 & 0.0600 & 0.1306 & 0.1906 \\ 0.1558 & 0.1322 & -0.2684 & 0.1243 & -0.1440 \\ -1.4166 & 0.1979 & -0.4600 & 1.0694 & 0.6094 \\ 1.3333 & 0 & 0 & -1.3333 & 0 \end{bmatrix},$$

$$B_\varrho^{\text{r}} = \begin{bmatrix} -0.1994 & -0.3032 & 0.1994 & 0.1141 & 0.1891 \\ 0.2025 & -0.1625 & -0.2025 & 0.2051 & -0.0426 \\ 0.5327 & -0.3635 & -0.5327 & -0.7807 & 1.1442 \\ -0.5 & 0.5 & 0 & 0.5 & -0.5 \end{bmatrix},$$

$$C_\rho^{\text{r}} = k \begin{bmatrix} -2.7549 & -1.4897 & 0.5098 & -0.75 \\ 1.4473 & -1.8694 & -0.8946 & -0.75 \\ 1.2151 & 1.3071 & 1.5698 & 1.5 \\ -2.7549 & -1.4897 & 0.5098 & 1.5 \\ -1.1226 & 0.7449 & -2.7549 & 3 \end{bmatrix},$$

$$C_\varrho^{\text{r}} = k \begin{bmatrix} 0.1084 & -1.9541 & 0.7832 & 2 \\ 1.5739 & -0.3690 & -1.1479 & -2 \\ 0.6588 & 1.1615 & 1.6823 & 2 \\ -2.2328 & -0.7926 & -0.5344 & -2 \\ -0.7672 & 0.7926 & -2.4656 & -2 \end{bmatrix}, \qquad (6.25)$$

while the matrix in (6.11) becomes

$$A_\rho^* = \begin{bmatrix} 2 & 1 & 0 & 1 & 0 \\ 1 & 1 & 1 & 0 & 0 \\ 0 & 1 & 1 & 0 & 1 \\ 1 & 0 & 0 & 2 & 1 \\ 0 & 0 & 1 & 1 & 2 \end{bmatrix}, \quad A_\varrho^* = \begin{bmatrix} 1 & 1 & 0 & 1 & 0 \\ 1 & 1 & 1 & 0 & 0 \\ 0 & 1 & 1 & 0 & 1 \\ 1 & 0 & 0 & 1 & 1 \\ 0 & 0 & 1 & 1 & 2 \end{bmatrix}.$$

6.5.1 Performance vs. Lifetime Trade-Offs

For a fixed topology, i.e., $\sigma(t) \equiv \rho$, a stabilizing τ_ρ yields (6.48). Notice that $\widetilde{K}$ and $\widetilde{\gamma}$ in (6.48) are functions of both τ_ρ and the control gain k. In the interest of brevity, suppose that $p \in [1, \infty)$ and $A_\rho^{\text{cl,r}}$ is a normal matrix for each $\rho \in \mathcal{P}$. For instance, the latter is fulfilled whenever the algebraic and geometric multiplicity of each eigenvalue of A_ρ^{cl} coincide (which is the case for L_ρ and L_ϱ considered above). Considering τ_ρ and k as variables, let us now determine dominant growth rates for K_ρ, γ_ρ, K_ρ^e, γ_ρ^e, K_ρ^d and γ_ρ^d that constitute $\widetilde{K}$ and $\widetilde{\gamma}$ in (6.48).

From [88, Corollary 5.2], (6.25), (6.13) and (6.14) we obtain

$$O(K_\rho) = kO(K_\rho^d) = kO\left(\left(\lambda_{\max}(P_\rho)\right)^{\frac{1}{p}} \sqrt{\frac{\lambda_{\max}(P_\rho)}{\lambda_{\min}(P_\rho)}}\right),$$

$$O(\gamma_\rho) = kO(\gamma_\rho^d) = kO\left(k \frac{\lambda_{\max}^2(P_\rho)}{\lambda_{\min}(P_\rho)}\right),$$

$$O(K_\rho^e) = \frac{\exp(k\tau_\rho)}{k(2 - \exp(k\tau_\rho))},$$

$$O(\gamma_\rho^e) = \frac{\exp(2k\tau_\rho)}{k(2 - \exp(k\tau_\rho))},$$

where P_ρ is the solution of the Lyapunov equation $P_\rho A_\rho^{\text{cl,r}} + \left(A_\rho^{\text{cl,r}}\right)^\top P_\rho = -\mathbf{I}_{n_x}$. Let us introduce $A_\rho^s := A_\rho^{\text{cl,r}} + \left(A_\rho^{\text{cl,r}}\right)^\top$. Employing [96, (70) and (88)], we obtain $\lambda_{\min}(P_\rho) \geq \frac{1}{2\|A_\rho^{\text{cl,r}}\|}$ and $\lambda_{\max}(P_\rho) \leq \frac{-2}{\lambda_{\max}(A_\rho^s)}$. According to [24, Problem VII.6.4], we know that $\lambda_{\max}(A_\rho^s) < 0$ since $A_\rho^{\text{cl,r}}$ is Hurwitz by construction. Moreover, $-\lambda_{\max}(A_\rho^s)$ equals the smallest singular value, denoted $\sigma_{\min}$, of A_ρ^s. The work in [145] provides $\sigma_{\min}(A_\rho^s) \geq \frac{|\det(A_\rho^s)|}{\sqrt{\text{trace}(A_\rho^{s\top} A_\rho^s)}}$. Using

$$\sqrt{\text{trace}(A_\rho^{s\top} A_\rho^s)} = \sqrt{\sum_{i=1}^{n_z}(\sigma_i(A_s))^2} \leq \sum_{i=1}^{n_z} \sigma_i(A_s) \leq n_{z_r}\|A_\rho^s\| \leq 2n_{z_r}\|A_\rho^{\text{cl,r}}\|$$

and the Marcus–de Oliviera conjecture from [24, p. 184], we infer that the lower bound of $\sigma_{\min}(A_\rho^s)$ is a polynomial in k of order n_{z_r}. Thus, $O(K_\rho)$, $O(\gamma_\rho)$, $O(K_\rho^d)$ and $O(\gamma_\rho^d)$ are polynomials in k. Since $O(K_\rho^e)$ and $O(\gamma_\rho^e)$ contain exponential terms, their impact on the growth rates of $\widetilde{K}$ and $\widetilde{\gamma}$ is predominant. Now, it is straightforward to show that the dominant growth rates of $\widetilde{K}$ and $\widetilde{\gamma}$ are given by

$$O(\widetilde{K}) = O(\widetilde{\gamma}) = \frac{1}{k^\alpha \left(\varkappa - \exp\left(k\tau_\rho\right)\right)}, \quad (6.26)$$

where $\alpha > 0$ while $\varkappa$ is a positive constant whose actual value is irrelevant for our purposes herein. However, we point out that $\varkappa$ is a function of L_ρ, agent dynamics and the selected decentralized control law. In order for (6.26) to be finite, the following needs to hold

$$\tau_\rho \in \left(0, \frac{\ln \varkappa}{k}\right), \quad (6.27)$$

which is in line with Theorem 6.1. Let us measure the MAN performance in terms of the time to reach (i.e., convergence rate) the MAN objective starting from some $(x(t_0), e(t_0))$ and in the presence of disturbance ω (refer to (6.48)).

From [178] and [175] we know there is an element of the set $\left(0, \frac{\ln \varkappa}{k}\right)$, denoted $\tau_\rho^{\max}$, such that any $\tau_\rho < \tau_\rho^{\max}$ stabilizes the control system of interest. Apparently, an increase in the control gain k decreases $\tau_\rho^{\max}$. From (6.48) and (6.26) one concludes that, for a fixed k, the greater τ_ρ (i.e., greater $\widetilde{K}$ and $\widetilde{\gamma}$) becomes, the smaller the rate of convergence toward the MAN objective, and the closed-loop system becomes more susceptible to disturbances/noise. On the other hand, the rate of convergence and disturbance sensitivity might be decreased by increasing k, which in turn requires a smaller τ_ρ (i.e., more frequent information exchange). Nevertheless, a decrease in k might impair the MAN performance and disturbance rejection due to k^α in (6.26). Therefore, one needs to carefully balance between performance and energy needs. It might even turn out that the hardware at hand cannot deliver information at a rate below $\tau_\rho^{\max}$. In that case, one should decrease k or augment $\varkappa$ by changing the underlying communication topology, control law or system dynamics. Alternatively, one could look for more advanced hardware. Lastly, in light of Theorem 6.5, the same inference about performance vs. resource consumption holds for time-varying topologies with switching signals satisfying the associated average dwell-time condition.

6.5.2 Experimental Setup

We select eZ430-RF2500 wireless sensor nodes because they offer a fairly short time window of activity for transmitting/receiving a message. In addition, these nodes are quite affordable and have modest energy requirements [82]. On the other hand, due to simplicity of the WSN nodes, the topology discovery algorithm [14] is not yet implemented and time-varying topologies are not examined experimentally. For numerical simulations involving switching topologies, albeit without agent partitions, refer to [175].

Each eZ430-RF2500 node (Figure 6.6) is a very low-power wireless platform built around an MSP430 microcontroller and CC2500 transceiver [174].

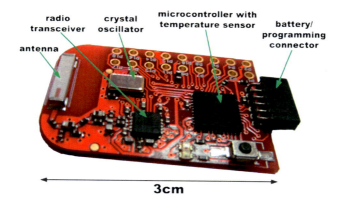

FIGURE 6.6
eZ430-RF2500 WSN node used in the experimental setup.

The ultra low-power MSP430 is a 16-bit microcontroller with 5 low-power operating modes. It is equipped with a digitally controlled oscillator (DCO) and internal very-low-power low-frequency oscillator (VLO) used in low-power operating modes. Besides the internal clock source, MSP430 also supports external crystal oscillators. Information about power consumption and oscillator characteristics are shown in Table 6.1. More information about typical drifts for commercial clock sources and procedures regarding how to suppress them can be found in [163].

The CC2500 is a 2.4-GHz RF transceiver implementing the SimpliciTI communication protocol with 250 kbps data rate. This proprietary stack of Texas Instruments is much simpler than the generally used IEEE 802.15.4/Zig-Bee protocol, requires less memory capacity and enables lower power consumption when the node is in the idle mode. It also has a much lower overhead in terms of additional headers per packet (only 14 extra bytes).

Current consumption	active mode (at 1 MHz and 3.3 V)	about 300 μA
	deepest low-power mode	about 500 nA
Oscillators	digitally controlled oscillator	+/-2% typical tolerance +/-5% maximal tolerance
	very-low-power low-frequency oscillator	frequency drift 0.5%/°C and 4%/V

TABLE 6.1
Current consumption and oscillator characteristics of MSP430 microcontroller.

6.5.3 Energy Consumption

To characterize the energy consumption of the eZ430-RF2500, the node was connected to a laboratory power source providing 3 V, and a 10 Ω resistor was connected in series to the node. The current in the circuit was measured by a multimeter (Fluke 45), while the voltage drop on the resistor was captured with a digital oscilloscope (Rigol DS1102E). Node current consumption in different operating modes is shown in Table 6.2. Figure 6.7 shows the voltage drop (proportional to the node current consumption) on the resistor, presenting a sequence in which the node is receiving and then transmitting a packet. As pointed out in the figure, the time interval required to merely receive or transmit is about 780 μs, but a significant amount of time is taken by switching between operating modes. The intrinsic platform-dependent functions (e.g., calibration, scanning the channel before sending, etc.) create overhead. Due to this overhead, it takes 2.2 ms to broadcast or receive a message. This represents the hardware-dependent lower bound on τ_ρ.

The energy consumption of the node, denoted E_{node}, is the sum of the energy utilized in all operating modes and energy utilized for all transitions between operating modes, i.e.,

$$E_{node} = \sum_{mode} P_{mode} \cdot t_{mode} + \sum_{trans} P_{trans} \cdot t_{trans}, \qquad (6.28)$$

where P_{mode} and t_{mode} represent power consumption of the node in a particular operation mode and the time spent in that mode, respectively. Taking into account the power consumption of a node from Table 6.2 and measuring the time intervals when staying in different modes and transitions (as in Figure 6.7), we estimate the long-term energy consumption of the node. In our experiment, each node was connected to a small-size battery providing 3 V and 1000 mAh. In the idle mode the radio was inactive while the microcontroller was active.

Component		Current consumption [mA]
Microcontroller	Transceiver	
on	on, RX	21.30
on	on, TX	25.11
on	off	3.00
off	off	0.001

TABLE 6.2
Current consumption of the eZ430-RF2500 node in different operating modes with a 3 V supply.

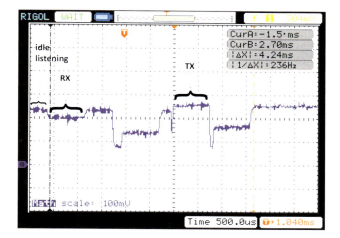

FIGURE 6.7
A sequence showing a node listening to the communication medium, then receiving a message and immediately transmitting another message. The packet has a 2 B payload (total 24 B on the physical layer due to the stack and radio overhead). Transitions between operating modes present a significant overhead in time and energy. Notice that the power consumption of the radio when merely listening is almost the same (and even slightly higher!) as when actually receiving a packet. This finding also advocates TDMA scheduling employing agent partitions.

6.5.4 Experimental Results

We select $k = 1$, $p = 2$, randomly pick initial states that are further apart (see Figure 6.8(c)), and use these initial states in all experiments for a fair comparison. The single-integrator problem was implemented for various broadcasting intervals τ_ρ ranging from the lower physical limit of 2.2 ms to the experimental upper limit on the stabilizing intervals of 0.3 s. We point out that the theoretical upper bound on the stabilizing intervals τ_ρ, obtained via Section 6.4, is 0.033 s. Hence, the theoretically predicted $\tau_\rho^{\max}$ is about 9 times more conservative than the experimental one.

As shown in Figure 6.4, each node cycles between the following modes: TX-idle-RX-idle-RX-idle-TX-..., and so on. Clearly, each cycle lasts for $T_\rho \tau_\rho$ seconds. The expected node lifetime, when powered by a 1000-mAh-capacity battery, is shown in Figure 6.8(a) for different τ_ρ. The smaller the τ_ρ is, the greater the average power consumed in one cycle becomes. If we want to increase the lifetime of our device by increasing τ_ρ, the closed-loop system takes longer to reach ϵ-vicinity of consensus (see Figure 6.8(b)). In addition, a greater τ_ρ causes greater sensitivity to noise as shown in Figure 6.8(b). The smaller noise/disturbance level is hardware dependent and arises from round-ing the exchanged data to two decimal places. The greater noise/disturbance level is emulated by adding random numbers to the exchanged data. Those random numbers are drawn from the uniform distribution over the interval $[-1, 1]$. Since the nodes in our experiments are within 10 m of each other, hardware-specific sporadic losses of messages are negligible [41]. Nevertheless, an increase in lost messages can easily be compensated for according to Remark 2.6.

As predicted by (6.26), the graphs in Figures 6.8(a) and 6.8(b) are rather nonlinear with vertical and horizontal asymptotes; hence, considerable decreases of intertransmission intervals may not significantly improve performance while the corresponding expected lifetimes may be significantly shortened (and vice versa). The rationale behind this observation is that, as τ_ρ decreases, practically the same information is being exchanged in several consecutive intervals. Basically, upcoming (but costly) transmissions do not bring significantly different information; hence, the "old" information (preserved via the zero-order-hold strategy) is an adequate replacement for the "new" information.

Figure 6.8(c) brings experimentally obtained states of the agents for the theoretical $\tau_\rho^{\max} = 0.033$ s. Apparently, from Figures 6.8(a) and 6.8(b), we infer that the theoretical $\tau_\rho^{\max} = 0.033$ s provides an acceptable compromise between the time to converge and node lifetime.

6.6 Conclusions and Perspectives

In this chapter, we devise a comprehensive self-triggered communication scheme for managing resources of MANs without compromising the original initial-condition-(in)dependent goals of the MANs. By extending the intervals of idleness of each agent, our communication scheme poses lesser demands on the agent resources (e.g., lowers the operating frequency and energy consumption). In addition, these resources can now be exploited for other activities (e.g., inter-network collaboration). However, extended intervals of idleness lead to degraded performance, which manifests in decreased disturbance/noise resilience and increased convergence times. On the other hand, shorter intertransmission intervals may significantly decrease network lifetime while performance may not improve considerably. In addition, our experimental findings are in good agreement with the theoretical analyses and suggest that our theoretical bounds are not overly conservative and provide a reasonable performance vs. energy trade-off.

In the future, a consideration of nonlinear agents is in order. Regarding the practical side of this work, the experiments indicate that it is important to keep agents' internal clocks synchronized. Evidently, the proposed scheme

Topology-Triggering of Multi-Agent Systems

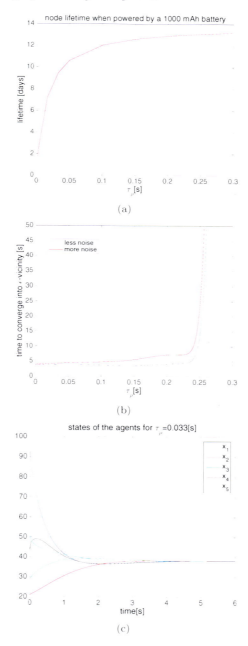

FIGURE 6.8
Experimental results that verify the theoretical exposition of Section 6.4: (a) expected lifetime of the battery, (b) time to converge into ϵ-vicinity of the consensus for $\epsilon = 0.4$, and (c) states of the agents for $\tau_\rho = 0.033$ s.

is suitable for clock synchronization during experiments; hence, it will be employed in the future.

6.7 Proofs and Derivations of Main Results

6.7.1 From Agent Dynamics to Closed-Loop Dynamics

Consider N heterogeneous linear systems, i.e., agents, given by

$$\dot{\xi}_i = A_i \xi_i + B_i u_i + \omega_i,$$
$$\zeta_i = C_i \xi_i, \quad (6.29)$$

where $\xi_i \in \mathbb{R}^{n_{\xi_i}}$ is the state, $u_i \in \mathbb{R}^{n_{u_i}}$ is the input, $\zeta_i \in \mathbb{R}^{n_\zeta}$ is the output of the ith system, $i \in \{1, 2, \ldots, N\}$, and $\omega_i \in \mathbb{R}^{n_\xi}$ reflects exogenous disturbances or unmodeled dynamics. In addition, A_i, B_i and C_i are matrices of appropriate dimensions. Since these agents are vertices of a communication graph, the set of all agents is denoted $\mathcal{V}$. Hence, $|\mathcal{V}| = N$. A common decentralized control policy is given by

$$u_i = -K_i \sum_{j \in \mathcal{N}_i} [(\zeta_i - \zeta_j) - (d_i - d_j)], \quad (6.30)$$

where K_i is a $n_{u_i} \times n_\zeta$ matrix, $\mathcal{N}_i$ denotes the set of neighbors of the ith agent and $d_i \in \mathbb{R}^{n_\zeta}$ is the bias term. Next, let us define the following stack vectors $\xi := (\xi_1, \xi_2, \ldots, \xi_N)$, $y := (\zeta_1, \zeta_2, \ldots, \zeta_N)$, $d := (d_1, d_2, \ldots, d_N)$ and $\omega := (\omega_1, \omega_2, \ldots, \omega_N)$. Knowing the Laplacian matrix L_ρ of a communication graph $\mathcal{G}$, the closed-loop dynamic equation of (6.29) given the control law (6.30) becomes

$$\dot{\xi} = A_\rho^{\text{cl}} \xi - B_\rho^{\text{cl}} d + \omega,$$
$$\zeta = C^{\text{cl}} \xi, \quad (6.31)$$

where $\rho \in \mathcal{P}$ denotes the subsystem associated with L_ρ, and

$$A_\rho^{\text{cl}} = [A_{\rho,ij}^{\text{cl}}], \quad A_{\rho,ij}^{\text{cl}} = \begin{cases} A_i - l_{\rho,ii} B_i K_i C_i, & i = j \\ -l_{\rho,ij} B_i K_i C_j, & \text{otherwise} \end{cases},$$
$$B_\rho^{\text{cl}} = [B_{\rho,ij}^{\text{cl}}], \quad B_{\rho,ij}^{\text{cl}} = -l_{\rho,ij} B_i K_i,$$
$$C^{\text{cl}} = \text{diag}(C_1, C_2, \ldots, C_N), \quad (6.32)$$

where $A_{\rho,ij}^{\text{cl}}$ and $B_{\rho,ij}^{\text{cl}}$ are matrix blocks whilst $\text{diag}(\cdot, \cdot, \ldots, \cdot)$ indicates a diagonal matrix.

Remark 6.6. *Using the Geršgorin circle theorem, the work in [50] provides sufficient conditions for A_ρ^{cl} to be Hurwitz. Applying these sufficient conditions to A_ρ^{cl} herein, we obtain that when*

(i) $A_i - l_{\rho,ii} B_i K_i C_i$ *is Hurwitz for all* $i \in \{1, 2, \cdots, N\}$, *and*

(ii) $\min_{\lambda_i \in \lambda(A_i - l_{\rho,ii} B_i K_i C_i)} |\lambda_i| \geq \sum_{j \in \mathcal{N}_i} \|B_i K_i C_j\|$,

where the set of all eigenvalues of a matrix is denoted $\lambda(\cdot)$, *are fulfilled, the matrix* A_ρ^{cl} *is Hurwitz. Thus, by changing* K_i's *for different topologies, one can ensure that* A_σ^{cl} *remains Hurwitz.*

When $\omega \equiv \mathbf{0}_{n_\omega}$, the equilibria of (6.31) satisfy

$$A_\rho^{\text{cl}} \xi = B_\rho^{\text{cl}} d. \qquad (6.33)$$

It is well known that the above matrix equality is solvable if and only if $B_\rho^{\text{cl}} d$ is in the column space of A_ρ^{cl}. If this condition is not met, simply select different d_i's. Notice that $d = \mathbf{0}_{n_d}$ or A_ρ^{cl} being Hurwitz immediately makes (6.33) solvable. Provided that (6.33) is solvable, we can find a particular solution ξ^{P} to (6.33). The corresponding output is $\zeta^{\text{P}} = C^{\text{cl}} \xi^{\text{P}}$. Now, the substitutions $x = \xi - \xi^{\text{P}}$ and $y = \zeta - \zeta^{\text{P}}$ transform (6.31) into the equivalent system (6.5). Notice that one can change ζ^{P} by changing d. For example, one can change formations by changing d.

6.7.2 Introducing Intermittent Data Exchange

The control law (6.30) can be modified as follows:

$$u_i = -K_i \sum_{j \in \mathcal{N}_i} [(\hat{\zeta}_i - \hat{\zeta}_j) - (d_i - d_j)], \qquad (6.34)$$

where signals $\hat{\zeta}_i : [t_0, \infty) \to \mathbb{R}^{n_x}$, $i \in \{1, \ldots, N\}$, are piece-wise constant and right-continuous functions with jumps accompanying the data exchange instants and $t_0 \in \mathbb{R}$ is the initial time. In other words, the control signal u_i is driven by sampled zero-order-hold versions of the actual signals $\zeta_i : [t_0, \infty) \to \mathbb{R}^{n_x}$, $i \in \{1, \ldots, N\}$. Now, let us introduce the error vector e as follows

$$e = \begin{bmatrix} e_1 \\ e_2 \\ \vdots \\ e_N \end{bmatrix} := \begin{bmatrix} \hat{\zeta}_1 - \zeta_1 \\ \hat{\zeta}_2 - \zeta_2 \\ \vdots \\ \hat{\zeta}_N - \zeta_N \end{bmatrix} = \hat{\zeta} - \zeta. \qquad (6.35)$$

The above expression uses $\hat{\zeta} := (\hat{\zeta}_1, \hat{\zeta}_2, \ldots, \hat{\zeta}_N)$. Taking e into account, the closed-loop dynamics (6.5) become (6.6). Since $\dot{\hat{\zeta}} = 0$ and $\dot{\zeta}^{\text{P}} = 0$, the corresponding error dynamics are (6.7).

6.7.3 Proof of Proposition 6.1

Proof. This proposition is proved by showing that

$$\| - C^{\text{cl}}(A_\rho^{\text{cl}} x + \omega) \|_{\mathcal{B}_{\mathring{y}_\rho}} \geq \| \overline{-C^{\text{cl}}(A_\rho^{\text{cl}} x + \omega)} \|_{\mathcal{B}_{\tilde{y}_\rho}}, \qquad (6.36)$$

where the sets $\mathcal{B}_{\mathring{y}_\rho}$ and $\mathcal{B}_{\tilde{y}_\rho}$ are obtained via (6.2). Apparently, $\mathcal{B}_{\tilde{y}_\rho}$ may be constructed from $\mathcal{B}_{\mathring{y}_\rho}$ by applying the $\bar{\cdot}$ operator to the elements of $\mathcal{B}_{\mathring{y}_\rho}$. In addition, notice that $n_{\mathring{y}_\rho} = n_{\tilde{y}_\rho}$. Now, the inequality (6.36) is obtained as follows. For each $i \in 1, \ldots, n_{\tilde{y}}$, the reverse triangle inequality yields

$$\left| \left(- C^{\text{cl}}(A_\rho^{\text{cl}} x + \omega) - b_{\mathring{y}_\rho} \right)_i \right| \geq \left| \left| \left(- C^{\text{cl}}(A_\rho^{\text{cl}} x + \omega) \right)_i \right| - \left| (b_{\mathring{y}_\rho})_i \right| \right|,$$

where $b_{\mathring{y}_\rho} \in \mathcal{B}_{\mathring{y}_\rho}$ and $(\cdot)_i$ denotes the ith component of a vector. Hence,

$$\| - C^{\text{cl}}(A_\rho^{\text{cl}} x + \omega) - b_{\mathring{y}_\rho} \| \geq \| \overline{-C^{\text{cl}}(A_\rho^{\text{cl}} x + \omega)} - \overline{b_{\mathring{y}_\rho}} \|,$$

yielding

$$\inf_{b_{\mathring{y}_\rho} \in \mathcal{B}_{\mathring{y}_\rho}} \| - C^{\text{cl}}(A_\rho^{\text{cl}} x + \omega) - b_{\mathring{y}_\rho} \| \geq \inf_{b_{\mathring{y}_\rho} \in \mathcal{B}_{\mathring{y}_\rho}} \| \overline{-C^{\text{cl}}(A_\rho^{\text{cl}} x + \omega)} - \overline{b_{\mathring{y}_\rho}} \|$$

$$= \inf_{b_{\tilde{y}_\rho} \in \mathcal{B}_{\tilde{y}_\rho}} \| \overline{-C^{\text{cl}}(A_\rho^{\text{cl}} x + \omega)} - b_{\tilde{y}_\rho} \|,$$

which is equivalent to (6.36). □

6.7.4 Proof of Theorem 6.2

Proof. According to [175, Section IV.D], the state x of system (6.6) is $\mathcal{L}_p$-detectable w.r.t. $\mathcal{B}$ from input (ω, e) and from any output $\mathring{y}_\rho$, i.e., there exist K_ρ^d and γ_ρ^d such that

$$\|x[t_0, t]\|_{p, \mathcal{B}} \leq K_\rho^d \|x(t_0)\|_\mathcal{B} + \gamma_\rho^d \|\mathring{y}_\rho[t_0, t]\|_{p, \mathcal{B}_y} + \gamma_\rho^d \|\omega[t_0, t]\|_p, \qquad (6.37)$$

for any $t \geq t_0$. In addition, from the assumptions of the theorem we have:

$$\|\tilde{y}_\rho[t_0, t]\|_{p, \mathcal{B}_{\tilde{y}_\rho}} \leq K_\rho \|x(t_0)\|_\mathcal{B} + \gamma_\rho \|(\omega, e)[t_0, t]\|_p, \qquad (6.38)$$

$$\|e[t_0, t]\|_p \leq K_\rho^e \|e(t_0)\| + \gamma_\rho^e \|\tilde{y}_\rho[t_0, t]\|_p, \qquad (6.39)$$

for all $t \geq t_0$. Notice that (6.38) involves $\|\tilde{y}_\rho[t_0, t]\|_{p, \mathcal{B}_{\tilde{y}_\rho}}$ while (6.39) involves $\|\tilde{y}_\rho[t_0, t]\|_p$. Therefore, in general, the small-gain theorem is not applicable to interconnections of systems that are $\mathcal{L}_p$-stable w.r.t. sets. However, our choice of the output $\tilde{y}_\rho$, stated in (6.12), yields $\mathcal{B}_{\tilde{y}_\rho} = \mathbf{0}_{n_{\tilde{y}_\rho}}$. In other words, $\|\tilde{y}_\rho[t_0, t]\|_p = \|\tilde{y}_\rho[t_0, t]\|_{p, \mathcal{B}_{\tilde{y}_\rho}}$. Next, notice that $\|(\omega, e)[t_0, t]\|_p \leq \|\omega[t_0, t]\|_p +$

$\|e[t_0,t]\|_p$. We now apply the small-gain theorem to (6.38)–(6.39) obtaining

$$\|\tilde{y}_\rho[t_0,t]\|_p \leq \frac{1}{1-\gamma_\rho\gamma_\rho^e}\left[K_\rho\|x(t_0)\|_\mathcal{B} + \gamma_\rho K_\rho^e\|e(t_0)\| + \gamma_\rho\|\omega[t_0,t]\|_p\right], \quad (6.40)$$

$$\|e[t_0,t]\|_p \leq \frac{1}{1-\gamma_\rho\gamma_\rho^e}\left[\gamma_\rho^e K_\rho\|x(t_0)\|_\mathcal{B} + K_\rho^e\|e(t_0)\| + \gamma_\rho\gamma_\rho^e\|\omega[t_0,t]\|_p\right]. \quad (6.41)$$

Merging (6.37) and (6.40), we obtain:

$$\|x[t_0,t]\|_{p,\mathcal{B}} \leq \left(\frac{K_\rho \gamma_\rho^d}{1-\gamma_\rho\gamma_\rho^e} + K_\rho^d\right)\|x(t_0)\|_\mathcal{B} + \frac{K_\rho^e \gamma_\rho \gamma_\rho^d}{1-\gamma_\rho\gamma_\rho^e}\|e(t_0)\|$$

$$+ \left(\frac{\gamma_\rho \gamma_\rho^d}{1-\gamma_\rho\gamma_\rho^e} + \gamma_\rho^d\right)\|\omega[t_0,t]\|_p. \quad (6.42)$$

Now, we use the following equality:

$$\|(x,e)[t_0,t]\|_{p,(\mathcal{B},\mathbf{0}_{n_e})} = \|(x,\mathbf{0}_{n_e})[t_0,t] + (\mathbf{0}_{n_x},e)[t_0,t]\|_{p,(\mathcal{B},\mathbf{0}_{n_e})}. \quad (6.43)$$

It is easily shown (simply follow the proof for the classical Minkowski inequality [15, Chapter 6] and the fact that $\|\cdot\|_\mathcal{B}$ is a seminorm) that

$$\|(x,\mathbf{0}_{n_e})[t_0,t] + (\mathbf{0}_{n_x},e)[t_0,t]\|_{p,(\mathcal{B},\mathbf{0}_{n_e})} \leq \|(x,\mathbf{0}_{n_e})[t_0,t]\|_{p,(\mathcal{B},\mathbf{0}_{n_e})} +$$
$$+ \|(\mathbf{0}_{n_x},e)[t_0,t]\|_{p,(\mathcal{B},\mathbf{0}_{n_e})}$$
$$= \|x[t_0,t]\|_{p,\mathcal{B}} + \|e[t_0,t]\|_p \quad (6.44)$$

holds. In fact, one can think of (6.44) as a variant of the classical Minkowski inequality. Combining (6.41), (6.42), (6.43) and (6.44) yields

$$\|(x,e)[t_0,t]\|_{p,(\mathcal{B},\mathbf{0}_{n_e})} \leq K_1\|x(t_0)\|_\mathcal{B} + K_2\|e(t_0)\| + \tilde{\gamma}\|\omega[t_0,t]\|_p, \quad (6.45)$$

where $K_1 = \frac{\gamma_\rho^e K_\rho + \gamma_\rho^d K_\rho}{1-\gamma_\rho\gamma_\rho^e} + K_\rho^d$, $K_2 = \frac{K_\rho^e + K_\rho^e \gamma_\rho \gamma_\rho^d}{1-\gamma_\rho\gamma_\rho^e}$ and $\tilde{\gamma} = \frac{\gamma_\rho \gamma_\rho^e + \gamma_\rho \gamma_\rho^d}{1-\gamma_\rho\gamma_\rho^e} + \gamma_\rho^d$. We proceed by obtaining the following expression:

$$\|x(t_0)\|_\mathcal{B} + \|e(t_0)\| = \inf_{b\in\mathcal{B}}\|x(t_0)-b\| + \|e(t_0)\|$$
$$\leq \inf_{b\in\mathcal{B}}\left(\sum_{i=1}^{n_x}|x_i(t_0)-b_i|\right) + \sum_{i=1}^{n_e}|e_i(t_0)-0|$$
$$= \inf_{b\in\mathcal{B}}\left(\sum_{i=1}^{n_x}|x_i(t_0)-b_i| + \sum_{i=1}^{n_e}|e_i(t_0)-0|\right)$$
$$\leq \sqrt{n_x+n_e}\inf_{b\in\mathcal{B}}\left(\|(x(t_0),e(t_0))-(b,\mathbf{0}_{n_e})\|\right)$$
$$= \sqrt{n_x+n_e}\|(x(t_0),e(t_0))\|_{(\mathcal{B},\mathbf{0}_{n_e})}. \quad (6.46)$$

In the above derivation, we use the following inequalities

$$\sqrt{\sum_{i=1}^{n_\vartheta}|\vartheta_i|^2} \leq \sum_{i=1}^{n_\vartheta}|\vartheta_i| \leq \sqrt{n_\vartheta}\sqrt{\sum_{i=1}^{n_\vartheta}|\vartheta_i|^2}, \qquad (6.47)$$

where $\vartheta = (\vartheta_1,\ldots,\vartheta_{n_\vartheta}) \in \mathbb{R}^{n_\vartheta}$. Finally, putting together (6.45) and (6.46), we obtain

$$\|(x,e)[t_0,t]\|_{p,(\mathcal{B},\mathbf{0}_{n_e})} \leq \widetilde{K}\|(x(t_0),e(t_0))\|_{(\mathcal{B},\mathbf{0}_{n_e})} + \widetilde{\gamma}\|\omega[t_0,t]\|_p, \qquad (6.48)$$

where $\widetilde{K} := \sqrt{n_x+n_e}\max\{K_1,K_2\}$. $\square$

6.7.5 Proof of Theorem 6.3

Proof. Let us show that (6.16) and (6.7) satisfy the assumptions of [168, Theorem 2.5]. In other words, we show that there exist nonnegative constants L_1, L_2, L_3 and L_4 such that

$$\left.\begin{array}{r}\|A_\rho^{\mathrm{cl,r}}z_r + B_\rho^{\mathrm{cl,r}}e\| \leq L_1(\|z_r\|+\|e\|) \\ \|C^{\mathrm{cl}}A_\rho^{\mathrm{cl}}S_\rho^{-1}(z_r,z_q) + C^{\mathrm{cl}}B_\rho^{\mathrm{cl}}e\| \leq L_2(\|z_r\|+\|e\|) \\ \|z_r^+(t)\| \leq L_3\|z_r(t)\| \\ \|e^+(t)\| \leq L_4\|e(t)\|\end{array}\right\} \qquad (6.49)$$

for all $z_r \in \mathbb{R}^{n_x-\mathcal{A}}$, all $e \in \mathbb{R}^{n_e}$ and each $\rho \in \mathcal{P}$. Notice that z_r does not experience jumps when new information arrives; hence, one can take $L_3 = 1$. From (6.9) it follows that the last inequality is satisfied with $L_4 = 1$. It is straightforward to show that $L_1 = \max\{\|A_\rho^{\mathrm{cl,r}}\|,\|B_\rho^{\mathrm{cl,r}}\|\}$ and $L_2 = \max\{\|C^{\mathrm{cl}}A_\rho^{\mathrm{cl}}S_\rho^{-1}\|,\|C^{\mathrm{cl}}B_\rho^{\mathrm{cl}}\|\}$ satisfy the above inequalities. The UGES property of (z_r,e) follows from [168, Theorem 2.5], i.e., there exist $k,l>0$ such that $\|(z_r,e)(t)\| \leq k\exp(-l(t-t_0))\|(z_r,e)(t_0)\|$ for all $t \geq t_0$ and for any $(z_r,e)(t_0)$.

Lastly, for all $t \geq t_0$ and for any $(x,e)(t_0)$ we have

$$\|(x,e)(t)\|_{(\mathcal{B},\mathbf{0}_{n_e})} \overset{\text{seminorm}}{\leq} \|(x(t),\mathbf{0}_{n_e})\|_{(\mathcal{B},\mathbf{0}_{n_e})} + \|(\mathbf{0}_{n_x},e(t))\|_{(\mathcal{B},\mathbf{0}_{n_e})}$$

$$\leq \|x(t)\|_\mathcal{B} + \|e(t)\| \overset{(6.15)}{\leq} \|S_\rho^{-1}\|\|z_r(t)\| + \|e(t)\|$$

$$\leq \max\{\|S_\rho^{-1}\|,1\}(\|z_r(t)\|+\|e(t)\|)$$

$$\overset{(6.47)}{\leq} \max\{\|S_\rho^{-1}\|,1\}\sqrt{n_{z_r}+n_e}\|(z_r,e)(t)\|$$

$$\leq k\max\{\|S_\rho^{-1}\|,1\}\sqrt{n_{z_r}+n_e}\exp(-l(t-t_0))\|(z_r,e)(t_0)\|$$

$$\leq k\max\{\|S_\rho^{-1}\|,1\}\sqrt{n_{z_r}+n_e}\exp(-l(t-t_0))(\|z_r(t_0)\|+\|e(t_0)\|)$$

$$\leq k\max\{\|S_\rho^{-1}\|,1\}\sqrt{n_{z_r}+n_e}\exp(-l(t-t_0))(\|S_\rho\|\|P\|\|x(t_0)\|_\mathcal{B}+\|e(t_0)\|)$$

$$\overset{(6.46)}{\leq} k_1\exp(-l(t-t_0))\|(x,e)(t_0)\|_{(\mathcal{B},\mathbf{0}_{n_e})},$$

where P is the oblique projector onto $\mathcal{B}^c$ along $\mathcal{B}$ (refer to [175, Section IV.C] for more) and $k_1 := k\sqrt{n_{z_r} + n_e}\sqrt{n_x + n_e} \max\{\|S_\rho^{-1}\|, 1\} \max\{\|S_\rho\|\|P\|, 1\}$. □

6.7.6 Proof of Theorem 6.4

Proof. This proof follows the proof of [105, Theorem 3.2]. Pick an arbitrary $T > 0$, let $t_0 := 0$, and denote the switching times on the interval $(0, T)$ by $t_1, \ldots, t_{N_\sigma(T,0)}$. Consider the function $W(t) := \exp(2\lambda_0 t)V_{\sigma(t)}(t, \chi(t))$. On each interval $[t_j, t_{j+1})$ we have

$$D^+_{\sigma(t_j)} W = 2\lambda_0 W + \exp(2\lambda_0 t) D^+_{\sigma(t_j)} V_{\sigma(t_j)}(t, \chi) \leq 0$$

due to (6.21). Due to (6.19), when state jumps occur we have $V_{\sigma(t_j)}(t^+, \chi^+) \leq V_{\sigma(t_j)}(t, \chi)$ no matter whether the jump times coincide with the switching times or not. Therefore, W is nonincreasing between two switching times. This, together with (6.19) and (6.22), yields

$$W(t_{j+1}) = \exp(2\lambda_0 t_{j+1}) V_{\sigma(t_{j+1})}(t_{j+1}, \chi(t_{j+1}))$$
$$\leq \mu \exp(2\lambda_0 t_{j+1}) V_{\sigma(t_j)}(t_{j+1}, \chi(t_{j+1}))$$
$$\leq \mu \exp(2\lambda_0 t_{j+1}^-) V_{\sigma(t_j)}(t_{j+1}^-, \chi^-(t_{j+1})) = \mu W(t_{j+1}^-) \leq \mu W(t_j).$$

In the above expressions, the time instant just before t is denoted t^-. When the functions of interest are continuous in t, then $t^- = t$, but we still write t^- instead of t for clarity. In addition, the left limit of a solution $\chi(t)$ at instant t is denoted $\chi^-(t)$. Iterating the last inequality from $j = 0$ to $j = N_\sigma(T, 0) - 1$, we obtain

$$W(T^-) \leq W(t_{N_\sigma(T,0)}) \leq \mu^{N_\sigma(T,0)} W(0).$$

Using the definition of W, the above inequality and (6.19) we have

$$\exp(2\lambda_0 T) V_{\sigma(T^-)}(T, \chi(T)) \leq \exp(2\lambda_0 T^-) V_{\sigma(T^-)}(T^-, \chi(T^-)) \leq$$
$$\leq \mu^{N_\sigma(T,0)} V_{\sigma(0)}(0, \chi(0)).$$

Now suppose that σ has the average dwell-time property (6.4). Hence, we can write

$$V_{\sigma(T^-)}(T, \chi(T)) \leq \exp(-2\lambda_0 T + (N_0 + \frac{T}{\tau_a})\ln\mu) V_{\sigma(0)}(0, \chi(0)) =$$
$$= \exp(N_0 \ln\mu) \exp((\frac{\ln\mu}{\tau_a} - 2\lambda)T) V_{\sigma(0)}(0, \chi(0)).$$

From the above inequality, we infer that if τ_a satisfies (6.23), then $V_{\sigma(T^-)}(T, \chi(T))$ converges to zero exponentially as $T \to \infty$, i.e., it is upper-bounded by $\mu^{N_0} \exp(-2\lambda T) V_{\sigma(0)}(0, \chi(0))$ for some $\lambda \in (0, \lambda_0)$. Using (6.20), we obtain $\|\chi(T)\| \leq \sqrt{\frac{c_2}{c_1} \mu^{N_0} \exp(-2\lambda T)} \|\chi(0)\|$. This proves GES.

Notice that the value of the initial time t_0 was fixed to 0 for convenience. In fact, the switched system of interest is GES for any t_0, i.e., the switched systems is UGES. □

7

Cooperative Control in Degraded Communication Environments

CONTENTS

7.1	Motivation, Applications and Related Works	178
7.2	Impulsive Delayed Systems	179
7.3	Problem Statement: Stabilizing Transmission Intervals and Delays	180
7.4	Computing Maximally Allowable Transfer Intervals	182
	7.4.1 Interconnecting the Nominal and Error System	183
	7.4.2 MASs with Nontrivial Sets $\mathcal{B}$	183
	7.4.3 Computing Transmission Intervals τ	184
7.5	Example: Consensus Control with Experimental Validation	185
7.6	Conclusions and Perspectives	190
7.7	Proofs of Main Results	193
	7.7.1 Proof of Theorem 7.1	193
	7.7.2 Proof of Corollary 7.1	194

This chapter computes transmission rates and delays that provably stabilize Multi-Agent Systems (MASs) in the presence of disturbances and noise. Namely, given the existing information delay among the agents and the underlying communication topology, we determine rates at which information between the agents need to be exchanged such that the MAS of interest is $\mathcal{L}_p$-stable with bias, where this bias accounts for noisy data. In order to consider MASs characterized by sets of equilibrium points, the notions of $\mathcal{L}_p$-stability (with bias) and $\mathcal{L}_p$-detectability with respect to a set are employed. Using Lyapunov–Razumikhin type of arguments, we are able to consider delays greater than the transmission intervals. Our methodology is applicable to general (not merely to single- and double-integrator) heterogeneous linear agents, directed topologies and output feedback. The computed transmission rates are experimentally verified employing a group of off-the-shelf quadcopters.

7.1 Motivation, Applications and Related Works

Recent years have witnessed an increasing interest in *decentralized control* of Multi-Agent Systems (MASs) [139, 159, 49, 197, 201, 43, 110, 57, 143]. Decentralized control is characterized by *local* interactions between *neighbors*. In comparison with centralized control, decentralized control avoids a single point of failure, which in turn increases robustness of MASs, allows for inexpensive and simple agents, and lowers the implementation cost. In addition, decentralized control scales better as the number of agents increases and is sometimes an intrinsic property of MASs. The problem of synchronizing agents' outputs is a typical problem solved in a decentralized fashion (e.g., [159] and [201]). The goal of output synchronization is to achieve a desired *collective behavior* of MASs. Examples are formation control, flocking, consensus control, etc.

Information exchange among neighbors is instrumental for coordination as discussed in all aforementioned references. However, degraded communication environments are commonly encountered in real-life applications. Namely, the exchanged information is often sampled, delayed, corrupted and prone to communication channel dropouts. Furthermore, the agent models utilized to devise cooperative control laws might not be accurate or external disturbances might be present. Hence, it is of interest to determine how robust a control law is with respect to realistic data exchange and to quantify this robustness. Herein, the impact of external disturbances is quantified through $\mathcal{L}_p$-gains while the influence of corrupted data is quantified through the bias term in $\mathcal{L}_p$-stability with bias. As expected, different transmission rates and communication delays as well as different controller parameters yield different robustness levels. In addition, since many MASs possess a set of equilibria (rather than a sole equilibrium point), we employ the notions of $\mathcal{L}_p$-stability (with bias) and $\mathcal{L}_p$-detectability with respect to a set (see Chapter 6 and [175, 182]).

The remainder of the chapter is organized as follows. Section 7.2 provides the notation and terminology used herein. Section 7.3 states the robustness problem of linear MASs with respect to realistic communication. A methodology to solve this robustness problem is presented in Section 7.4 and experimentally verified in Section 7.5. Conclusions and future directions are in Section 7.6. The proofs are provided in Section 7.7.

FIGURE 7.1
A snapshot of our experimental setup with three quadcopters.

7.2 Impulsive Delayed Systems

In this chapter, we consider impulsive delayed systems

$$\Sigma \begin{cases} \chi(t^+) = h_\chi(\chi(t), \chi(t-d)) & t \in \mathcal{T} \\ \dot\chi(t) = f_\chi(\chi(t), \chi(t-d), \omega) \\ y = \ell_\chi(\chi, \omega) \end{cases} \text{otherwise} , \quad (7.1)$$

where $\chi \in \mathbb{R}^{n_\chi}$ is the state, $\omega \in \mathbb{R}^{n_\omega}$ is the input, $y \in \mathbb{R}^{n_y}$ is the output and $d \geq 0$ is the time delay. The functions f_χ and h_χ are regular enough to guarantee forward completeness of solutions which, given initial condition $\psi_\chi \in PC([t_0 - d, t_0], \mathbb{R}^{n_\chi})$ and initial time t_0, are given by right-continuous functions $t \mapsto \chi(t) \in PC([t_0 - d, \infty], \mathbb{R}^{n_\chi})$. Jumps of the state $\chi(t^+)$ occur at each $t \in \mathcal{T} := \{t_1, t_2, \ldots\}$, where $t_i < t_{i+1}$, $i \in \mathbb{N}_0$. For a comprehensive discussion regarding the solutions to (7.1) considered herein, refer to [13, Chapter 2 & 3]. Even though the considered solutions to (7.1) allow for jumps at t_0, we exclude such jumps in favor of notational convenience.

In the following definitions, we use the set

$$\mathcal{B}_y := \{y \in \mathbb{R}^{n_y} | \exists b \in \mathcal{B} \text{ such that } y = \ell_\chi(b, \mathbf{0}_{n_\omega})\}, \quad (7.2)$$

where $\mathcal{B} \subseteq \mathbb{R}^{n_\chi}$.

Definition 7.1 ($\mathcal{L}_p$-Stability with Bias b and w.r.t. a set). Let $p \in [1, \infty]$. The system Σ is $\mathcal{L}_p$-stable w.r.t. a set $\mathcal{B}$ and with bias $b(t) \equiv b \geq 0$ from ω to y with gain $\gamma \geq 0$ if there exists $K \geq 0$ such that, for each $t_0 \in \mathbb{R}$ and each $\psi_\chi \in PC([t_0 - d, t_0], \mathbb{R}^{n_\chi})$, each solution to Σ from ψ_χ at $t = t_0$ satisfies $\|y[t_0, t]\|_{p, \mathcal{B}_y} \leq K\|\psi_\chi\|_{d, \mathcal{B}} + \gamma\|\omega[t_0, t]\|_p + \|b[t_0, t]\|_p$ for each $t \geq t_0$.

Definition 7.2 ($\mathcal{L}_p$-Detectability w.r.t. a set). *Let $p \in [1, \infty]$. The state χ of Σ is $\mathcal{L}_p$-detectable w.r.t. a set $\mathcal{B}$ from (y, ω) with gain $\gamma_d \geq 0$ if there exists $K_d \geq 0$ such that, for each $t_0 \in \mathbb{R}$ and each $\psi_\chi \in PC([t_0 - d, t_0], \mathbb{R}^{n_\chi})$, each solution to Σ from ψ_χ at $t = t_0$ satisfies $\|\chi[t_0, t]\|_{p, \mathcal{B}} \leq K_d \|\psi_\chi\|_{d, \mathcal{B}} + \gamma_d \|y[t_0, t]\|_{p, \mathcal{B}_y} + \gamma_d \|\omega[t_0, t]\|_p$ for each $t \geq t_0$.*

7.3 Problem Statement: Stabilizing Transmission Intervals and Delays

Consider N heterogeneous linear agents given by

$$\dot{\xi}_i = A_i \xi_i + B_i u_i + \omega_i,$$
$$\zeta_i = C_i \xi_i, \qquad (7.3)$$

where $\xi_i \in \mathbb{R}^{n_{\xi_i}}$ is the state, $u_i \in \mathbb{R}^{n_{u_i}}$ is the input, $\zeta_i \in \mathbb{R}^{n_\zeta}$ is the output of the ith agent, $i \in \{1, 2, \ldots, N\}$, and $\omega_i \in \mathbb{R}^{n_{\xi_i}}$ reflects modeling uncertainties and/or exogenous disturbances. The matrices A_i, B_i and C_i are of appropriate dimensions. A common decentralized policy is

$$u_i = -K_i \sum_{j \in \mathcal{N}_i} (\zeta_i - \zeta_j), \qquad (7.4)$$

where K_i is an $n_{u_i} \times n_\zeta$ matrix, and $\mathcal{N}_i$ denotes the set of neighbors of the ith agent according to Section 6.2.2.

When considering real-life applications, the controllers (7.4) are typically fed by estimates of signals ζ_i's and ζ_j's. Basically, (7.4) becomes

$$u_i = -K_i \sum_{j \in \mathcal{N}_i} (\hat{\zeta}_i - \hat{\zeta}_j), \qquad (7.5)$$

where estimates $\hat{\zeta}_i \in ([t_0, \infty), \mathbb{R}^\zeta)$, $i \in \{1, \ldots, N\}$, are piece-wise continuous functions with jump times contained in $\mathcal{T}$. Observe that the ZOH estimation strategy yields $\hat{\zeta}_i$'s, which are merely delayed, sampled (i.e., piece-wise constant) and noisy versions of ζ_i's. For simplicity, we assume that all communication links introduce the same propagation delay equal to some $d \geq 0$. In addition, we neglect message collisions and consider that all agents' outputs are received at time instants $t_i \in \mathcal{T}$ as indicated in Figure 7.2. If one is concerned with packet collisions, the scheduling protocols among agents' outputs from Chapter 6 readily complement the methodology presented herein.

Next, we define the following stack vectors $\xi := (\xi_1, \ldots, \xi_N)$, $y := (\zeta_1, \ldots, \zeta_N)$, $\hat{\zeta} := (\hat{\zeta}_1, \ldots, \hat{\zeta}_N)$, and $\omega := (\omega_1, \ldots, \omega_N)$. The discrepancy between $\hat{\zeta}$ and the delayed signal ζ is captured by the error vector e, i.e.,

$$e(t) := \hat{\zeta}(t) - \zeta(t - d). \qquad (7.6)$$

Cooperative Control in Degraded Communication Environments

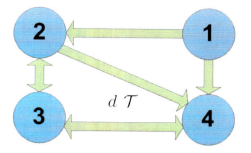

FIGURE 7.2
An illustration of communication links among four agents characterized with the same propagation delays and sampling instants.

Utilizing the Laplacian matrix L of the underlying communication graph $\mathcal{G}$, the closed-loop dynamic equation becomes

$$\dot{\xi}(t) = A^{cl}\xi(t) + A^{cld}\xi(t-d) + A^{cle}e(t) + \omega(t), \qquad (7.7)$$
$$\zeta = C^{cl}\xi, \qquad (7.8)$$

with

$$A^{cl} = \mathrm{diag}(A_1,\ldots,A_N), \qquad A^{cld} = [A^{cld}_{ij}], \qquad A^{cld}_{ij} = -l_{ij}B_iK_iC_j,$$
$$A^{cle} = [A^{cle}_{ij}], \qquad A^{cle}_{ij} = -l_{ij}B_iK_i, \qquad C^{cl} = \mathrm{diag}(C_1,\ldots,C_N),$$

where A^{cld}_{ij} and A^{cle}_{ij} are matrix blocks whilst $\mathrm{diag}(\cdot,\cdot,\ldots,\cdot)$ indicates a diagonal matrix. Our methodology allows for employment of model-based estimators [48, 181], that is,

$$\dot{\hat{\zeta}}(t) = E\hat{\zeta}(t-d), \qquad (7.9)$$

for each $t \in [t_0,\infty) \setminus \mathcal{T}$, where E is a block diagonal matrix. Notice that E is implemented locally due to the decentralized nature of the control problem. Basically, each agent runs an estimator regarding itself (e.g., for $\hat{\zeta}_i$) and an estimator regarding each of its neighbors (i.e., for each $\hat{\zeta}_j$, $j \in \mathcal{N}_i$). Notice that the obtained estimates are utilized in (7.5). In order for the estimates regarding each $\hat{\zeta}_i$, $i \in \{1,\ldots,N\}$, to match, the estimators are provided with same initial conditions while the estimators' dynamics are identical and given by the respective block matrices of E. Using (7.6), (7.8) and (7.9), we arrive at

$$\dot{e}(t) = \dot{\hat{\zeta}}(t) - \dot{\zeta}(t-d) = E\Big(e(t-d) + C^{cl}\xi(t-2d)\Big) - C^{cl}\dot{\xi}(t-d), \qquad (7.10)$$

for each $t \in [t_0,\infty) \setminus \mathcal{T}$ while for each $t_i \in \mathcal{T}$ we have

$$e(t_i^+) = \nu(t_i), \qquad (7.11)$$

where $\nu(t_i)$ models any discrepancy (e.g., due to measurement noise and/or channel distortions) between the received values and their actual values at time $t_i - d$ (when the agents' outputs were sampled). We consider bounded ν, i.e.,

$$\sup_{t \in \mathbb{R}} \|\nu(t)\| = K_\nu \geq 0.$$

Let $\mathcal{B} := \mathrm{Ker}(A^{\mathrm{cl}} + A^{\mathrm{cld}})$ denote the set of equilibrium points of the MAS (7.3)–(7.5) in the absence of realistic control system phenomena (e.g., delays and exogenous disturbances as well as intermittent and distorted information). Since $\mathcal{B}$ specifies the desired MAS behavior (as the control law (7.4) is designed to steer MAS toward $\mathcal{B}$), it is natural to analyze robustness of the MAS (7.3)–(7.5) w.r.t. $\mathcal{B}$. Refer to Chapter 6 for a comprehensive discussion regarding stability w.r.t. sets.

Intuitively, the *nominal system* (defined explicitly in Section 7.4.1) pertaining to MAS (7.3)–(7.5) might be robustly stable w.r.t. $\mathcal{B}$ (in the $\mathcal{L}_p$-sense according to (7.14)) only for some values of d. We refer to such delays as *admissible delays*. The condition (7.14) typically entails the existence of a directed spanning tree [159], but one can envision (less challenging/appealing) MAS scenarios in which this directed spanning tree requirement is redundant. Furthermore, given an admissible delay d, the maximal transmission interval τ, i.e., $t_{i+1} - t_i \leq \tau$ for each $i \in \mathbb{N}_0$, which renders $\mathcal{L}_p$-stability of MAS (7.3)–(7.5) w.r.t. $\mathcal{B}$ is called Maximally Allowable Transfer Interval (MATI) and is denoted $\bar{\tau}$. In order to exclude ill-disposed choices, such as $t_{i+1} - t_i = \frac{\tau}{i^2}$, that lead to Zeno behavior, we impose an arbitrary small lower bound on $t_{i+1} - t_i$, that is, $0 < \varepsilon \leq t_{i+1} - t_i$ for each $i \in \mathbb{N}_0$. The existence of such ε follows from Remark 7.3. We are now ready to state the main problem considered herein.

Problem 7.1. *Given an admissible delay d existent in the MAS (7.3)–(7.5) with estimator (7.9), determine the MATI $\bar{\tau}$ to update the control law (7.5) with novel neighbors' outputs such that the MAS of interest is $\mathcal{L}_p$-stable from ω to ξ w.r.t. $\mathcal{B}$ for some $p \in [1, \infty]$.*

Remark 7.1. *Using our framework, it is straightforward to compute $\bar{\tau}$ resulting in $\mathcal{L}_p$-stability from ω to ξ w.r.t. $\mathcal{B}$ with a prespecified $\mathcal{L}_p$-gain for some $p \in [1, \infty]$. In order not to the conceal the main points of this chapter, these details are left out herein and the reader is referred to Chapter 2.*

7.4 Computing Maximally Allowable Transfer Intervals

In order to solve Problem 7.1, we bring together the concepts of $\mathcal{L}_p$-stability and $\mathcal{L}_p$-detectability w.r.t. a set (refer to Chapter 6 and [175] for more) as well as $\mathcal{L}_p$-stability and $\mathcal{L}_p$-detectability for delayed (impulsive) systems (see Chapter 2).

7.4.1 Interconnecting the Nominal and Error System

As in Chapter 2, the principal idea is to consider two interconnected systems and invoke the small-gain theorem [88]. Basically, we want to interconnect dynamics (7.7) and dynamics (7.10)–(7.11). To that end, let us expand the right side of (7.10):

$$\dot{e}(t) = E\Big(e(t-d) + C^{\text{cl}}\xi(t-2d)\Big) - \\ - C^{\text{cl}}\Big(A^{\text{cle}}e(t-d) + A^{\text{cl}}\xi(t-d) + A^{\text{cld}}\xi(t-2d) + \omega(t-d)\Big), \quad (7.12)$$

and select

$$\tilde{\zeta} := -C^{\text{cl}}A^{\text{cl}}\xi(t-d) + \Big(EC^{\text{cl}} - C^{\text{cl}}A^{\text{cld}}\Big)\xi(t-2d) - C^{\text{cl}}\omega(t-d) \quad (7.13)$$

to be the output of the *nominal system* Σ_n given by (7.7). Notice that the input to Σ_n is (e, ω). Now, let us interconnect with Σ_n the *error system* Σ_e given by (7.10) and (7.11) with the input and output being $\tilde{\zeta}$ and e, respectively. Notice that Σ_n is a delayed system whilst Σ_e is an impulsive delayed system from Section 7.2. According to Problem 7.1, the delay d is admissible so that the nominal system Σ_n is $\mathcal{L}_p$-stable w.r.t. $\mathcal{B}$ for some $p \in [1, \infty]$ and the corresponding $\mathcal{L}_p$-gain is denoted γ_n, i.e., there exist $K_n, \gamma_n \geq 0$ such that

$$\|\tilde{\zeta}[t_0, t]\|_{p, \mathcal{B}_{\tilde{\zeta}}} \leq K_n \|\psi_\xi\|_{d, \mathcal{B}} + \gamma_n \|(e, \omega)[t_0, t]\|_p, \quad (7.14)$$

for all $t \geq t_0$. In order to invoke the small-gain theorem and infer $\mathcal{L}_p$-stability, the transmission interval τ has to render the $\mathcal{L}_p$-gain γ_e of Σ_e such that $\gamma_n \gamma_e < 1$. Recall that the maximal such τ is in fact $\bar{\tau}$. When $\mathcal{B} = \mathbf{0}_{n_\xi}$, a procedure for attaining such a period τ is provided in Chapter 2. Nevertheless, Chapter 2 does not cover cases when $\mathcal{B} \neq \mathbf{0}_{n_\xi}$.

7.4.2 MASs with Nontrivial Sets $\mathcal{B}$

MASs with nontrivial sets $\mathcal{B}$, that is, $\mathcal{B} \neq \mathbf{0}_{n_\xi}$, are designed to primarily look for an agreement while the actual value of the agreement/consensus point depends on the agents initial conditions. However, available tools for computing $\mathcal{L}_p$-stability and $\mathcal{L}_p$-detectability gains of delayed dynamics, such as the software HINFN [124], cannot be utilized as they are. Nevertheless, Chapter 6 delineates how to compute the respective gains in the quotient space $\mathbb{R}^{n_\xi} \setminus \mathcal{B}$, where the existing tools can be utilized. In order to exploit Chapter 6, we impose the following two assumptions on the MAS (7.3)–(7.5):

Assumption 7.1. *The algebraic multiplicity of the zero eigenvalue of* $A^{\text{cl}} + A^{\text{cld}}$, *denoted* $\mathcal{A}$, *equals its geometric multiplicity, denoted* $\mathcal{G}$.

Due to Assumption 7.1, we know that there exists a Real Jordan Form basis such that the last $\mathcal{A}$ rows and columns of the operator $A^{\text{cl}} + A^{\text{cld}}$ written in this basis contain all zero entries. For more, refer to Chapter 6. In order to streamline the exposition of this chapter, we consider such Real Jordan Form bases herein.

Assumption 7.2. *There exists a Real Jordan Form basis of $A^{\text{cl}} + A^{\text{cld}}$ such that the last $\mathcal{A}$ rows and columns of operators A^{cl} and A^{cld} written in this basis contain all zero entries as well.*

Remark 7.2. *It is certainly of interest to investigate which class of $A^{\text{cl}} + A^{\text{cld}}$ yields Assumption 7.2. However, since about two dozen examples that we have considered so far readily fulfill Assumption 7.2, we have deferred this investigation for the future.*

Assumption 7.3. *The estimator E yields $\text{Ker}(A^{\text{cl}} + A^{\text{cld}}) \subseteq \text{Ker}(-C^{\text{cl}}A^{\text{cl}} + EC^{\text{cl}} - C^{\text{cl}}A^{\text{cld}})$.*

Moreover, Chapter 6 states that, given the Assumptions 7.1 and 7.2, one can utilize the same procedure to infer $\mathcal{L}_p$-detectability w.r.t. $\mathcal{B}$ of the state ξ of Σ_n from $(e, \omega, \tilde{\zeta})$. Let γ_d denote the corresponding $\mathcal{L}_p$-detectability gain.

7.4.3 Computing Transmission Intervals τ

From Lemma 2.1 and Theorem 2.1, we know that $\gamma_e = \frac{2}{\lambda}\sqrt{M}$, $K_e = 2\sqrt{M}\left(1 + \|E - C^{\text{cl}}A^{\text{cle}}\|\frac{2}{\lambda}\left(e^{\frac{d\lambda}{2}} - 1\right)\right)\left(\frac{1}{p\lambda}\right)^{\frac{1}{p}}$ and bias $b = \frac{\tilde{K}_\nu\sqrt{M}}{e^{\frac{\lambda\varepsilon}{2}} - 1}$ for any $p \in [1, \infty]$, where constants $\lambda > 0$ and $M > 1$ satisfy inequalities

(I) $\tau(\lambda + r + \lambda_1 M e^{-\lambda\tau}) < \ln M$, and

(II) $\tau(\lambda + r + \frac{\lambda_1}{\lambda_2} e^{\lambda d}) < -\ln \lambda_2$,

with $r > 0$ being an arbitrary constant, $\lambda_1 := \frac{\|E - C^{\text{cl}}A^{\text{cle}}\|^2}{r}$ and $\lambda_2 \in (0, 1)$. It is worth mentioning that Lyapunov–Razumikhin techniques are employed to reach conditions (I) and (II). The expression for λ_1 is obtained by considering the Try-Once-Discard (TOD) protocol in Theorem 2.2. Namely, L in Theorem 2.2 equals $\|E - C^{\text{cl}}A^{\text{cle}}\|$ according to (7.12). Altogether, we have reached the following result:

Theorem 7.1. *Assume that an admissible delay d for the MAS (7.3)–(7.5) is given so that (7.14) holds for some $p \in [1, \infty]$. If the transmission interval τ satisfies (I) and (II) for some $\lambda > 0$ and $M > 1$ such that $\frac{2}{\lambda}\sqrt{M}\gamma_n < 1$, the interconnection of the nominal system Σ_n and the error system Σ_e is $\mathcal{L}_p$-stable from ω to $(\tilde{\zeta}, e)$ w.r.t. $(\mathcal{B}, \mathbf{0}_{n_e})$ (and with bias when $K_\nu \neq 0$).*

Corollary 7.1. *Assume that the conditions of Theorem 7.1 holds and that ξ is $\mathcal{L}_p$-detectable from $(e, \omega, \tilde{\zeta})$ w.r.t. $\mathcal{B}$. Then the MAS (7.3)–(7.5) is $\mathcal{L}_p$-stable (with bias when $K_\nu \neq 0$) w.r.t. $(\mathcal{B}, \mathbf{0}_{n_e})$ from ω to (ξ, e).*

Section 7.5 delineates once more all the pieces put together in order to attain Theorem 7.1 (i.e., Corollary 7.1). However, the following remarks regarding the relevance of Theorem 7.1 to practical implementations are in order.

Remark 7.3. *The left-hand sides of conditions (I) and (II) are nonnegative continuous functions of $\tau \geq 0$ and tend to ∞ as $\tau \to \infty$. Also, these left-hand sides equal zero for $\tau = 0$. Note that both sides of (I) and (II) are continuous in λ, M, λ_1, λ_2 and d. Hence, for every $\lambda > 0$, $\lambda_1 \geq 0$, $M > 1$, $\lambda_2 \in (0,1)$ and $d \geq 0$, there exists $\tau > 0$ such that (I) and (II) are satisfied. Finally, since $\frac{2}{\lambda}\sqrt{M}$ is continuous in λ and M, we infer that for every finite $\gamma_n > 0$ there exists $\tau > 0$ such that $\frac{2}{\lambda}\sqrt{M}\gamma_n < 1$. In other words, for each admissible d the unwanted Zeno behavior is avoided and the proposed methodology does not yield continuous feedback that might be impossible to implement. Notice that each τ yielding $\frac{2}{\lambda}\sqrt{M}\gamma_n < 1$ is a candidate for $\bar{\tau}$. Depending on r, λ_2, λ and M, the maximal such τ is in fact MATI $\bar{\tau}$.*

Remarks 2.5 and 2.6 are applicable to Theorem 7.1 as well.

7.5 Example: Consensus Control with Experimental Validation

The agents in our experimental setup are commercially available AR.Drone Parrot quadcopter platforms [146]. In order to account for the specifics of our experimental settings (e.g., baseline controller, custom battery supplies, latencies due to the OptiTrack motion caption system [140] and our ROS code [151]), we decided to perform system identification of these quadcopters on our own (rather than employing models or model parameters obtained by other researchers). As we are interested in coordinating the quadcopters in the plane, we identify the quadcopters' transfer functions in x- and y-axis. The experimental data for identification is collected for each axis independently. Potential coupling among x- and y-axis or slight model differences between quadcopters are taken into account using ω_i's in our MAS model (7.3)–(7.5). Using the MATLAB® System Identification Toolbox, for both axes the obtained transfer function is

$$G_{x_i}(s) = \frac{X_i(s)}{U_{x_i}(s)} = \frac{K_p}{s(1+T_p s)} e^{-ds},$$

$$G_{y_i}(s) = \frac{Y_i(s)}{U_{y_i}(s)} = \frac{K_p}{s(1+T_p s)} e^{-ds}, \quad (7.15)$$

where $K_p = 5.2$, $T_p = 0.38$ and $d = 0.104$ s for all $i \in \{1,2,3,4\}$. The inputs $U_{x_i}(s)$ and $U_{y_i}(s)$ are the Laplace transforms of the normalized pitch and roll angles of the ith agent, respectively, whilst the outputs $X_i(s)$ and

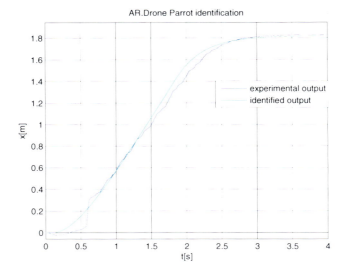

FIGURE 7.3
Experimental verification of the identified transfer function for an AR.Drone Parrot quadcopter in the x-axis. These responses are obtained for the pulse input with duration of 1.75 s and amplitude of 0.2. This 0.2 is the normalized roll angle that is fed to the low-level controller and corresponds to the actual pitch angle of $9°$. Since our agents are homogeneous, we write $x[m]$ instead of $x_i[m]$.

$Y_i(s)$ are the Laplace transforms of the x- and y-coordinate of the ith agent, respectively. This second-order transfer function with delay is a reasonable trade-off between the model complexity and accuracy. In addition, since our framework allows for modeling uncertainties through w_i's, additional model accuracy is not essential. A fragment of comparison among the identified model and the actual model is provided in Figure 7.3. In order to avoid notation ambiguity, we restrict our attention to $G_{x_i}(s)$ (i.e., to x-axis direction) from now on. The analysis with respect to y-axis is the same.

Let us now write (7.15) in the form of MAS (7.3)–(7.5) with delays. Basically, the delay d from (7.15) becomes the communication link delay (see Figure 7.2) while the remainder of the transfer function can be written in the state-space form (7.3) as

$$\dot{\xi}_i = \begin{bmatrix} 0 & 1 \\ 0 & -T_p \end{bmatrix} \xi_i + \begin{bmatrix} 0 \\ K_p \end{bmatrix} u_i + w_i.$$

Since the consensus controller (7.4) is a time-invariant static controller (i.e., without memory/states on its own), the choice for d to represent the communication delay is justified. The first component of ξ_i corresponds to the position of the ith agent while the second component corresponds to its velocity in the x-axis direction. The more intuitive notation for the first component of ξ_i is x_i while for the second component of ξ_i is v_i^x, which is the notation found in the plots. We weight the position and velocity in the output ζ_i as follows

$$\zeta_i = \begin{bmatrix} 0.05 & 0.025 \end{bmatrix} \xi_i,$$

in order to put (twice) more emphasis on the position discrepancy among the agents because the velocity measurements are rather noisy (see Figures 7.4 and 7.5). The selected graph topology is depicted in Figure 7.2 and the graph Laplacian matrix is

$$L = \begin{bmatrix} 0 & 0 & 0 & 0 \\ -1 & 2 & -1 & 0 \\ 0 & -1 & 2 & -1 \\ -1 & -1 & -1 & 3 \end{bmatrix}.$$

As given by L, we consider one leader and three followers. Due to the space constraints of our laboratory, the leader is virtual (i.e., generated on a computer). This chapter includes two sets of experiments. The first set is obtained using $K := K_1 = \ldots = K_4 = 2$ while the second set of experiments uses $K := K_1 = \ldots = K_4 = 4$.

For simplicity, we set $E = \mathbf{0}_{n_\zeta \times n_\zeta}$, which corresponds to the ZOH estimation strategy. Observe that this choice of E meets Assumption 7.3. In addition, one can readily verify that Assumption 7.1 holds for both $K = 2$ and $K = 4$ with $\mathcal{A} = \mathcal{G} = 1$. A Real Jordan Form basis of $A^{\text{cl}} + A^{\text{cld}}$ for $K = 2$ is given by the columns of the following matrix

$$T_2 = \begin{bmatrix} 0 & 0 & 0 & 0 & 0 & 0 & -2.6316 & 1 \\ 0 & 0 & 0 & 0 & 0 & 0 & 1 & 0 \\ 0.3718 & 0.7091 & -0.8738 & 1.5873 & -0.3570 & 0.6601 & -2.6316 & 1 \\ -1 & 0 & 1 & 0 & 1 & 0 & 1 & 0 \\ -0.3718 & -0.7091 & -1.2357 & 2.2448 & 0.5049 & -0.9336 & -2.6316 & 1 \\ 1 & 0 & 1.4142 & 0 & -1.4142 & 0 & 1 & 0 \\ 0 & 0 & -0.8738 & 1.5873 & -0.3570 & 0.6601 & -2.6316 & 1 \\ 0 & 0 & 1 & 0 & 1 & 0 & 1 & 0 \end{bmatrix},$$

while for $K = 4$ is given by the columns of

$$T_4 = \begin{bmatrix} 0 & 0 & 0 & 0 & 0 & 0 & -2.6316 & 1 \\ 0 & 0 & 0 & 0 & 0 & 0 & 1 & 0 \\ 0.3109 & 0.4731 & -0.5619 & 1.1514 & -0.3035 & 0.4353 & -2.6316 & 1 \\ -1 & 0 & 1 & 0 & 1 & 0 & 1 & 0 \\ -0.3109 & -0.4731 & -0.7946 & 1.6283 & 0.4292 & -0.6156 & -2.6316 & 1 \\ 1 & 0 & 1.4142 & 0 & -1.4142 & 0 & 1 & 0 \\ 0 & 0 & -0.5619 & 1.1514 & -0.3035 & 0.4353 & -2.6316 & 1 \\ 0 & 0 & 1 & 0 & 1 & 0 & 1 & 0 \end{bmatrix}.$$

The last column of the above two matrices spans $\mathcal{B}$, i.e.,

$$\mathcal{B} := \rho(1, 0, 1, 0, 1, 0, 1, 0), \quad \rho \in \mathbb{R}.$$

Assumption 7.2 is readily verified as well. Apparently, the above $\mathcal{B}$ is not trivial as discussed in Section 7.4.2. Hence, we need to work in the quotient space $\mathbb{R}^{n_\varepsilon} \setminus \mathcal{B}$. To that end, we utilize the above Real Jordan Form bases and apply the procedure from Chapter 6 to write the closed-loop dynamics (7.7) with the output $\tilde{\zeta}$, given by (7.13), in the reduced form. Basically, we introduce the substitution (i.e., change of coordinates) $z = (T_K)^{-1}\xi$, where T_K is given above for the selected K's, prune the last component of z, obtaining z_r, so that the reduced system for $K = 2$ becomes:

$$\dot{z}_r(t) = \begin{bmatrix} -0.38 & 0 & 0 & 0 & 0 & 0 & 0 \\ -1.211 & 0 & 0 & 0 & 0 & 0 & 0 \\ 0 & 0 & -0.38 & 0 & 0 & 0 & 0 \\ 0 & 0 & 0.4208 & 0 & 0 & 0 & 0 \\ 0 & 0 & 0 & 0 & -0.38 & 0 & 0 \\ 0 & 0 & 0 & 0 & 1.3093 & 0 & 0 \\ 0 & 0 & 0 & 0 & 0 & 0 & -0.38 \end{bmatrix} z_r(t)$$

$$+ \begin{bmatrix} -0.2 & 1.1062 & 0 & 0 & 0 & 0 & 0 \\ 0.1049 & -0.58 & 0 & 0 & 0 & 0 & 0 \\ 0 & 0 & 0.1138 & -0.4835 & 0 & 0 & 0 \\ 0 & 0 & 0.0627 & -0.2662 & 0 & 0 & 0 \\ 0 & 0 & 0 & 0 & -0.2538 & -1.172 & 0 \\ 0 & 0 & 0 & 0 & -0.1373 & -0.6338 & 0 \\ 0 & 0 & 0 & 0 & 0 & 0 & 0 \end{bmatrix} z_r(t-d)$$

$$+ \begin{bmatrix} 0 & 31.2 & 0 & -31.2 & 0 & 0 \\ 0 & -16.3592 & 0 & 16.3592 & 0 & 0 \\ 5.2 & -2.1539 & -2.1539 & -0.8922 & 0 & -0.8536 \\ 2.8624 & -1.1857 & -1.1857 & -0.4911 & -0.5377 & -0.4699 \\ 5.2000 & 12.5539 & 12.5539 & -30.3078 & 0 & -0.1464 \\ 2.8122 & 6.7894 & 6.7894 & -16.3910 & -0.2218 & -0.0792 \\ 0 & 0 & 0 & 0 & 0 & 1 \end{bmatrix}$$

$$\begin{bmatrix} 0 & -1 & 0 & 0 & 0 & 1 \\ 1.4103 & 0.5243 & 0 & 0 & -1.4103 & -0.5243 \\ 0 & 0.3536 & 0 & 0.3536 & 0 & 0.1464 \\ 0.2227 & 0.1946 & 0.2227 & 0.1946 & 0.0923 & 0.0806 \\ 0 & -0.3536 & 0 & -0.3536 & 0 & 0.8536 \\ -0.5356 & -0.1912 & -0.5356 & -0.1912 & 1.293 & 0.4616 \\ 0 & 0 & 0 & 0 & 0 & 0 \end{bmatrix} (e, \omega)(t),$$

where the output of the reduced system Σ_n is the following one

$$\tilde{\zeta}_r(t) = \underbrace{\begin{bmatrix} 0 & 0 & 0 & 0 & 0 & 0 & -0.0405 \\ 0.0405 & 0 & -0.0405 & 0 & -0.0405 & 0 & -0.0405 \\ -0.0405 & 0 & -0.0573 & 0 & 0.0573 & 0 & -0.0405 \\ 0 & 0 & -0.0405 & 0 & -0.0405 & 0 & -0.0405 \end{bmatrix}}_{C_r} z_r(t)$$

$$+ \underbrace{\begin{bmatrix} 0 & 0 & 0 & 0 & 0 & 0 & 0 \\ -0.005 & 0.0277 & -0.0028 & 0.0121 & 0.0063 & 0.0293 & 0 \\ 0.005 & -0.0277 & -0.004 & 0.0171 & -0.009 & -0.0414 & 0 \\ 0 & 0 & -0.0028 & 0.0121 & 0.0063 & 0.0293 & 0 \end{bmatrix}}_{C_r^d} z_r(t-d)$$

$$+ \underbrace{\begin{bmatrix} 0 & 0 & 0 & 0 & -0.05 & -0.025 & 0 & 0 & 0 & 0 & 0 & 0 \\ 0 & 0 & 0 & 0 & 0 & 0 & -0.05 & -0.025 & 0 & 0 & 0 & 0 \\ 0 & 0 & 0 & 0 & 0 & 0 & 0 & 0 & -0.05 & -0.025 & 0 & 0 \\ 0 & 0 & 0 & 0 & 0 & 0 & 0 & 0 & 0 & 0 & -0.05 & -0.025 \end{bmatrix}}_{D_r} (e, \omega)(t),$$

delayed for additional d time units (compare with (7.13)). We point out that the $\mathcal{L}_p$-gain corresponding to this additional delay d is unity [87]. Likewise, for $K = 4$ we obtain:

$$\dot{z}_r(t) = \begin{bmatrix} -0.38 & 0 & 0 & 0 & 0 & 0 & 0 \\ -1.8639 & 0 & 0 & 0 & 0 & 0 & 0 \\ 0 & 0 & -0.38 & 0 & 0 & 0 & 0 \\ 0 & 0 & 0.6831 & 0 & 0 & 0 & 0 \\ 0 & 0 & 0 & 0 & -0.38 & 0 & 0 \\ 0 & 0 & 0 & 0 & 2.0322 & 0 & 0 \\ 0 & 0 & 0 & 0 & 0 & 0 & -0.38 \end{bmatrix} z_r(t)$$

$$+ \begin{bmatrix} -0.59 & 1.4762 & 0 & 0 & 0 & 0 & 0 \\ 0.3877 & -0.97 & 0 & 0 & 0 & 0 & 0 \\ 0 & 0 & 0.0377 & -0.7015 & 0 & 0 & 0 \\ 0 & 0 & 0.0184 & -0.3423 & 0 & 0 & 0 \\ 0 & 0 & 0 & 0 & -0.6977 & -1.5458 & 0 \\ 0 & 0 & 0 & 0 & -0.4864 & -1.0777 & 0 \\ 0 & 0 & 0 & 0 & 0 & 0 & 0 \end{bmatrix} z_r(t-d)$$

$$+ \begin{bmatrix} 0 & 62.4 & 0 & -62.4 & 0 & 0 \\ 0 & -41.0032 & 0 & 41.0032 & 0 & 0 \\ 10.4 & -4.3078 & -4.3078 & -1.7844 & 0 & -0.8536 \\ 5.0751 & -2.1022 & -2.1022 & -0.8707 & -0.7413 & -0.4165 \\ 10.4 & 25.1078 & 25.1078 & -60.6156 & 0 & -0.1464 \\ 7.2509 & 17.5051 & 17.5051 & -42.2611 & -0.3364 & -0.1021 \\ 0 & 0 & 0 & 0 & 0 & 1 \\ 2.1136 & 0.6571 & 0 & 0 & -2.1136 & -0.6571 \\ 0 & 0.3536 & 0 & 0.3536 & 0 & 0.1464 \\ 0.3071 & 0.1725 & 0.3071 & 0.1725 & 0.1272 & 0.0715 \\ 0 & -0.3536 & 0 & -0.3536 & 0 & 0.8536 \\ -0.8122 & -0.2465 & -0.8122 & -0.2465 & 1.9607 & 0.5951 \\ 0 & 0 & 0 & 0 & 0 & 0 \end{bmatrix} (e, \omega)(t),$$

$$\tilde{\zeta}_r(t) = \underbrace{\begin{bmatrix} 0 & 0 & 0 & 0 & 0 & 0 & -0.0405 \\ 0.0405 & 0 & -0.0405 & 0 & -0.0405 & 0 & -0.0405 \\ -0.0405 & 0 & -0.0573 & 0 & 0.0573 & 0 & -0.0405 \\ 0 & 0 & -0.0405 & 0 & -0.0405 & 0 & -0.0405 \end{bmatrix}}_{C_r} z_r(t)$$

$$+ \underbrace{\begin{bmatrix} 0 & 0 & 0 & 0 & 0 & 0 & 0 \\ -0.0147 & 0.0369 & -0.0009 & 0.0175 & 0.0174 & 0.0386 & 0 \\ 0.0147 & -0.0369 & -0.0013 & 0.0248 & -0.0247 & -0.0547 & 0 \\ 0 & 0 & -0.0009 & 0.0175 & 0.0174 & 0.0386 & 0 \end{bmatrix}}_{C_r^d} z_r(t-d)$$

$$+ \underbrace{\begin{bmatrix} 0 & 0 & 0 & 0 & -0.05 & -0.025 & 0 & 0 & 0 & 0 & 0 & 0 \\ 0 & 0 & 0 & 0 & 0 & 0 & -0.05 & -0.025 & 0 & 0 & 0 & 0 \\ 0 & 0 & 0 & 0 & 0 & 0 & 0 & 0 & -0.05 & -0.025 & 0 & 0 \\ 0 & 0 & 0 & 0 & 0 & 0 & 0 & 0 & 0 & 0 & -0.05 & -0.025 \end{bmatrix}}_{D_r} (e, \omega)(t).$$

We are now ready to apply the software HINFN [124] in order to compute γ_n and γ_d for $p = 2$. Altogether, for $K = 2$ we obtain $\gamma_n = 3.1588$ and $\gamma_d = 62.0433$, while for $K = 4$ we obtain $\gamma_n = 4.9444$ and $\gamma_d = 66.5614$. Apparently, $d = 0.108$ s is an admissible delay. As far as the computation of γ_e is concerned, for $K = 2$ we have $\|C^{\text{cl}} A^{\text{cle}}\| = 0.9881$, while for $K = 4$ we have $\|C^{\text{cl}} A^{\text{cle}}\| = 1.9762$. Using Theorem 7.1, for $K = 2$ we obtain $\bar{\tau} = 0.0835$ s when $r = 5$, $\lambda = 14.363$, $M = 5.168$ and $\lambda_2 = 0.08$. Additionally, for $K = 4$ we obtain $\bar{\tau} = 0.042$ s when $r = 8$, $\lambda = 17.2$, $M = 3$ and $\lambda_2 = 0.13$. It is worth mentioning that, in general, the ratio among two MATIs is not almost the inverse of the gain ratio. This is corroborated by the following: $K = 6$ results in $\bar{\tau} = 0.0262$ s, and $K = 8$ leads to $\bar{\tau} = 0.0176$ s. Apparently, an increase in K decreases $\bar{\tau}$.

Next, we experimentally verify the computed MATIs $\bar{\tau} = 0.0835$ s and $\bar{\tau} = 0.042$ s for $K = 2$ and $K = 4$, respectively, which corresponds to 12 Hz and

		$\dfrac{\\|\xi[t_0,t_{\text{end}}]\\|_{2,\mathcal{B}}}{(t_{\text{end}}-t_0)}$
	50 Hz	0.054
K=2	12 Hz	0.042
	9 Hz	0.078
	50 Hz	0.052
K=4	24 Hz	0.057
	20 Hz	0.065

TABLE 7.1
Comparison of the normalized $\|\xi[t_0, t_{\text{end}}]\|_{2,\mathcal{B}}$ for different gains K and sampling frequencies. The consensus controller is turned on at t_0 while it is switched off at t_{end}. The difference $t_{\text{end}} - t_0$ is about 40 s for each experiment.

24 Hz, respectively. In order to test the claimed robustness of the experimental MAS (see Corollary 7.1) and to examine how different gains K and different sampling frequencies affect the system performance, we generate an exogenous disturbance (e.g., wind) that moves the virtual agent with speed 0.5 m/s along the square of side length 0.5 m. This disturbance is generated in our computer in order to have its repeatability while carrying out the experiments, which in turn yields a fair comparison. This disturbance was reproduced several times during each experimental trial. Also, we have introduced offsets among agent positions in order to avoid agent collisions. Of course, these offsets do not alter the MATIs as clarified in Chapter 6. Experimental data is presented in Table 7.1 as well as in Figures 7.4 and 7.5.

From Table 7.1 we infer that faster sampling improves the system performance (with the only exception being the case $K = 2$, $f = 12$), which is expected. Notice that the delay d is larger than the sampling intervals. Such delays are referred to as *large delays*. In addition, the computed MATIs appear to be rather accurate and useful for our experimental setup (see Figures 7.4 and 7.5). Hence, the obtained experimental results regarding the conservativeness of our methodology are encouraging. Nevertheless, further research efforts are needed in this regard.

7.6 Conclusions and Perspectives

This chapter investigates multi-agent control in degraded communication environments. Our theoretical framework allows for sampled, delayed and corrupted data as well as for general heterogeneous linear agents, model-based estimators, output feedback and directed (not necessarily balanced) topologies. We seek Maximally Allowable Transmission Intervals (MATIs) that yield

Cooperative Control in Degraded Communication Environments

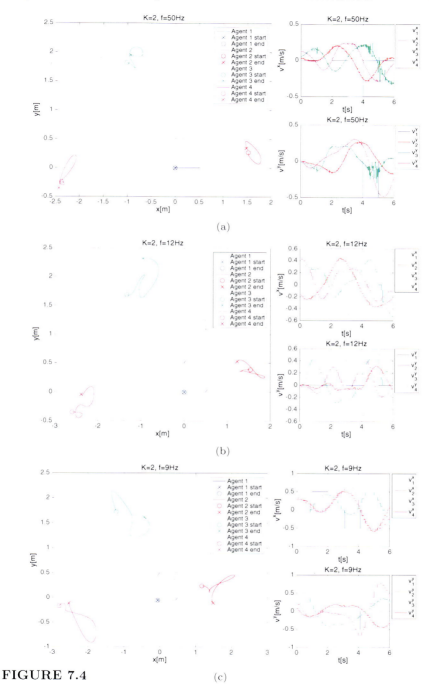

FIGURE 7.4
Experimental data for one square maneuver and $K = 2$. The computed marginal frequency is 12 Hz while the experimentally obtained one is 9 Hz.

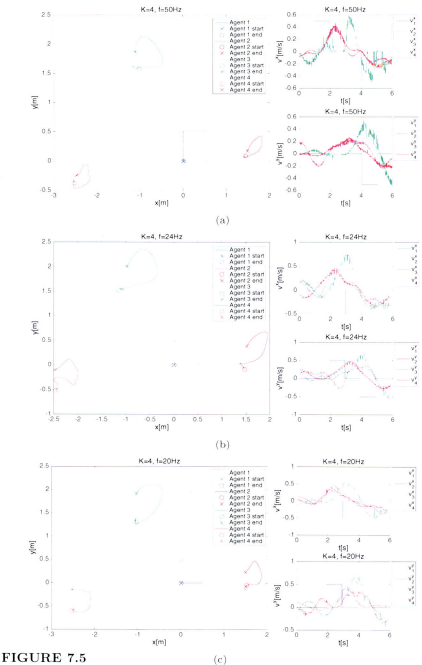

FIGURE 7.5
Experimental data for one square maneuver and $K = 4$. The computed marginal frequency is 24 Hz while the experimentally obtained one is 20 Hz.

a certain performance in terms of $\mathcal{L}_p$-gains. In addition, a salient feature of our framework is the consideration of so-called *large delays*. Essentially, we combine results from previous chapters regarding $\mathcal{L}_p$-stability of delay systems and multi-agent systems under intermittent information, and experimentally verify the obtained MATIs. The experiments suggest that our methodology yields MATIs which are rather accurate and useful for practical purposes.

Future research avenues include nonuniform and time-varying delays (using the work presented in Chapter 2) as well as time-varying topologies. It is also of interest to further study the impact that different parameters (e.g., different sampling frequencies, delays, controller gains or noise levels) have on the control performance. Lastly, in order to further reduce communication efforts, one can consider the event- and self-triggered communication paradigms.

7.7 Proofs of Main Results

7.7.1 Proof of Theorem 7.1

Proof. From the paragraph directly above Theorem 7.1, we have $\mathcal{L}_p$-stability from $\tilde{\zeta}$ to e with bias, i.e.,

$$\|e[t_0, t]\|_p \leq K_e \|\psi_e\|_d + \gamma_e \|\tilde{\zeta}[t_0, t]\|_p + \|b[t_0, t]\|_p, \tag{7.16}$$

for any $t \geq t_0$ and any $p \in [1, \infty]$. Expressions for K_e, γ_e and b are provided directly above Theorem 7.1.

Let us now infer $\mathcal{L}_p$-stability with bias from ω to $(\tilde{\zeta}, e)$ via the small-gain theorem [88]. Notice that (7.14) involves $\|\tilde{\zeta}[t_0, t]\|_{p, \mathcal{B}_{\tilde{\zeta}}}$ while (7.16) involves $\|\tilde{\zeta}[t_0, t]\|_p$. Therefore, in general, the small-gain theorem is not applicable to interconnections of systems that are $\mathcal{L}_p$-stable w.r.t. sets. However, our choice of the output $\tilde{\zeta}$, stated in (7.13), yields $\mathcal{B}_{\tilde{\zeta}} = \mathbf{0}_{n_{\tilde{\zeta}}}$ due to Assumption 7.3. In other words, $\|\tilde{\zeta}[t_0, t]\|_p = \|\tilde{\zeta}[t_0, t]\|_{p, \mathcal{B}_{\tilde{\zeta}}}$. Next, notice that $\|(\omega, e)[t_0, t]\|_p \leq \|\omega[t_0, t]\|_p + \|e[t_0, t]\|_p$. Combining (7.14) with (7.16) yields

$$\|e[t_0, t]\|_p \leq \frac{K_e}{1 - \gamma_e \gamma_n} \|\psi_e\|_d + \frac{\gamma_e K_n}{1 - \gamma_e \gamma_n} \|\psi_\xi\|_{d, \mathcal{B}} + $$
$$+ \frac{\gamma_e \gamma_n}{1 - \gamma_e \gamma_n} \|\omega[t_0, t]\|_p + \frac{1}{1 - \gamma_e \gamma_n} \|b[t_0, t]\|_p, \tag{7.17}$$
$$\|\tilde{\zeta}[t_0, t]\|_p \leq \frac{K_n}{1 - \gamma_e \gamma_n} \|\psi_\xi\|_{d, \mathcal{B}} + \frac{\gamma_n K_e}{1 - \gamma_e \gamma_n} \|\psi_e\|_d + $$
$$+ \frac{\gamma_n}{1 - \gamma_e \gamma_n} \|\omega[t_0, t]\|_p + \frac{\gamma_n}{1 - \gamma_e \gamma_n} \|b[t_0, t]\|_p. \tag{7.18}$$

Using the Minkowski inequality [15, Chapter 6], we obtain:

$$\|(\tilde{\zeta}, e)[t_0, t]\|_p = \|(\tilde{\zeta}, \mathbf{0}_{n_e})[t_0, t] + (\mathbf{0}_{n_{\tilde{\zeta}}}, e)[t_0, t]\|_p \leq \|\tilde{\zeta}[t_0, t]\|_p + \|e[t_0, t]\|_p. \tag{7.19}$$

Combining (7.17), (7.18) and (7.19), along with $\mathcal{B}_{\tilde{\zeta}} = \mathbf{0}_{n_{\tilde{\zeta}}}$, yields

$$\|(\tilde{\zeta}, e)[t_0, t]\|_{p,(\mathcal{B}_{\tilde{\zeta}}, \mathbf{0}_{n_e})} \leq K_1 \|\psi_\xi\|_{d,\mathcal{B}} + K_2 \|\psi_e\|_d + \gamma_1 \|\omega[t_0, t]\|_p + \|b_1[t_0, t]\|_p, \tag{7.20}$$

for all $t \geq t_0$, where $K_1 = \frac{\gamma_e K_n + K_n}{1 - \gamma \gamma_e}$, $K_2 = \frac{K_e + \gamma_n K_e}{1 - \gamma \gamma_e}$, $\gamma_1 = \frac{\gamma_n \gamma_e + \gamma_n}{1 - \gamma \gamma_e}$ and $b_1 = \frac{1 + \gamma_n}{1 - \gamma_e \gamma_n} b$. We proceed by using the following expression:

$$\|\psi_\xi\|_{d,\mathcal{B}} + \|\psi_e\|_d \leq 2 \max\{\|\psi_\xi\|_{d,\mathcal{B}}, \|\psi_e\|_d\} \leq 2\|(\psi_\xi, \psi_e)\|_{d,(\mathcal{B}, \mathbf{0}_{n_e})}. \tag{7.21}$$

Finally, putting together (7.20) and (7.21), we obtain

$$\|(\tilde{\zeta}, e)[t_0, t]\|_{p,(\mathcal{B}_{\tilde{\zeta}}, \mathbf{0}_{n_e})} \leq K_3 \|(\psi_\xi, \psi_e)\|_{d,(\mathcal{B}, \mathbf{0}_{n_e})} + \gamma_1 \|\omega[t_0, t]\|_p + \|b_1[t_0, t]\|_p, \tag{7.22}$$

for all $t \geq t_0$, where $K_3 := 2\max\{K_1, K_2\}$. $\square$

7.7.2 Proof of Corollary 7.1

Proof. The $\mathcal{L}_p$-detectability of ξ w.r.t. $\mathcal{B}$ from $(e, \omega, \tilde{\zeta})$ implies that there exist $K_d, \gamma_d \geq 0$ such that

$$\begin{aligned}\|\xi[t_0, t]\|_{p,\mathcal{B}} &\leq K_d \|\psi_\xi\|_{d,\mathcal{B}} + \gamma_d \|\tilde{\zeta}[t_0, t]\|_{p,\mathcal{B}_{\tilde{\zeta}}} + \gamma_d \|(e, \omega)[t_0, t]\|_p \\ &\leq K_d \|\psi_\xi\|_{d,\mathcal{B}} + \gamma_d \|\tilde{\zeta}[t_0, t]\|_{p,\mathcal{B}_{\tilde{\zeta}}} + \gamma_d \|e[t_0, t]\|_p + \gamma_d \|\omega[t_0, t]\|_p\end{aligned} \tag{7.23}$$

for all $t \geq t_0$. Plugging (7.14) into (7.23) leads to

$$\begin{aligned}\|\xi[t_0, t]\|_{p,\mathcal{B}} &\leq K_d \|\psi_\xi\|_{d,\mathcal{B}} + \gamma_d K_n \|\psi_\xi\|_{d,\mathcal{B}} + \gamma_d \gamma_n \|e[t, t_0]\|_p + \\ &\quad + \gamma_d \gamma_n \|\omega[t_0, t]\|_p + \gamma_d \|e[t_0, t]\|_p + \gamma_d \|\omega[t_0, t]\|_p \\ &\leq K_d \|\psi_\xi\|_{d,\mathcal{B}} + \gamma_d K_n \|\psi_\xi\|_{d,\mathcal{B}} + (\gamma_d \gamma_n + \gamma_d)\|e[t, t_0]\|_p + \\ &\quad + (\gamma_d \gamma_n + \gamma_d)\|\omega[t_0, t]\|_p\end{aligned}$$

for all $t \geq t_0$. Observe that (simply follow the proof for the classical Minkowski inequality [15, Chapter 6] and the fact that $\|\cdot\|_{\mathcal{B}}$ is a seminorm)

$$\begin{aligned}\|(\xi, e)[t_0, t]\|_{p,(\mathcal{B}, \mathbf{0}_{n_e})} &= \|(\xi, \mathbf{0}_{n_e})[t_0, t] + (\mathbf{0}_{n_\xi}, e)[t_0, t]\|_{p,(\mathcal{B}, \mathbf{0}_{n_e})} \leq \\ &\leq \|(\xi, \mathbf{0}_{n_e})[t_0, t]\|_{p,(\mathcal{B}, \mathbf{0}_{n_e})} + \|(\mathbf{0}_{n_\xi}, e)[t_0, t]\|_{p,(\mathcal{B}, \mathbf{0}_{n_e})} \\ &= \|\xi[t_0, t]\|_{p,\mathcal{B}} + \|e[t_0, t]\|_p\end{aligned} \tag{7.24}$$

holds. In fact, one can think of (7.24) as a variant of the classical Minkowski inequality. Finally, we include (7.17) in (7.23) and add the obtained inequality to (7.17), which establishes $\mathcal{L}_p$-stability w.r.t. $(\mathcal{B}, \mathbf{0}_{n_e})$ with bias from ω to (ξ, e) following the same steps as in the proof of Theorem 7.1. □

8

Optimal Intermittent Feedback via Least Square Policy Iteration

CONTENTS

8.1	Motivation, Applications and Related Works	197
8.2	Problem Statement: Cost-Minimizing Transmission Policies	199
8.3	Computing Maximally Allowable Transfer Intervals	201
	8.3.1 Stabilizing Interbroadcasting Intervals	201
	8.3.2 (Sub)optimal Interbroadcasting Intervals	204
8.4	Example: Consensus Control (Revisited)	207
8.5	Conclusions and Perspectives	209

This chapter investigates how often information between neighbors in cooperative MASs needs to be exchanged in order to meet a desired performance. To that end, we formulate a DP problem for each agent and employ an online Least Square Policy Iteration (LSPI) method to solve these problems. Basically, LSPI computes aperiodic broadcasting instants for each individual agent such that the cost functions of interest are minimized. Cost functions studied herein capture trade-offs between the MAS local control performance in the presence of exogenous disturbances and energy consumption of each agent. Agent energy consumption is critical for prolonging the MAS mission and is comprised of both control (e.g., acceleration, velocity, etc.) and communication effort. The obtained broadcasting intervals adapt to the most recent information (e.g., delayed and noisy agents' inputs and/or outputs) received from neighbors and provably stabilize the MAS. We point out that LSPI is a model-free Reinforcement Learning (RL) algorithm capable of dealing with continuous states and actions. Chebyshev polynomials are utilized as the approximator in LSPI while Kalman Filtering (KF) handles sampled, corrupted and delayed data. The proposed methodology is exemplified on a consensus control problem with general linear agent dynamics.

8.1 Motivation, Applications and Related Works

Decentralized control of MASs has been one of the most active research areas in the last decade [139, 159, 49, 197, 201, 110, 57]. Decentralized control is characterized by local interactions between neighbors. As discussed in all aforementioned references, information exchange among neighbors is instrumental for MAS coordination, especially in the presence of exogenous disturbances and noisy data. Owing to digital technology and limited communication channel capacities, continuous data flows among neighbors are often not achievable. Instead, data among agents are exchanged intermittently (i.e., not necessarily in a periodic fashion). In addition, realistic communication channels introduce delays, distort the transmitted data and give rise to packet dropouts. In consequence of these network-induced phenomena, the MAS control performance is impaired and MASs can even become unstable. At the same time, agents' sensing and broadcasting units consume energy while in operation. Since agents typically have scarce energy supplies at disposal (e.g., batteries), it may erroneously seem that sensing and broadcasting should be executed as sparsely as possible in order to prolong the MAS mission (i.e., agents' lifetime), which is the driving force in some of the related works as pointed out in [68] and [179]. Namely, compared to scenarios with "frequent" sampling and broadcasting, "infrequent" data exchange results in greater settling times (i.e., slower convergence to a vicinity of MAS equilibria according to Chapter 6), which in turn may exhaust the energy supplies more than the "frequent" counterpart due to greater control effort in the long run (e.g., acceleration or velocity). This observation underlies the work in the present chapter.

We quantify the repercussions of intermittent feedback on the MAS control performance and MAS lifetime. An optimization problem is posed for each agent involving only locally available data. Each cost function captures the local control performance vs. agent lifetime trade-offs. In theory, DP solves a wide spectrum of optimization problems (including the optimization problems considered herein) providing an optimal solution [20, 21]. In practice, straightforward implementations of DP algorithms are deemed computationally intractable for most of the applications due to the infamous *curses of dimensionality* [150]. Therefore the need for efficient Approximate Dynamic Programming (ADP) or RL methods (refer to [22, 150] and [166, 32], respectively). The RL method utilized herein is the online model-free LSPI reported in [142, 141, 97]. Notice that the local DP problems are coupled, which in turn yields nonautonomous underlying dynamics so that the corresponding cost-to-go is non-stationary [86]. In addition, the distribution of exogenous disturbances is often unknown or unpredictable (e.g., winds, sea currents, waves, etc.). The behavior of neighbors may not be predictable due to the presence of other agents, disturbances and noise. Hence the need for an online model-free RL method. Besides, online RL methods do not require the training phase (unlike offline RL methods). Our LSPI approximation architecture for both continuous state and action pairs is based on Chebyshev polynomials. In addition, we use KF to alleviate the problem of delayed, sampled and noisy data [2].

The present chapter is a continuation of Chapter 4, where we consider similar cost functions for plant-controller settings without delays. In addition, the ADP method in Chapter 4 is an offline perceptron-based value iteration supplied with the exact control system model. Building upon Chapters 6 and 7, this chapter first devises stabilizing upper bounds for intervals between two consecutive broadcasting instants of each individual agent, thereby giving rise to asynchronous communication. Afterward, these stabilizing upper bounds are exploited in LSPI to ensure MAS stability. To the best of our knowledge, the optimal MAS intermittent feedback problem is yet to be addressed. Arguably, the line of research found in [86] and [92] is the most similar to our work, even though [86] and [92] employ different RL approaches and aim at different goals. In addition to the references provided herein, the reader interested in ADP and RL approaches for MASs should as well consult the references from [86] and [92].

The remainder of the chapter is organized as follows. Section 8.2 formulates the optimal intermittent feedback problem for MASs. In Section 8.3 we bring together $\mathcal{L}_p$-stability of MASs and LSPI in order to solve the problem of interest. A numerical example is provided in Section 8.4. Conclusions and future research avenues are in Section 8.5.

8.2 Problem Statement: Cost-Minimizing Transmission Policies

Consider N heterogeneous linear agents given by

$$\dot{\xi}_i = A_i \xi_i + B_i u_i + \omega_i,$$
$$\zeta_i = C_i \xi_i, \qquad (8.1)$$

where $\xi_i \in \mathbb{R}^{n_{\xi_i}}$ is the state, $u_i \in \mathbb{R}^{n_{u_i}}$ is the input, $\zeta_i \in \mathbb{R}^{n_\zeta}$ is the output of the ith agent, $i \in \{1, 2, \ldots, N\}$, and $\omega_i \in \mathbb{R}^{n_{\xi_i}}$ reflects modeling uncertainties and/or exogenous disturbances. The matrices A_i, B_i and C_i are of appropriate dimensions. A common decentralized policy is

$$u_i(t) = -K_i \sum_{j \in \mathcal{N}_i} (\zeta_i(t) - \zeta_j(t)), \qquad (8.2)$$

where K_i is an $n_{u_i} \times n_\zeta$ gain matrix [139, 159]. In real-life applications, the transmitted information about signals $\zeta_i(t)$ and $\zeta_j(t)$ in (8.2) arrive at the receivers with some delay. For simplicity, suppose that all communication links introduce the same propagation delay $d \geq 0$ so that $\zeta_i(t)$ and $\zeta_j(t)$ in (8.2) can be replaced with $\zeta_i(t-d)$ and $\zeta_j(t-d)$, respectively.

Remark 8.1. *The delay constancy, and hence delay uniformity (i.e., the delays in all communication channels are equal), can be enforced through protocols with constant end-to-end time delays or protocols with guaranteed and known upper bounds on the time delay (e.g., Time Division Multiple Access (TDMA) and Controller Area Network (CAN) protocols) and appropriate buffering at the receiver ends (refer to [73] and the references therein). In case time-varying and/or nonuniform delays are inevitable, the present approach can readily be augmented with the ideas from Chapter 2 at the expense of more complex notation and mathematics.*

Next, we define the following stack vectors $\xi := (\xi_1, \ldots, \xi_N)$, $\zeta := (\zeta_1, \ldots, \zeta_N)$ and $\omega := (\omega_1, \ldots, \omega_N)$. We assume that $\omega \in \mathcal{L}_p$ (even though ω may not be known entirely). Utilizing the Laplacian matrix L of the underlying communication graph $\mathcal{G}$, the closed-loop dynamic equation becomes

$$\dot{\xi}(t) = A^{\text{cl}} \xi(t) + A^{\text{cld}} \xi(t-d) + \omega(t), \tag{8.3}$$

$$\zeta = C^{\text{cl}} \xi, \tag{8.4}$$

with

$$A^{\text{cl}} = \text{diag}(A_1, \ldots, A_N), \qquad A^{\text{cld}} = [A_{ij}^{\text{cld}}],$$
$$A_{ij}^{\text{cld}} = -l_{ij} B_i K_i C_j, \qquad C^{\text{cl}} = \text{diag}(C_1, \ldots, C_N),$$

where A_{ij}^{cld} are matrix blocks whilst $\text{diag}(\cdot, \cdot, \ldots, \cdot)$ indicates a diagonal matrix.

Next to being delayed, the signals $\zeta_i(t)$ and $\zeta_j(t)$ forwarded to the controllers (8.2) are typically sampled and noisy versions of the neighbors' outputs. We denote these signals $\hat{\zeta}_i(t)$ and $\hat{\zeta}_j(t)$, respectively. In order not to conceal the main points of this chapter, we assume Zero-Order-Hold (ZOH) sampling even though model-based estimators can be employed in order to extend the stabilizing interbroadcasting intervals (see Chapter 6). In addition to broadcasting sampled versions of ζ_i's at time instant belonging to $\mathcal{T}$ (see Figure 7.2), the agents broadcast sampled versions of u_i's as well because these input data are required for KF. Noise related to ζ_i is $\nu_i(t) \in \mathbb{R}^{n_\zeta}$ whilst the noise pertaining to u_i is $\upsilon_i(t) \in \mathbb{R}^{n_{u_i}}$. For each $i \in \{1, \ldots, N\}$, we assume $\nu_i(t) \in \mathcal{L}_\infty$ and $\upsilon_i(t) \in \mathcal{L}_\infty$.

In what follows, we need to differentiate among the broadcasting instants in $\mathcal{T}$ pertaining to each agent. Consequently, let $t_i^j \in \mathcal{T}$, $i \in \mathbb{N}_0$, label the broadcasting instants pertaining to the jth agent. The interbroadcasting intervals τ_i^j's are defined by $\tau_{i-1}^j = t_i^j - t_{i-1}^j$, $i \in \mathbb{N}$. We allow some broadcasting instants of different agents to coincide. Also, agents' broadcasting instants are not orchestrated by some scheduling protocol. In other words, communication between agents is asynchronous. Furthermore, let us define local discrepancy vectors x_i, $i \in \{1, \ldots, N\}$, as follows: $x_i := (\ldots, \zeta_i - \zeta_j, \ldots)$, where $j \in \mathcal{N}_i$.

According to Chapter 6, the vast majority of MASs (8.1)–(8.2) are designed to steer ξ toward $\mathcal{B} := \mathrm{Ker}(A^{\mathrm{cl}} + A^{\mathrm{cld}})$ in an effort to attain $\|x_i(t)\| \to 0$ as $t \to \infty$ for all $i \in \{1, \ldots, N\}$. In this chapter, we use x_i's to measure the MAS local performance while the energy effort comprises both control and broadcasting effort. We are now ready to state the main problem solved herein.

Problem 8.1. *For each $j \in \{1, \ldots, N\}$, minimize the following cost function that captures performance vs. energy trade-offs*

$$\mathbb{E}_{\omega}\left\{\sum_{i=1}^{\infty}(\gamma_j)^i\Big[\underbrace{\int_{t_{i-1}^j}^{t_i^j}(x_j^\top P_j x_j + u_j^\top R_j u_j)\mathrm{d}t + S_j}_{r_j(x_j, u_j, \tau_i^j)}\Big]\right\} \qquad (8.5)$$

for the jth agent of MAS (8.1)–(8.2) over all sampling policies τ_i^j and for all initial conditions $x_j(t_0) \in \mathbb{R}^{n_{x_j}}$.

In Problem 8.1, the discount factor $\gamma_j \in (0,1)$ makes the sum (8.5) finite provided that the reward $r_j(x_j, u_j, \tau_i^j)$ is uniformly bounded over all τ_i^j. The matrices P_j and R_j are positive semidefinite, and a nonnegative constant S_j represents the cost incurred for sampling and transmitting ζ_j and u_j. In (8.5), the conditional expectation over a stochastic signal ω is denoted $\mathbb{E}_{\omega}$.

8.3 Computing Maximally Allowable Transfer Intervals

Section 8.3.1 computes stabilizing upper bounds on τ_i^j, $j \in \{1, \ldots, N\}$, $i \in \mathbb{N}_0$, such that the MAS (8.1)–(8.2) is $\mathcal{L}_p$-stable from ω to (ξ, e) w.r.t. $\mathcal{B}$ and with bias (see Definition 7.1). Subsequently, these upper bounds on τ_i^j's are utilized in the online LSPI procedure of Section 8.3.2 while obtaining (sub)optimal τ_i^j's in the sense of Problem 8.1. Accordingly, our solutions of the optimal problems (8.5) yield control system stability in addition to (sub)optimality, which is a perennial problem when employing RL or ADP (see [206] or any RL/ADP reference herein for more). In addition, we employ KF in Section 8.3.2 to handle partially observable DP problems.

8.3.1 Stabilizing Interbroadcasting Intervals

Before proceeding further, let us point out that the analysis of the present subsection, and therefore of the entire chapter, is applicable to MASs with general heterogeneous continuous-time linear agents (not merely to single- or double-integrator dynamics) with exogenous disturbances, directed communication topologies and output feedback (inherited from Chapter 7). Furthermore, when communication delays are greater than the sampling period (the so-called large delays), no other work takes into account the concurrent adverse effects of the aforementioned realistic data exchange phenomena such as sampled, corrupted and delayed data as well as lossy communication channels (also inherited from Chapter 7). Consequently, even though there exist a number of works investigating self- or time-triggered control strategies for MASs (refer to Chapter 7 for a list of such approaches), no other approach is applicable to settings as general as those considered herein. However, when some of the settings are relaxed, alternative approaches for computing stabilizing upper bounds on τ_i^j, $j \in \{1, \ldots, N\}$, $i \in \mathbb{N}_0$, may be employed as well. Basically, the LSPI pursuit for (sub)optimal τ_i^j's in Section 8.3.2 is decoupled from the methodology for computing stabilizing upper bounds on τ_i^j's. Lastly, since Chapter 7 does not yield asynchronous broadcasting, the present subsection extends Chapter 7 in this regard.

The discrepancy between the vector of received information $\hat{\zeta} := (\hat{\zeta}_1, \ldots, \hat{\zeta}_N)$ that actually feeds the control laws (8.2), and the delayed signal $\zeta(t)$ is captured by the error vector e, i.e.,

$$e(t) = (e_1(t), \ldots, e_N(t)) := \hat{\zeta}(t) - \zeta(t - d). \tag{8.6}$$

Thus, the closed-loop dynamic equations become

$$\dot{\xi}(t) = A^{\mathrm{cl}}\xi(t) + A^{\mathrm{cld}}\xi(t-d) + A^{\mathrm{cle}}e(t) + \omega(t), \tag{8.7}$$
$$\zeta = C^{\mathrm{cl}}\xi, \tag{8.8}$$

with

$$A^{\mathrm{cle}} = [A_{ij}^{\mathrm{cle}}], \qquad A_{ij}^{\mathrm{cle}} = -l_{ij}B_i K_i.$$

Due to ZOH sampling, that is $\dot{\hat{\zeta}} = \mathbf{0}_{n_\zeta}$ for each $t \in [t_0, \infty) \setminus \mathcal{T}$, we have

$$\dot{e}(t) = -\dot{\zeta}(t-d) = -C^{\mathrm{cl}}\dot{\xi}(t-d), \tag{8.9}$$

while for each $t_i^j \in \mathcal{T}$ we have

$$e_k(t_i^{j+}) = e_k(t_i^j), \qquad k \in \{1, \ldots, N\}, k \neq j,$$
$$e_j(t_i^{j+}) = \nu_j(t_i^j). \tag{8.10}$$

The principal idea is to interconnect two thoughtfully constructed systems and invoke the small-gain theorem [88]. Basically, we want to interconnect dynamics (8.7) and dynamics (8.9)–(8.10). To that end, let us expand the right side of (8.9):

$$\dot{e}(t) = -C^{\mathrm{cl}}\big[A^{\mathrm{cle}}e(t-d) + A^{\mathrm{cl}}\xi(t-d) + A^{\mathrm{cld}}\xi(t-2d) + \omega(t-d)\big], \tag{8.11}$$

and select

$$\tilde{\zeta} := -C^{\mathrm{cl}}\big[A^{\mathrm{cl}}\xi(t-d) + A^{\mathrm{cld}}\xi(t-2d) + \omega(t-d)\big] \qquad (8.12)$$

to be the output of the *nominal system* Σ_n given by (8.7). Notice that the input to Σ_n is (e,ω). Now, let us interconnect with Σ_n the *error system* Σ_e given by (8.9)–(8.10) with the input and output being $\tilde{\zeta}$ and e, respectively. Notice that Σ_n is a delay system whilst Σ_e is an impulsive delay system from Section 7.2. For the given d, we assume that the nominal system Σ_n is $\mathcal{L}_p$-stable w.r.t. $\mathcal{B}$ for some $p \in [1,\infty]$ and denote the corresponding $\mathcal{L}_p$-gain γ_n. Equivalently, there exist $K_n, \gamma_n \geq 0$ such that

$$\|\tilde{\zeta}[t_0,t]\|_{p,\mathcal{B}_{\tilde{\zeta}}} \leq K_n \|\psi_\xi\|_{d,\mathcal{B}} + \gamma_n \|(e,\omega)[t_0,t]\|_p, \qquad (8.13)$$

for all $t \geq t_0$. The condition (8.13) typically entails the existence of a directed spanning tree [159], but one can envision less challenging/appealing MAS scenarios in which this directed spanning tree requirement is redundant. A procedure for computing γ_n is provided in Chapter 7.

In order to invoke the small-gain theorem, the interbroadcasting intervals τ_i^j's have to render the $\mathcal{L}_p$-gain γ_e of Σ_e such that $\gamma_n \gamma_e < 1$. From Lemma 2.1 and Theorem 2.1, we know that $\gamma_e = \frac{2}{\lambda}\sqrt{M}$ and bias for any $p \in [1,\infty]$, where constants $\lambda > 0$ and $M > 1$ satisfy inequalities

(I) $\tau_i^j\big(\lambda + r + \lambda_1 M e^{-\lambda \tau_i^j}\big) < \ln M$, and

(II) $\tau_i^j\big(\lambda + r + \frac{\lambda_1}{\lambda_2}e^{\lambda d}\big) < -\ln \lambda_2$,

with $r > 0$ being an arbitrary constant, $\lambda_1 := \frac{N\|C^{\mathrm{cl}}A^{\mathrm{cle}}\|^2}{r}$ and $\lambda_2 = \frac{N-1}{N}$.

Theorem 8.1. *Suppose the communication link delay d for the MAS (8.1)–(8.2) yields (8.13) for some $p \in [1,\infty]$. If the interbroadcasting intervals τ_i^j, $i \in \mathbb{N}_0$, $j \in \{1,\ldots,N\}$, satisfy (I) and (II) for some $\lambda > 0$ and $M > 1$ such that $\frac{2}{\lambda}\sqrt{M}\gamma_n < 1$, then the MAS (8.1)–(8.2) is $\mathcal{L}_p$-stable from ω to $(\tilde{\zeta},e)$ w.r.t. $(\mathcal{B},\mathbf{0}_{n_e})$ and with bias.*

Proof. The main idea behind this proof is already outlined in the second part of Remark 2.6. Consider the Round Robin (RR) protocol from Chapter 2, and select any RR broadcasting pattern (i.e., any permutation $\{1,\ldots,N\}$). Essentially, the broadcasting instants of the first agent in the RR sequence are followed by the broadcasting instants of the second agent in this RR sequence, which in turn are followed by the broadcasting instants of the third agent in this RR sequence, etc. The broadcasting instants of the Nth agent in the RR sequence are followed by the broadcasting instants of the first agent in the RR sequence and so on (see Figure 8.1). Since the conditions of Theorem 8.1 correspond to any RR broadcasting sequence (i.e., conditions (I) and (II) pertain to a generic RR protocol through λ_1 and λ_2), let τ_{RR} denote the obtained stabilizing interbroadcasting interval. Arbitrarily select

FIGURE 8.1
Illustration of the RR protocol.

one stabilizing interbroadcasting interval of RR, denote it τ_{RR}, and take all τ_i^j, $i \in \mathbb{N}_0, j \in \{1, \ldots, N\}$, to be less than or equal to τ_{RR}. Note that this results in asynchronous broadcasting between the agents. In addition, observe that each agent broadcasts at most every $N\tau_{RR}$ units of time in the RR protocol whilst each agent broadcasts at most every τ_{RR} units in the asynchronous scheme from Section 8.2. Clearly, one can always find the selected RR broadcasting pattern among the resulting asynchronous broadcasting instants that complies with τ_{RR}. Thus, the resulting asynchronous communication possesses the same $\mathcal{L}_p$-gain γ_e as the original RR protocol, which concludes the proof. □

The largest τ_i^j satisfying Theorem 8.1 is denoted $\bar{\tau}$ and represents the upper bound of stabilizing interbroadcasting intervals. Essentially, as long as $\tau_i^j \leq \bar{\tau}$, $i \in \mathbb{N}_0, j \in \{1, \ldots, N\}$, MAS stability is ensured. Owing to Remark 2.3, it follows that $\bar{\tau} > 0$. In other words, the unwanted Zeno behavior is avoided.

Corollary 8.1. *Suppose the conditions of Theorem 8.1 hold and ξ is $\mathcal{L}_p$-detectable from $(e, \omega, \tilde{\zeta})$ w.r.t. $\mathcal{B}$. Then the MAS (8.1)–(8.2) is $\mathcal{L}_p$-stable with bias w.r.t. $(\mathcal{B}, \mathbf{0}_{n_e})$ from ω to (ξ, e).*

Proof. Refer to the proof of Theorem 6.2. □

Notice that Remark 2.6 holds for Theorem 8.1 as well.

Remark 8.2. *We point out that we are not interested in stochastic differential equations or stochastic stability (see [89] or Chapter 5), but rather in $\mathcal{L}_p$-stability. Hence, even though we allow for stochastic disturbances ω herein, Theorem 8.1 holds solely for disturbance sample paths (i.e., trajectories or realizations) with finite $\mathcal{L}_p$-norm. Following [150, Section 8.2] and [32, Chapter 5], this sample path stability approach is congruent with our online LSPI method. Refer to Remark 8.4 for more regarding the relation between $\mathcal{L}_p$-stability and stochastic signals considered herein.*

8.3.2 (Sub)optimal Interbroadcasting Intervals

In order to utilize the results of Section 8.3.1, one needs $\mathcal{L}_p$ disturbances and $\mathcal{L}_\infty$ noise, which in turn excludes the Gaussian distribution. Since KF is the optimal linear filter for linear non-Gaussian systems [69], we focus on KF herein. In addition, even though nonlinear filters (such as KF banks [204] or particle filters) might outperform KF, their employment is not always justifiable due to higher computational and implementation complexity (on the top of the LSPI complexity, which is discussed in Remark 8.3). Based on the Cramér-Rao lower bound for linear non-Gaussian systems, a procedure for determining whether it is advantageous to employ nonlinear filters over KF is found in [69]. For instance, in order to facilitate the process of drawing samples from a distribution as well as to obtain $\mathcal{L}_p$ disturbances and $\mathcal{L}_\infty$ noise, Section 8.4 considers truncated Gaussian distributions [31] over the interval $[\mu - 10\sigma, \mu + 10\sigma]$, where μ and σ are the mean and standard deviation, respectively. Hence, it is legitimate to use KF in Section 8.4, rather than nonlinear filters. Of course, in case more general disturbances and noise distributions are encountered, one is free to employ nonlinear filters (preceded with pertaining theoretical analyses, of course), since our framework is not limited merely to KF.

In order to provide LSPI with up-to-date and more accurate (i.e., less noisy) values of x_j, $j \in \{1, \ldots, N\}$, in the aforementioned settings, we employ the concepts from [2]. Basically, each agent runs one KF for itself and one KF for each of its neighbors. Similar ideas are found in [20, Chapter 5] while handling partially observable DP problems. Results for inferring detectability and stabilizability of time-varying discrete systems, which are requirements for KF to work properly, are found in [3] and [46] (see Section 8.4 for more). In what follows, for notational convenience we drop the agent identifier j while solving (8.5). Recall that N LSPI algorithms are executed in parallel (one on each agent).

Adaptive optimal algorithms from the literature are typically based on solving the Bellman equation [17]. In the present work, we use an online version of LSPI, which interacts with the system at time instants t_i (i.e., iteration step i), $i \in \mathbb{N}_0$, through the following three variables $x(t_i), \tau(t_i)$ and $r(t_i)$ found in (8.5). The first variable (i.e., the LSPI state) is the local discrepancy vector $x(t_i) \in \mathcal{X}$, which represents the local agent knowledge at time t_i obtained via KF. The second variable is the LSPI action $\tau(t_i) \in \mathcal{A}$. The third variable is the LSPI reward function $r(t_i)$ defined in (8.5) and measures the local performance after taking the action $\tau(t_i)$ in state $x(t_i)$. In other words, we are solving the sequential decision-making problem (8.5) and LSPI aims to provide optimal actions $\tau(t_i)$ for each state $x(t_i)$.

Using policy iteration, policies $h_\kappa(x(t_i))$ are evaluated by iteratively constructing an estimate of the LSPI state and action value function called Q-function. The use of Q-functions alleviates the problem of computing the expectation in (8.5) as discussed in [150, Chapter 8.2]. Following [142], the LSPI state-action approximate value function is defined by

$$\hat{Q}(x(t_i), \tau(t_i)) = \Phi^\top(x(t_i), \tau(t_i))\alpha_\kappa, \qquad (8.14)$$

where

$$\Phi(x(t_i), \tau(t_i)) = \psi(\tau(t_i)) \otimes \phi(x(t_i))$$

is the Kronecker product of the basis function vectors $\psi(\tau(t_i))$ and $\phi(x(t_i))$ formed with Chebyshev polynomials while α_κ is the approximation parameter vector that is being learned. Clearly, since Chebyshev polynomials are defined on the interval $[-1, 1]$, the state space $\mathcal{X}$ needs to be compact. Observe that $\mathcal{A}$ is a compact set due to the upper bound $\bar{\tau}$. Consequently, a compact $\mathcal{X}$ yields finiteness of $r(t_i)$ so that the cost function in Problem 8.1 is well defined. The control action $\tau(t_i) \in \mathcal{A}$ is given by

$$\tau(t_i) = h_\kappa\big(x(t_i)\big),$$

where

$$h_\kappa\big(x(t_i)\big) = \begin{cases} \text{u.r.a.} \in \mathcal{A} & \text{every } \varepsilon \text{ iterations,} \\ h_\kappa\big(x(t_i)\big) & \text{otherwise,} \end{cases} \quad (8.15)$$

where "u.r.a." stands for "uniformly chosen random action" and yields the algorithm's exploration phase every ε steps while $h_\kappa(x(t_i))$ is the policy obtained according to

$$h_\kappa(x(t_i)) \in \arg\min_u \hat{Q}\big(x(t_i), \tau(t_i)\big). \quad (8.16)$$

Since LSPI solves optimal problems, the algorithm needs to have an exploration phase in order to avoid local minima and to cope with non-stationary cost-to-go functions [86]. The parameter vector α_κ is updated every $\kappa \geq 1$ steps from the projected Bellman equation for model-free policy iteration that, according to [97], can be written as

$$\Gamma_i \alpha_\kappa = \gamma \Lambda_i \alpha_\kappa + z_i, \quad (8.17)$$

where γ originates from (8.5) and

$$\begin{aligned} &\Gamma_0 = \beta_\Gamma I, \quad \Lambda_0 = \mathbf{0}, \quad z_0 = \mathbf{0}, \\ &\Gamma_i = \Gamma_{i-1} + \phi\big(x(t_i), \tau(t_i)\big)\phi\big(x(t_{i-1}), \tau(t_{i-1})\big)^\top, \\ &\Lambda_i = \Lambda_{i-1} + \phi\big(x(t_i), \tau(t_i)\big)\phi\big(x(t_i), h(x(t_{i+1}))\big)^\top, \\ &z_i = z_{i-1} + \phi\big(x(t_i), \tau(t_i)\big)r(t_i), \end{aligned} \quad (8.18)$$

where Γ_i, Λ_i and z_i represent the policy evaluation mapping and projection operator of the Bellman equation [32] and are updated at every iteration step i. The updated vector α_κ via (8.17) is used to improve the Q-function estimate (8.14). Afterward, new and improved policies (in the sense of Problem 8.1) are obtained from (8.16). The interested reader is referred to [142] and [97] for more details regarding the LSPI variant employed herein.

Remark 8.3. *Some of the benefits of LSPI include implementation simplicity due to the linear-in-parameter approximators, which in turn strips the algorithm down to a set of linear matrix equations. Accordingly, the principal* numerical concern lies in the need for matrix inversion, which is a well-understood nuisance circumvented, among others, via the LU factorization [54]. Of course, one can always additionally decrease the numerical complexity of LSPI by employing lower-order Chebyshev polynomials at the expense of greater suboptimality (while stability is still guaranteed in light of Section 8.3.1). Furthermore, the approximation in LSPI focuses on the value-function representation, thereby eliminating the "actor" part in the actor-critic framework of related approximate policy-iteration algorithms (see [86] and the references therein). Consequently, LSPI eliminates one potential source of inaccuracy and decreases computational complexity with respect to the actor-critic framework. As noted earlier, LSPI is a model-free algorithm, that is, it does not need a model for policy iteration nor does it need to learn one. Additionally, LSPI is able to learn (sub)optimal policies using a small number of samples when compared to conventional learning approaches such as the classical Q-learning. For a comprehensive discussion, refer to [97].

Remark 8.4. ($\mathcal{L}_p$-stability and stochastic inputs) KF requires the disturbances ω_i's to be Gaussian white processes. As the Gaussian white processes are not Lebesgue integrable, one cannot invoke Theorem 8.1 immediately. However, the Gaussian white process is an idealization convenient for a number of analytical purposes, but does not exist as a physical phenomenon (see [147, Chapter 6], [56, Chapter 10] and [204]). Instead, band-limited approximations/realizations are encountered in practice. For instance, all sample paths of ω_i's in Section 8.4 possess finite $\mathcal{L}_p$-norm owing to MATLAB® realizations of white noise (e.g., finite noise bandwidth and nonzero sampling rate). When it comes to prospective experimental verifications, such as those in Chapters 6 and 7, similar inferences could be drawn (refer to [56] and [204]).

Remark 8.5. The investigation regarding (sub)optimality of the obtained interbroadcasting intervals is the principal direction of our future work. While we already have reassuring results regarding (sub)optimality of LSPI in plant-controller settings, the (sub)optimality repercussions of the KF-LSPI and LSPI-LSPI interplay in MAS settings is still unclear. Basically, the generality of our MAS settings (e.g., partially observable and coupled DP problems with external disturbances) hinders a fruitful comprehensive investigation at the moment.

8.4 Example: Consensus Control (Revisited)

Consider a group of four agents with identical dynamics

$$\dot{\xi}_i = \begin{bmatrix} 0 & 1 \\ 0 & -T_p \end{bmatrix} \xi_i + \begin{bmatrix} 0 \\ K_p \end{bmatrix} u_i + \omega_i,$$
$$\zeta_i = \begin{bmatrix} 0.05 & 0.025 \end{bmatrix} \xi_i,$$

where $K_p = 5.2$ and $T_p = 0.38$, with the communication topology shown in Figure 7.2 and communication link delay $d = 0.104$ s. Thus, the corresponding graph Laplacian matrix is

$$L = \begin{bmatrix} 0 & 0 & 0 & 0 \\ -1 & 2 & -1 & 0 \\ 0 & -1 & 2 & -1 \\ -1 & -1 & -1 & 3 \end{bmatrix}.$$

Apparently, the first agent is the leader. We point out that the above agent dynamics is the result of AR.Drone Parrot quadcopter identification in the x-axis direction (see Chapter 7). Accordingly, the first component of ξ_i corresponds to the position of the ith agent while the second component corresponds to its velocity. The more intuitive notation for the first component of ξ_i is x_i while for the second component of ξ_i is v_i, which is the notation found in Figure 8.2. In (8.2), select $K_1 = \ldots = K_4 = 0.5$. Using Theorem 8.1, the resulting $\bar{\tau}$ equals 0.04 s. Apparently, the delay is greater than the broadcasting interval (i.e., the so-called large delay).

From [3, Corollary 3.4], we infer that τ_i^j, $j \in \{1, \ldots, N\}$, $i \in \mathbb{N}_0$, bounded away from zero, say by an arbitrary $\underline{\tau} > 0$, suffice to fulfill the detectability requirement for KF convergence. Basically, one can find K_k for any $\tau_i^j \in \mathcal{A} := [\underline{\tau}, \bar{\tau}]$ such that each state matrix norm in [3, Corollary 3.4] is upper bounded by the same constant strictly less than unity. We choose $\underline{\tau} = 10^{-5}$ s. In real-life applications, observe that a lower bound on τ_i^j intrinsically stems from hardware limitations. Next, owing to duality between detectability and stabilizability, the stabilizability requirement for KF convergence is obtained via [3, Corollary 3.4] as well.

The selected tuning parameters for LSPI are $\kappa = 2$, $\varepsilon = 50$, and the orders of Chebyshev polynomials for both $\mathcal{X}$ and $\mathcal{A}$ are 15, 15, 25 and 33 of the first, second, third and fourth agent, respectively. In addition, each component of x_j, $j \in \{1, \ldots, N\}$, belongs to the interval $[-30, 30]$, which in turn defines $\mathcal{X}$. The cost function parameters are $\gamma_1 = \ldots = \gamma_4 = 0.99$, $P_2 = P_3 = 5I_2$, $P_4 = 5I_3$, $R_1 = \ldots = R_4 = 5$ and $S_1 = \ldots = S_4 = 20$. Notice that, since the first agent is the leader, its local discrepancy vector x_1 is not defined; hence, P_1 does not exist. The output and input measurement noise adheres to the truncated zero-mean normal distribution with variance 0.3, which in turn renders $\nu_i(t), \upsilon_i(t) \in \mathcal{L}_\infty$. The external disturbances ω_i's also adhere to the truncated zero-mean normal distribution with covariance diag$(0, 1)$. Refer to [2] regarding continuous-time to discrete-time modeling transition required in order to employ KF. The interbroadcasting intervals are initialized to $\bar{\tau}$. The corresponding numerical results with the agents' initial conditions $(5,5)$, $(-3,-3)$, $(10,10)$ and $(4,4)$ are provided in Figure 8.2. It is worth mentioning that the execution time of one LSPI iteration with the above parameters is less than 1 ms performed on a computer with an i7-3770@3.4GHz processor and 8GB of memory.

A remark is in order regarding the choice of P_j, R_j and S_j. On the one hand, as we decrease S_j and keep P_j and R_j fixed, the obtained interbroadcasting intervals τ_i^j approach $\underline{\tau}$. On the other hand, as S_j becomes greater, τ_i^j approach $\bar{\tau}$. Notice that this remark is expected and underlines the principal motivation for this work—overly sparse information exchange is not preferable a priori and needs to be applied thoughtfully, especially when the relative cost of communication with respect to the cost of control performance (e.g., stabilization or tracking performance) and control effort (e.g., movement) is low. However, the related literature does not take into account this cost viewpoint, but solely focuses on minimizing the communication effort.

Recall that each agent learns its own approximation parameter vector α_κ^j, which in turn renders the corresponding policy $h_\kappa^j(x_j(t_i^j))$ and actions τ_i^j's. Therefore, convergence of α_κ^j's indicates a successful completion of the learning process (see Figure 8.3), while τ_i^j's do not necessarily display smooth behaviors (see Figure 8.2(c)). Also, keep in mind that the exploration phase takes place every ε steps. Furthermore, observe that α_κ^j's are different for each agent even though homogeneous agents are encountered (which is the motivation behind utilizing homogenous agents in this numerical example). The evidence that the underlying cost-to-go is non-stationary [86] is provided in Figure 8.3 (cf. [141, Figure 9]). Basically, α_κ^j adapts to changes in the underlying nonautonomous dynamics.

8.5 Conclusions and Perspectives

This chapter studies the optimal intermittent feedback problem in MASs with respect to a goal function that captures local MAS performance vs. agent lifetime trade-offs. Starting off with provably stabilizing upper-bounds on agents' interbroadcasting intervals, we bring together KF and an online model-free LSPI method to tackle coupled partially observable DP problems. Our framework is rather general since it is applicable to general heterogeneous linear agents with output feedback in the presence of disturbances, to realistic communication environments (e.g., intermittency, noise, delays and dropouts) as well as to directed and unbalanced communication topologies. Lastly, a consensus control problem is provided to demonstrate the benefits of the proposed methodology.

210 *Networked Control Systems with Intermittent Feedback*

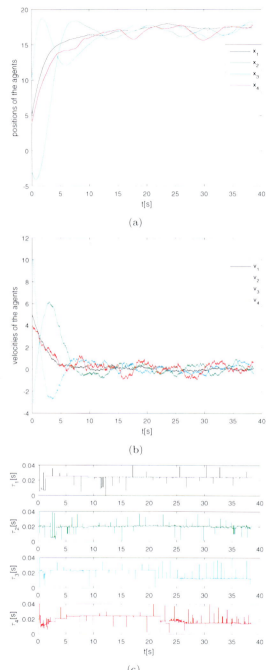

FIGURE 8.2
Numerically obtained data: (a) agents' positions, (b) agents' velocities, and (c) the corresponding (sub)optimal interbroadcasting intervals for each agent.

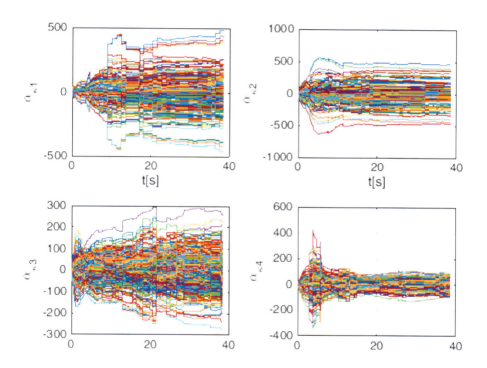

FIGURE 8.3
The process of learning α_κ for each agent.

In the future, the main goal is to estimate how suboptimal the methodology developed in this chapter is. In addition, it is of interest to investigate time-varying topologies and nonlinear agent dynamics. Lastly, experimental verification of the presented theory is in order.

Bibliography

[1] I. F. Akyildiz, W. Su, Y. Sankarasubramaniam, and E. Cayirci. Wireless sensor networks: A survey. *Computer Networks*, 38(4):393–422, 2002.

[2] H. L. Alexander. State estimation for distributed systems with sensing delay. In *SPIE Proceedings Vol. 1470. Data Structures and Target Classification*, pages 103–111, 1991.

[3] B. D. O. Anderson and J. B. Moore. Detectability and stabilizability of time-varying discrete-time linear systems. *SIAM Journal on Control and Optimization*, 19(1):20–32, 1981.

[4] A. Anokhin, L. Berezansky, and E. Braverman. Exponential stability of linear delay impulsive differential equations. *Journal of Mathematical Analysis and Applications*, 193(3):923–941, 1995.

[5] A. Anta and P. Tabuada. Isochronous manifolds in self-triggered control. In *Proc. IEEE Conference for Decision and Control*, pages 3194–3199, Shanghai, China, December 2009.

[6] A. Anta and P. Tabuada. To sample or not to sample: Self-triggered control for nonlinear systems. *IEEE Trans. on Automatic Control*, 55(9):2030–2042, 2010.

[7] D. Antunes, W. P. M. H. Heemels, and P. Tabuada. Dynamic programming formulation of periodic event-triggered control: Performance guarantees and co-design. In *Decision and Control (CDC), 2012 IEEE 51st Annual Conference on*, pages 7212–7217, 2012.

[8] D. Antunes, J. Hespanha, and C. Silvestre. Control of impulsive renewal systems: Application to direct design in networked control. In *Decision and Control, 2009 Held Jointly with the 2009 28th Chinese Control Conference. CDC/CCC 2009. Proceedings of the 48th IEEE Conference on*, pages 6882–6887, December 2009.

[9] J. Araújo, M. Mazo, A. Anta, P. Tabuada, and K. H. Johansson. System architectures, protocols and algorithms for aperiodic wireless control systems. *IEEE Transactions on Industrial Informatics*, 10(1):175–184, 2014.

[10] K. E. Arzén. A simple event-based PID controller. In *Preprints IFAC World Conf.*, volume 18, pages 423–428, 1999.

[11] K. Åström and B. Bernhardsson. Comparison of Riemann and Lebesgue sampling for first order stochastic systems. In *Proceedings of the 41st IEEE Conference on Decision and Control*, pages 2011–2016, December 2002.

[12] D. Bainov and P. Simeonov. *Systems with Impulse Effects: Stability, Theory and Applications*. Ellis Horwood Limited, 1989.

[13] G. H. Ballinger. *Qualitative Theory of Impulsive Delay Differential Equations*. PhD thesis, Univ. of Waterloo, Canada, 1999.

[14] L. Bao and J. Garcia-Luna-Acaves. Link-state routing in networks with unidirectional links. In *Eight International Conference in Computer Communications and Networks*, pages 358–363, 1999.

[15] R. G. Bartle. *The Elements of Integration and Lebesgue Measure*. Wiley-Interscience, 1st edition, 1995.

[16] G. Bekey and R. Tomović. On the optimum sampling rate for discrete-time modeling of continuous-time systems. *IEEE Transactions on Automatic Control*, 11(2):284–287, 1966.

[17] R. E. Bellman. *Dynamic Programming*. Princeton University Press, Princeton, New Jersey, 1957.

[18] A. Bemporad, M. Heemels, and M. Vejdemo-Johansson. *Networked Control Systems*, volume 406 of *Lecture Notes in Control and Information Sciences*. Springer-Verlag London, 2010.

[19] S. C. Bengea and R. A. DeCarlo. Optimal control of switching systems. *Automatica*, 41(1):11–27, 2005.

[20] D. P. Bertsekas. *Dynamic Programming and Optimal Control, Vol. I*. Athena Scientific, Belmont, Massachusetts, 3rd edition, 2005.

[21] D. P. Bertsekas. *Dynamic Programming and Optimal Control, Vol. II*. Athena Scientific, Belmont, Massachusetts, 3rd edition, 2007.

[22] D. P. Bertsekas and J. N. Tsitsiklis. *Neuro-Dynamic Programming*. Athena Scientific, Belmont, Massachusetts, 1996.

[23] B. M. Bethke. *Kernel-Based Approximate Dynamic Programming Using Bellman Residual Elimination*. PhD thesis, MIT, Cambridge, Massachusetts, February 2010.

[24] R. Bhatia. *Matrix Analysis*. Springer-Verlag, New York, 1997.

[25] R. Blind and F. Allgöwer. Analysis of Networked Event-Based Control with a Shared Communication Medium: Part I—Pure ALOHA. In *IFAC World Congress*, 2011.

[26] R. Blind and F. Allgöwer. Analysis of Networked Event-Based Control with a Shared Communication Medium: Part II—Slotted ALOHA. In *IFAC World Congress*, 2011.

[27] R. Blind and F. Allgöwer. On the optimal sending rate for networked control systems with a shared communication medium. In *Proc. 50th IEEE Conference on Decision and Control (CDC'11)*, 2011.

[28] D. P. Borgers and W. P. M. H. Heemels. Stability analysis of large-scale networked control systems with local networks: A hybrid small-gain approach. In *Hybrid Systems: Computation and Control (HSCC)*, pages 103–112. ACM, 2014.

[29] S. Boyd, L. El-Ghaoui, E. Feron, and V. Balakrishnan. *Linear Matrix Inequalities in Systems and Control Theory*. SIAM, Philadelphia, 1994.

[30] S. Boyd and L. Vandenberghe. *Convex Optimization*. Cambridge University Press, 2004.

[31] J. M. Brankart. *Optimal Nonlinear Niltering and Smoothing Assuming Truncated Gaussian Probability Distributions in a Reduced Dimension Space of Adaptive Size*. technical report, MEOM-LEGI, Grenoble, 2006.

[32] L. Busoniu, R. Babuska, B. D. Schutter, and D. Ernst. *Reinforcement Learning and Dynamic Programming Using Function Approximators*. Automation and Control Engineering Series. CRC Press, 2010.

[33] A. Cervin and T. Henningsson. Scheduling of event-triggered controllers on a shared network. In *IEEE Conference on Decision and Control*, pages 3601–3606, December 2008.

[34] A. Cervin and T. Henningsson. A simple model for the interference between event-based control loops using a shared medium. In *IEEE Conference on Decision and Control*, pages 3240–3245, Atlanta, GA, 2010.

[35] D. Christmann, R. Gotzhein, S. Siegmund, and F. Wirth. Realization of Try-Once-Discard in wireless multihop networks. *IEEE Transactions on Industrial Informatics*, 10(1):17–26, 2014.

[36] D. Ciscato and L. Martiani. On increasing sampling efficiency by adaptive sampling. *IEEE Transactions on Automatic Control*, 12(3):318, 1967.

[37] R. Cogill, S. Lall, and J. Hespanha. A constant factor approximation algorithm for event-based sampling. In *American Control Conference, 2007. ACC '07*, pages 305–311, July 2007.

[38] D. F. Coutinho and C. E. de Souza. Delay-dependent robust stability and $\mathcal{L}_2$-gain analysis of a class of nonlinear time-delay systems. *Automatica*, 44(8):2006–2018, 2008.

[39] P. J. Davis and P. Rabinowitz. *Methods of Numerical Integration*. Dover Publications, Inc., Mineola, NY, 2nd edition, 2007.

[40] C. C. de Wit, H. Khennouf, C. Samson, and O. Sordalen. *Nonlinear control design for mobile robots*. World Scientific Publishing, Singapore, 1993.

[41] R. S. Dilmaghani, H. Bobarshad, M. Ghavami, S. Choobkar, and C. Wolfe. Wireless sensor networks for monitoring physiological signals of multiple patients. *IEEE Trans. on Biomedical Circuits and Systems*, 5(4):347–356, 2011.

[42] D. Dimarogonas and K. Johansson. Event-triggered control for multi-agent systems. In *Proceedings of the IEEE Conference on Decision and Control Held Jointly with the Chinese Control Conference*, pages 7131–7136, December 2009.

[43] D. V. Dimarogonas, E. Frazzoli, and K. H. Johansson. Distributed event-triggered control for multi-agent systems. *IEEE Trans. on Automatic Control*, 57(5):1291–1297, 2012.

[44] V. S. Dolk, D. P. Borgers, and W. P. M. H. Heemels. Dynamic event-triggered control: Tradeoffs between transmission intervals and performance. In *Proceedings of the IEEE Conference on Decision and Control*, pages 2764–2769, Los Angeles, CA, December 2014.

[45] A. Doucet and A. Johansen. *A Tutorial on Particle Filtering and Smoothing: Fifteen Years Later*. Technical report, Department of Statistics, University of British Columbia, December 2008.

[46] J. C. Engwerda. Stabilizability and detectability of discrete-time time-varying systems. *IEEE Trans. on Automatic Control*, 35(4):425–429, 1990.

[47] T. Estrada and P. J. Antsaklis. Stability of model-based networked control systems with intermittent feedback. In *Proceedings of the 17th IFAC World Congress on Automatic Control*, pages 12581–12586, July 2008.

[48] T. Estrada and P. J. Antsaklis. Model-based control with intermittent feedback: Bridging the gap between continuous and instantaneous feedback. *Int. J. of Control*, 83(12):2588–2605, 2010.

[49] L. Fang and P. J. Antsaklis. On communication requirements for multi-agent consensus seeking. In P. Antsaklis and P. Tabuada, editors,

Networked Embedded Sensing and Control, Proceedings of Workshop NESC05: University of Notre Dame, USA, October 17–18, 2005, volume 331 of *Lecture Notes in Control and Information Sciences*, pages 133–147. Springer, 2006.

[50] M. Franceschelli, A. Gasparri, A. Giua, and G. Ulivi. Decentralized stabilization of heterogeneous linear multi-agent systems. In *IEEE Int. Conf. on Robotics and Automation*, pages 3556–3561, May 2010.

[51] R. A. Freeman. On the relationship between induced l^∞-gain and various notions of dissipativity. In *Proceedings of the IEEE Conference on Decision and Control*, pages 3430–3434, 2004.

[52] E. Fridman. *Introduction to Time-Delay Systems: Analysis and Control*. Systems & Control: Foundations & Applications. Springer International Publishing, 2014.

[53] R. Goebel, R. G. Sanfelice, and A. R. Teel. Hybrid dynamical systems. *IEEE Control Systems Magazine*, 29(2):28–93, 2009.

[54] G. H. Golub and C. F. V. Loan. *Matrix Computations*. John Hopkins University Press, Baltimore and London, 3rd edition, 1996.

[55] G. J. Gordon. *Stable Function Approximation in Dynamic Programming*. Technical report, School of Computer Science, Carnegie Mellon University, 1995.

[56] J. A. Gubner. *Probability and Random Processes for Electrical and Computer Engineers*. Cambridge University Press, New York, 2006.

[57] M. Guinaldo, D. V. Dimarogonas, K. H. Johansson, J. S. Moreno, and S. Dormido. Distributed event-based control strategies for interconnected linear systems. *IET Control Theory & Applications*, 7(6):877–886, 2013.

[58] S. Gupta. Increasing the sampling efficiency for a control system. *IEEE Transactions on Automatic Control*, 8(3):263–264, 1963.

[59] V. Gupta, A. Dana, J. Hespanha, R. Murray, and B. Hassibi. Data transmission over networks for estimation and control. *Automatic Control, IEEE Transactions on*, 54(8):1807–1819, 2009.

[60] J. Gutiérrez-Gutiérrez and P. Crespo. Asymptotically equivalent sequences of matrices and Hermitian block Toeplitz matrices with continuous symbols: Applications to MIMO systems. *IEEE Trans. on Information Theory*, 54(12):5671–5680, 2008.

[61] J. K. Hale. *Theory of Functional Differential Equations*, volume 3 of *Applied Mathematical Sciences*. Springer-Verlag, New York, 1977.

[62] A. Haug. *A Tutorial on Bayesian Estimation and Tracking Techniques Applicable to Nonlinear and Non-Gaussian Processes*. Technical report, Coorporation MITRE, January 2005.

[63] S. Haykin. *Neural Networks and Learning Machines*. Prentice Hall, 3rd edition, 2008.

[64] S. Hedlund and A. Rantzer. Optimal control of hybrid systems. In *Conference on Decision and Control*, pages 3972–3977, 1999.

[65] W. Heemels, J. H. Sandee, and P. V. D. Bosch. Analysis of event-driven controllers for linear systems. *International Journal of Control*, 81(4):571–590, 2008.

[66] W. P. M. H. Heemels, M. C. F. Donkers, and A. Teel. Periodic event-triggered control for linear systems. *IEEE Transactions on Automatic Control*, 58(4):847–861, 2013.

[67] W. P. M. H. Heemels, K. H. Johansson, and P. Tabuada. An introduction to event-triggered and self-triggered control. In *IEEE Conference on Decision and Control*, pages 3270–3285, Maui, Hawaii, 2012.

[68] W. P. M. H. Heemels, A. R. Teel, N. V. de Wouw, and D. Nešić. Networked Control Systems with communication constraints: Tradeoffs between transmission intervals, delays and performance. *IEEE Tran. on Automatic Control*, 55(8):1781–1796, 2010.

[69] G. Hendeby. *Fundamental Estimation and Detection Limits in Linear Non-Gaussian Systems*. PhD thesis, Linköping University, Sweden, 2005.

[70] T. Henningsson and A. Cervin. Scheduling of event-triggered controllers on a shared network. In *Proc. of IEEE Conference on Decision and Control*, pages 3601–3606, Cancun, Mexico, 2008.

[71] T. Henningsson, E. Johannesson, and A. Cervin. Sporadic event-based control of first-order linear stochastic systems. *Automatica*, 44(11):2890–2895, 2008.

[72] O. Hernandez Lerma and J. Lasserre. Further criteria for positive Harris recurrence of Markov chains. *Proceedings of the American Mathematical Society*, 129(5):1521–1524, 2001.

[73] J. Hespanha, P. Naghshtabrizi, and X. Yonggang. A survey of recent results in Networked Control Systems. *Proceedings of the IEEE*, 95(1):138–162, 2007.

[74] J. P. Hespanha and A. S. Morse. Stability of switched systems with average dwell-time. In *IEEE Conference on Decision and Control*, pages 2655–2660, December 1999.

[75] S. Hirche, C. Chen, and M. Buss. Performance oriented control over networks: Switching controllers and switched time delay. *Asian Journal of Control*, 10(1):24–33, 2008.

[76] S. Hirche, T. Matiakis, and M. Buss. A distributed controller approach for delay-independent stability of networked control systems. *Automatica*, 45(5):1828–1836, 2009.

[77] H. Hjalmarsson. From experiment design to closed-loop control. *Automatica*, 41(3):393–438, 2005.

[78] J. Imae. L2-gain computation for nonlinear systems using optimal control algorithms. In *Proceedings of the IEEE Conference on Decision and Control*, pages 547–551, Kobe, Japan, December 1996.

[79] J. Imae and G. Wanyoike. H_∞ norm computation for LTV systems using nonlinear optimal control algorithms. *International Journal of Control*, 63(1):161–182, 1996.

[80] O. Imer. *Optimal Estimation and Control under Communication Network Constraints*. PhD thesis, University of Illinois at Urbana-Champaign, Illinois, 2005.

[81] M. James and J. Baras. Robust H_∞ output feedback control for nonlinear systems. *IEEE Trans. on Automatic Control*, 40(6):1007–1017, 1995.

[82] V. Jeličić, M. Magno, D. Brunelli, V. Bilas, and L. Benini. An energy efficient multimodal wireless video sensor network with eZ430-RF2500 modules. In *Proc. 5th Internat. Conf. on Pervasive Computing and Appl. (ICPCA)*, pages 161–166, 2010.

[83] V. Jeličić, D. Tolić, and V. Bilas. Consensus-based decentralized resource sharing between co-located wireless sensor networks. In *IEEE Ninth International Conference on Intelligent Sensors, Sensor Networks and Information Processing (ISSNIP)*, April 2014.

[84] Z. Jiang, A. R. Teel, and L. Praly. Small-gain theorem for ISS systems and applications. *Mathematics of Control, Signals and Systems*, 7:95–120, 1994.

[85] W. D. Jones. Keeping cars from crashing. *IEEE Spectrum*, 38(9):40–45, 2001.

[86] R. Kamalapurkar, H. Dinh, P. Walters, and W. E. Dixon. Approximate optimal cooperative decentralized control for consensus in a topological network of agents with uncertain nonlinear dynamics. In *Proceedings of the American Control Conference*, pages 1320–1325, 2013.

[87] C.-Y. Kao and A. Rantzer. Stability analysis of systems with uncertain time-varying delays. *Automatica*, 43(6):959–970, 2007.

[88] H. Khalil. *Nonlinear Systems*. Prentice Hall, 3rd edition, 2002.

[89] R. Khasminskii and G. N. Milstein. *Stochastic Stability of Differential Equations*. Stochastic Modelling and Applied Probability. Springer Berlin Heidelberg, 2011.

[90] W.-J. Kim, K. Ji, and A. Ambike. Real-time operating environment for networked control systems. *IEEE Transactions on Automation Science and Engineering*, 3(3):287–296, 2006.

[91] D. B. Kingston, W. Ren, and R. W. Beard. Consensus algorithms are input-to-state stable. In *Proceedings of the American Control Conference*, pages 1686–1690, 2005.

[92] J. Klotz, R. Kamalapurkar, and W. E. Dixon. Concurrent learning-based network synchronization. In *Proceedings of the American Control Conference*, pages 796–801, 2014.

[93] G. Kortuem, F. Kawsar, D. Fitton, and V. Sundramoorthy. Smart objects as building blocks for the internet of things. *Internet Computing, IEEE*, 14(1):44–51, 2010.

[94] F. Kozin. A survey of stability of stochastic systems. *Automatica*, 5(1):95–112, 1969.

[95] A. Kruszewski, W. J. Jiang, E. Fridman, J. P. Richard, and A. Toguyeni. A switched system approach to exponential stabilization through communication network. *IEEE Trans. on Automatic Control*, 20(4):887–900, 2012.

[96] W. H. Kwon, Y. S. Moon, and S. C. Ahn. Bounds in Algebraic Riccati and Lyapunov Equations: A Survey and Some New Results. *International Journal of Control*, 64(3):377–389, 1996.

[97] M. G. Lagoudakis and R. Parr. Least-squares policy iteration. *Journal of Machine Learning Research*, 4:1107–1149, 2003.

[98] I. D. Landau and Z. Gianluca. *Digital Control Systems: Design, Identification and Implementation*. Communications and Control Engineering. Springer-Verlag, London, 2006.

[99] P. Le-Huy and S. Roy. Low-power wake-up radio for wireless sensor networks. *Mobile Networks and Applications*, 15(2):226–236, 2010.

[100] J. M. Lee, N. S. Kaisare, and J. H. Lee. Choice of approximator and design of penalty function for an approximate dynamic programming based control approach. *Journal of Process Control*, 16(2):135–156, 2006.

Bibliography

[101] D. Lehmann and J. Lunze. Event-based output-feedback control. In *Control Automation (MED), 2011 19th Mediterranean Conference on*, pages 982–987, June 2011.

[102] D. Lehmann, J. Lunze, and K. H. Johansson. Comparison between sampled-data control, deadband control and model-based event-triggered control. In *4th IFAC Conference on Analysis and Design of Hybrid Systems*, pages 7–12, Eindhoven, Netherlands, 2012.

[103] M. Lemmon. *Event-triggered Feedback in Control, Estimation, and Optimization*, volume 405 of *Lecture Notes in Control and Information Sciences*. Springer Verlag, 2010.

[104] C. Li, G. Feng, and X. Liao. Stabilization of nonlinear systems via periodically intermittent control. *IEEE Transactions on Circuits and Systems, II: Express Briefs*, 54(11):1019–1023, 2007.

[105] D. Liberzon. *Switching in Systems and Control*. Birkhauser, Boston, 2003.

[106] A. Liff and J. Wolf. On the optimum sampling rate for discrete-time modeling of continuous-time systems. *IEEE Transactions on Automatic Control*, 11(2):288–290, 1966.

[107] L. Lin. Self-improving reactive agents based on reinforcement learning, planning and teaching. *Machine Learning*, 8(3–4):293–322, 1992.

[108] Y. Lin, E. D. Sontag, and Y. Wang. A smooth converse Lyapunov theorem for robust stability. *SIAM Journal on Control and Optimization*, 34:124–160, 1996.

[109] X. Liu, X. Shen, and Y. Zhang. A comparison principle and stability for large-scale impulsive delay differential systems. *Anziam Journal: The Australian & New Zealand Industrial and Applied Mathematics Journal*, 47(2):203–235, 2005.

[110] Y. Liu and Y. Jia. H_∞ consensus control for multi-agent systems with linear coupling dynamics and communication delays. *International Journal of Systems Science*, 43(1):50–62, 2012.

[111] S. Longo, T. Su, G. Herrmann, and P. Barber. *Optimal and Robust Scheduling for Networked Control Systems*. Automation and Control Engineering. Taylor & Francis, 2013.

[112] J. Lunze and D. Lehmann. A state-feedback approach to event-based control. *Automatica*, 46(1):211–215, 2010.

[113] M. H. Mamduhi, M. Kneissl, and S. Hirche. A Decentralized Event-triggered MAC for Output Feedback Networked Control Systems. In *Decision and Control, 2016, Proceedings of the 55th IEEE Conference on*, 2016.

[114] M. H. Mamduhi, A. Molin, and S. Hirche. On the stability of prioritized error-based scheduling for resource-constrained networked control systems. In *Distributed Estimation and Control in Networked Systems (NecSys), 4th IFAC Workshop on*, pages 356–362, 2013.

[115] M. H. Mamduhi, D. Tolić, and S. Hirche. Decentralized event-based scheduling for shared-resource networked control systems. In *14th European Control Conf.*, pages 941–947, 2015.

[116] M. H. Mamduhi, D. Tolić, and S. Hirche. Robust event-based data scheduling for resource constrained networked control systems. In *IEEE American Control Conf.*, pages 4695–4701, 2015.

[117] M. H. Mamduhi, D. Tolić, A. Molin, and S. Hirche. Event-triggered scheduling for stochastic multi-loop networked control systems with packet dropouts. In *Proceedings of the IEEE Conference on Decision and Control*, pages 2776–2782, Los Angeles, CA, December 2014.

[118] N. Martins. Finite gain $\mathcal{L}_p$ stability requires analog control. *Systems and Control Letters*, 55(11):949–954, 2006.

[119] F. Mazenc, M. Malisoff, and T. N. Dinh. Robustness of nonlinear systems with respect to delay and sampling of the controls. *Automatica*, 49(6):1925–1931, 2013.

[120] M. Mazo and P. Tabuada. On event-triggered and self-triggered control over sensor/actuator networks. In *Decision and Control, 2008. CDC 2008. 47th IEEE Conference on*, pages 435–440, December 2008.

[121] A. D. McKernan and G. W. Irwin. Event-based sampling for wireless network control systems with QoS. In *American Control Conference (ACC), 2010*, pages 1841–1846, June 2010.

[122] C. D. Meyer. *Matrix Analysis and Applied Linear Algebra*. SIAM, 2000.

[123] S. Meyn and R. Tweedie. *Markov Chains and Stochastic Stability*. Springer London, 1996.

[124] W. Michiels and S. Gumussoy. Characterization and computation of H-infinity norms of time-delay systems. *IEEE Transactions on Automatic Control*, 31(4):2093–2115, 2010.

[125] D. Miorandi, S. Sicari, F. D. Pellegrini, and I. Chlamtac. Internet of things: Vision, applications and research challenges. *Ad Hoc Networks*, 10(7):1497–1516, 2012.

[126] J. Mitchell and J. McDaniel. Adaptive sampling technique. *IEEE Transactions on Automatic Control*, 14(2):200–201, 1969.

[127] A. Molin and S. Hirche. Optimal event-triggered control under costly observations. *Proceedings of the 19th International Symposium on Mathematical Theory of Networks and Systems*, 2010.

[128] A. Molin and S. Hirche. A bi-level approach for the design of event-triggered control systems over a shared network. *Discrete Event Dynamic Systems*, 24(2):153–171, 2014.

[129] A. Molin and S. Hirche. Price-based adaptive scheduling in multi-loop control systems with resource constraints. *Automatic Control, IEEE Trans. on*, 59(12):3282–3295, 2014.

[130] R. Murray, K. Astrom, S. Boyd, R. Brockett, and G. Stein. Future directions in control in an information-rich world. *Control Systems, IEEE*, 23(2):20–33, 2003.

[131] D. Nešić and A. R. Teel. Input-output stability properties of Networked Control Systems. *IEEE Transactions on Automatic Control*, 49(10):1650–1667, October 2004.

[132] D. Nešić, A. R. Teel, and P. V. Kokotović. Sufficient conditions for stabilization of sampled-data nonlinear systems via discrete-time approximations. *Systems and Control Letters*, 38(4-5):259–270, 1999.

[133] S.-I. Niculescu. *Delay Effects on Stability: A Robust Control Approach*. Number 269 in *Lecture Notes in Control and Information Sciences*. Springer-Verlag, London, 2001.

[134] C. Nowzari and J. Cortés. Self-triggered coordination of robotic networks for optimal deployment. In *Proceedings of the American Control Conference*, pages 1039–1044, San Francisco, CA, June–July 2011.

[135] C. Nowzari and J. Cortés. Team-triggered coordination of networked systems. In *Proceedings of the American Control Conference*, pages 3827–3832, June 2013.

[136] R. Obermaisser. *Event-Triggered and Time-Triggered Control Paradigms*, volume 22 of *Real-Time Systems Series*. Springer, 2005.

[137] R. Obermaisser. *Time-Triggered Communication*. CRC Press, 2011.

[138] B. O'Donoghue, Y. Wang, and S. Boyd. Min-max approximate dynamic programming. In *IEEE Multi-Conference on Systems and Control*, pages 424–431, Denver, CO, September 2011.

[139] R. Olfati-Saber and R. M. Murray. Consensus problems in networks of agents with switching topology and time-delays. *IEEE Trans. on Automatic Control*, 49(9):1520–1533, 2004.

[140] OptiTrack. https://www.naturalpoint.com/optitrack/, 2014.

[141] I. Palunko, P. Donner, M. Buss, and S. Hirche. Cooperative suspended object manipulation using reinforcement learning and energy-based control. In *IEEE/RSJ Int. Conf. on Intelligent Robots and Systems*, pages 885–891, 2014.

[142] I. Palunko, A. Faust, P. Cruz, L. Tapia, and R. Fierro. A reinforcement learning approach towards autonomous suspended load manipulation using aerial robots. In *IEEE Int. Conf. on Robotics and Automation*, pages 4896–4901, 2013.

[143] C. D. Persis and P. Frasca. Robust self-triggered coordination with ternary controllers. *IEEE Trans. on Automatic Control*, 58(12):3024–3038, 2013.

[144] C. D. Persis, R. Sailer, and F. Wirth. Parsimonious event-triggered distributed control: A zeno free approach. *Automatica*, 49(7):2116–2124, 2013.

[145] G. Piazza and T. Politi. An upper bound for the condition number of a matrix in spectral norm. *Journal of Computational and Appl. Math.*, 143(1):141–144, 2002.

[146] S. Piskorski and N. Brulez. *AR.Drone Developer Guide*. Technical report, Parrot, February 2011.

[147] H. V. Poor. *An Introduction to Signal Detection and Estimation*. Springer-Verlag, New York, 2nd edition, 1994.

[148] R. Poovendran. Cyber-physical systems: Close encounters between two worlds [point of view]. *Proceedings of the IEEE*, 98(8):1363–1366, 2010.

[149] R. Postoyan, M. C. Bragagnolo, E. Galbrun, J. Daafouz, D. Nešić, and E. B. Castelan. Nonlinear event-triggered tracking control of a mobile robot: design, analysis and experimental results. In *NOLCOS (IFAC Symposium on Nonlinear Control), Invited Paper*, pages 318–323, 2013.

[150] W. B. Powell. *Approximate Dynamic Programming: Solving the Curses of Dimensionality*. Wiley Series in Probability and Statistics. John Wiley and Sons, Inc., Hoboken, NJ, 2007.

[151] M. Quigley, K. Conley, B. P. Gerkey, J. Faust, T. Foote, J. Leibs, R. Wheeler, and A. Y. Ng. ROS: An open-source Robot Operating System. In *ICRA Workshop on Open Source Software*, 2009.

[152] M. Rabi and K. Johansson. Scheduling packets for event-triggered control. In *Proc. of 10th European Control Conf*, pages 3779–3784, 2009.

[153] M. Rabi, G. V. Moustakides, and J. S. Baras. Adaptive sampling for linear state estimation. *SIAM J. Control Optim.*, 50(2):672–702, 2012.

[154] V. Raghunathan, C. Schurgers, S. Park, and M. Srivastava. Energy-aware wireless microsensor networks. *IEEE Signal Processing Magazine*, 19(2):40–50, 2002.

[155] C. Ramesh, H. Sandberg, and K. Johansson. Stability analysis of multiple state-based schedulers with CSMA. In *Decision and Control (CDC), 2012 IEEE 51st Annual Conference on*, pages 7205–7211, December 2012.

[156] C. Ramesh, H. Sandberg, and K. H. Johansson. Design of state-based schedulers for a network of control loops. *IEEE Transactions on Automatic Control*, 58(8):1962–1975, 2013.

[157] K. J. Åström and B. Bernhardsson. Comparison of periodic and event based sampling for first order stochastic systems. In *Proceedings of IFAC World Conf.*, pages 301–306, 1999.

[158] K. J. Åström and B. Wittenmark. *Computer Controlled Systems*. Prentice Hall, Englewood Cliffs, NJ, 1990.

[159] W. Ren and R. W. Beard. *Distributed Consensus in Multi-Vehicle Cooperative Control: Theory and Applications*. Communications and Control Engineering. Springer, London, 2008.

[160] W. J. Rugh. *Linear System Theory*. Prentice Hall, Englewood Cliffs, NJ, 2nd edition, 1996.

[161] A. Samuels. Some studies in machine learning using the game of checkers. *IBM Journal of Research and Development*, 3(3):210–229, 1959.

[162] D. Sauter, M. Sid, S. Aberkane, and D. Maquin. Co-design of safe networked control systems. *Annual Reviews in Control*, 37(2):321–332, 2013.

[163] T. Schmid, Z. Charbiwala, J. Friedman, Y. H. Cho, and M. B. Srivastava. Exploiting manufacturing variations for compensating environment-induced clock drift in time synchronization. In *Proc. ACM International conference on Measurement and modeling of computer systems (SIGMETRICS)*, pages 97–108, 2008.

[164] H. Smith. *An Introduction to Delay Differential Equations with Applications to the Life Sciences*, volume 57 of *Texts in Applied Mathematics*. Springer, New York, 2011.

[165] E. D. Sontag. Comments on integral variants of ISS. *Systems & Control Letters*, 34(1–2):93–100, 1998.

[166] R. Sutton and A. Barto. *Reinforcement Learning*. The MIT Press, Cambridge, Massachusetts, 1998.

[167] I. Szita. *Rewarding Excursions: Extending Reinforcement Learning to Complex Domains*. PhD thesis, Eötvös Loránd University, Budapest, Hungary, March 2007.

[168] M. Tabbara, D. Nešić, and A. R. Teel. Stability of wireless and wireline networked control systems. *IEEE Transactions on Automatic Control*, 52(9):1615–1630, 2007.

[169] P. Tabuada. Event-triggered real-time scheduling of stabilizing control tasks. *IEEE Trans. on Automatic Control*, 52(9):1680–1685, 2007.

[170] P. Tallapragada and N. Chopra. On event triggered trajectory tracking for control affine nonlinear systems. In *Proceedings of the IEEE Conference on Decision and Control*, pages 5377–5382, December 2011.

[171] A. R. Teel. Asymptotic convergence from $\mathcal{L}_p$ stability. *IEEE Transactions on Automatic Control*, 44(11):2169–2170, 1999.

[172] A. R. Teel, E. Panteley, and A. Loría. Integral characterizations of uniform asymptotic and exponential stability with applications. *Mathematics of Control, Signals and Systems*, 15(3):177–201, 2002.

[173] G. Tesauro. Neurogammon: a neural network backgammon program. In *IJNN Proceedings III*, pages 33–39, 1990.

[174] Texas Instruments. *eZ430-RF2500 Development Tool User's Guide*, 2009.

[175] D. Tolić. $\mathcal{L}_p$-stability with respect to sets applied towards self-triggered communication for single-integrator consensus. In *Proc. IEEE Conf. on Decision and Control*, pages 3409–3414, December 2013.

[176] D. Tolić and R. Fierro. *A Comparison of a Curve Fitting Tracking Filter and Conventional Filters under Intermittent Information*. Technical report, Department of Electrical and Computer Engineering, University of New Mexico, October 2010.

[177] D. Tolić and R. Fierro. Adaptive sampling for tracking in pursuit-evasion games. In *IEEE Multi-Conference on Systems and Control*, pages 179–184, Denver, CO, September 2011.

[178] D. Tolić and R. Fierro. Decentralized output synchronization of heterogeneous linear systems with fixed and switching topology via self-triggered communication. In *Proceedings of the American Control Conference*, pages 4648–4653, June 2013.

[179] D. Tolić, R. Fierro, and S. Ferrari. Optimal self-triggering for nonlinear systems via approximate dynamic programming. In *IEEE Multi-Conference on Systems and Control*, pages 879–884, October 2012.

[180] D. Tolić and S. Hirche. Stabilizing transmission intervals and delays for linear time-varying control systems: The large delay case. In *IEEE Mediterranean Conf. Contr. and Automat.*, pages 686–691, Palermo, Italy, June 2014.

[181] D. Tolić and S. Hirche. Stabilizing transmission intervals and delays for nonlinear networked control systems: The large delay case. In *Proceedings of the IEEE Conference on Decision and Control*, pages 1203–1208, Los Angeles, CA, December 2014.

[182] D. Tolić, V. Jeličić, and V. Bilas. Resource management in cooperative multi-agent networks through self-triggering. *IET Control Theory & Applications*, 9(6):915–928, 2014.

[183] D. Tolić, R. G. Sanfelice, and R. Fierro. Self-triggering in nonlinear systems: A small gain theorem approach. In *20th Mediterranean Conference on Control and Automation*, pages 935–941, July 2012.

[184] D. Tolić, R. G. Sanfelice, and R. Fierro. Input-output triggered control using Lp-stability over finite horizons. *Int. J. Robust and Nonlinear Control*, 25(14):2299–2327, 2015.

[185] R. Tomović and G. Bekey. Adaptive sampling based on amplitude sensitivity. *IEEE Transactions on Automatic Control*, 11(2):282–284, 1966.

[186] J. Tsinias. A theorem on global stabilization of nonlinear systems by linear feedback. *Systems & Control Letters*, 17(5):357–362, 1991.

[187] P. Vamplew and R. Ollington. Global versus local constructive function approximation for on-line reinforcement learning. In *Proceedings of the 18th Australian Joint Conference on Advances in Artificial Intelligence*, AI'05, pages 113–122, Berlin, Heidelberg, 2005. Springer-Verlag.

[188] P. Varutti, B. Kern, T. Faulwasser, and R. Findeisen. Event-based model predictive control for networked control systems. In *Decision and Control, 2009 Held Jointly with the 2009 28th Chinese Control Conference. CDC/CCC 2009. Proceedings of the 48th IEEE Conference on*, pages 567–572, December 2009.

[189] G. C. Walsh, H. Ye, and L. G. Bushnell. Stability analysis of Networked Control Systems. *IEEE Transactions on Control Systems Technology*, 10(3):438–446, 2002.

[190] P. Wan and M. Lemmon. An event-triggered distributed primal-dual algorithm for network utility maximization. In *Decision and Control, 2009 Held Jointly with the 2009 28th Chinese Control Conference. CDC/CCC 2009. Proceedings of the 48th IEEE Conference on*, pages 5863–5868, December 2009.

[191] X. Wang and M. Lemmon. Self-triggered feedback control systems with finite-gain l2 stability. *IEEE Trans. on Automatic Control*, 54(3):452–467, 2009.

[192] X. Wang and M. Lemmon. Self-triggering under state-independent disturbances. *IEEE Trans. on Automatic Control*, 55(6):1494–1500, 2010.

[193] X. Wang and M. Lemmon. Event-triggering in distributed networked control systems. *IEEE Trans. on Automatic Control*, 56(3):586–601, 2011.

[194] Y. Wei, J. Heidemann, and D. Estrin. An energy-efficient MAC protocol for wireless sensor networks. In *Twenty-First Annual Joint Conference of the IEEE Computer and Communications Societies INFOCOM*, volume 3, pages 1567–1576, 2002.

[195] R. L. Williams-II and D. A. Lawrence. *Linear State-Space Control Systems*. John Wiley & Sons, 2007.

[196] Y. Xia, M. Fu, and G. Liu. *Analysis and Synthesis of Networked Control Systems*, volume 409 of *Lecture Notes in Control and Information Sciences*. Springer Berlin Heidelberg, 2011.

[197] F. Xiao and L. Wang. Asynchronous consensus in continuous-time multi-agent systems with switching topology and time-varying delays. *IEEE Trans. on Automatic Control*, 53(8):1804–1816, 2008.

[198] H. Xu, A. Sahoo, and S. Jagannathan. Stochastic adaptive event-triggered control and network scheduling protocol co-design for distributed networked systems. *IET Control Theory Applications*, 8(18):2253–2265, 2014.

[199] X. Xuping and P. J. Antsaklis. Optimal control of switched systems based on parameterization of the switching instants. *IEEE Trans. on Automatic Control*, 49(1):2–16, 2004.

[200] H. Yu and P. J. Antsaklis. Event-triggered real-time scheduling for stabilization of passive and output feedback passive systems. In *American Control Conf.*, pages 1674–1679, San Francisco, CA, 2011.

[201] H. Yu and P. J. Antsaklis. *Output Synchronization of Multi-Agent Systems with Event-Driven Communication: Communication Delay and Signal Quantization*. Technical report, Department of Electrical Engineering, University of Notre Dame, July 2011.

[202] H. Yu and P. J. Antsaklis. Quantized output synchronization of networked passive systems with event-driven communication. In *Proceedings of the American Control Conference*, pages 5706–5711, June 2012.

[203] H. Yu and P. J. Antsaklis. Event-triggered output feedback control for networked control systems using passivity: Achieving $\mathcal{L}_2$ stability in the presence of communication delays and signal quantization. *Automatica*, 49(1):30–38, 2013.

[204] P. Zarchan and H. Musoff. *Fundamentals of Kalman filtering: A practical approach*. American Institute of Aeronautics and Astronautics, Reston, VA, 3rd edition, 2009.

[205] B. Zhang, A. Kruszewski, and J. P. Richard. A novel control design for delayed teleoperation based on delay-scheduled Lyapunov–Krasovskii functionals. *International Journal of Control*, 87(8):1694–1706, 2014.

[206] W. Zhang. *Controller Synthesis for Switched Systems Using Approximate Dynamic Programming*. PhD thesis, Purdue University, Indiana, December 2009.

[207] J. Zhao and D. J. Hill. Vector $\mathcal{L}_2$-gain and stability of feedback switched systems. *Automatica*, 45(7):1703–1707, 2009.

[208] J. Zhou and Q. Wu. Exponential stability of impulsive delayed linear differential equations. *IEEE Transactions on Circuits and Systems, II: Express Briefs*, 56(9):744–748, 2009.

Index

Q-function, 206
$\mathcal{L}_p$-detectability, 18
    with respect to set, 150, 180
$\mathcal{L}_p$-norm, 10
$\mathcal{L}_p$-stability
    with bias and w.r.t. set, 179
    with respect to set, 150
$\mathcal{L}_p$-stability with bias, 18
    over finite horizon, 58
ψ-irreducible Markov chain, 105
f-ergodicity, 106
f-norm ergodic theorem, 106

admissible delays, 21
aperiodic Markov chain, 105
Approximate Dynamic Programming (ADP), 86
average dwell time, 150

Bellman
    equation, 90, 205
    operator, 90
bi-character scheduler, 110

centralized scheduling policy, 109
Chebyshev polynomials, 206
complementary space, 158
contraction approximators, 92
Cramér–Rao lower bound, 205
CSMA with collision avoidance, 103

decentralized scheduler, 117
delayed NCSs, 16
directed spanning tree, 151
Dynamic Programming (DP), 86

emulation-based approach, 3
ergodicity, 105

event-based scheduling, 109
event-triggering, 2
expansion approximators, 92

Geršgorin theorem, 171

heterogeneous agents, 170

impulsive delayed
    LTI systems, 23
    systems, 17, 179
inclusive cycle, 151
input-output triggering, 52
Input-to-State Stability (ISS) with respect to set, 150
intermittent
    feedback, 2
    information, 2

Kalman filter, 204

Laplacian matrix, 151
large delays, 15
Least Square Policy Iteration (LSPI), 205
left-hand
    derivative, 78
    limit, 11
Lyapunov
    function, 40
Lyapunov Mean Square Stability, 106
Lyapunov Stability in Probability, 107

Markov chain, 104
Markov property, 104
matrix kernel, 11
Maximally Allowable Transmission Interval (MATI), 22

medium access control, 100
Minkowski inequality, 77
multi-loop NCS, 100

network-aware
    co-design, 3
    control design, 3
Networked Control Systems (NCSs), 1
neuro-dynamic programming, 87
nonuniform delays, 16
null space, 11

occupation time, 104
operational research, 87

packet dropouts, 116
policy iteration, 205
positive Harris recurrent, 106
probabilistic scheduler, 112

reachable set, 69
Real Jordan Form, 158, 183
reinforcement learning, 86
return time, 104
Riesz–Thorin interpolation theorem, 44
right-hand limit, 11

sampled-data control, 9
scheduling, 100
self-triggering, 2
shared communication channel, 99
small set, 105
stochastic stability, 105
strongly aperiodic Markov chain, 105
support of a function, 92
switched system, 58, 150
switching signal, 58, 150

TOD scheduler, 111
topology
    discovery, 155
    graph, 151
topology-triggering, 155
translation operator, 17

truncated Gaussian, 205

unidirectional link, 151
Uniform Global Asymptotic Stability (UGAS), 18
Uniform Global Exponential Stability (UGES), 18
    with respect to set, 150
Uniform Global Stability (UGS), 18
Uniformly Globally Exponentially Stable (UGES) protocols, 21
upper right/Dini derivative, 41, 160

Value Iteration, 91

Young's inequality, 44

Zeno behavior, 26